Pathogenic Bacteria

Edited by Sahra Kırmusaoğlu
and Sonia Bhonchal Bhardwaj

Published in London, United Kingdom

IntechOpen

Supporting open minds since 2005

Pathogenic Bacteria
http://dx.doi.org/10.5772/intechopen.78081
Edited by Sahra Kırmusaoğlu and Sonia Bhonchal Bhardwaj

Contributors

Sonia Bhonchal Bhardwaj, Jeannette Barba-León, Lilia Mercedes Mancilla-Becerra, Teresa Lías-Macías, Cristina Lizbeth Ramírez-Jiménez, Mónica Sparo, Gastón Delpech, Leonardo Garcia Allende, S M Kadri, Marija Petković, Arshi Taj, Ailbhe H Brady, Graciela Avila-Quezada, Pavel Gerardo Espino-Solís, Bilal Aslam, Muhammad Hidayat Rasool, Muhammad Asif Zahoor, Muhammad Waseem, Muhammad Shafique, Saima Muzammil, Abu Baker Siddique, Zeeshan Nawaz, Mohsin Khurshid, Maria Rsool, Colleen M Pike, Ramona Neunuebel, Rebecca Noll, Ogueri Nwaiwu, Chiugo Claret Aduba, Eniola Oluyemisi Oni, Nahla Eltai, Asma Al Thani, Hadi Yassine, Hashim Alnjoomi Alhussain, Sanjay Doiphode, Neha Farid, Asma Bashir, Kashif Ali, Kiran Fatima, Nazish Ali, Safia Rehman, Abdullah Mazhar, Iram Liaqat, Bushra Mazhar, Gyoguevara Patriota, Benno Ejnisman, Nivaldo Cardozo, Paulo Santoro Belangero, Luiz Marcelo Leite, Yutaka Tsutsumi, Rana Binyamin, Muhammad Imran Arshad, Muhammad Aamir Aslam, Naveed Shahzad, Zulqarnain Baloch

First published in London, United Kingdom, 2020 by IntechOpen
IntechOpen is the global imprint of INTECHOPEN LIMITED, registered in England and Wales, registration number: 11086078, 5 Princes Gate Court, London, SW7 2QJ, United Kingdom
Printed in Croatia

British Library Cataloguing-in-Publication Data
A catalogue record for this book is available from the British Library

Additional hard and PDF copies can be obtained from orders@intechopen.com

Pathogenic Bacteria
Edited by Sahra Kırmusaoğlu and Sonia Bhonchal Bhardwaj
p. cm.
Print ISBN 978-1-78985-987-4
Online ISBN 978-1-78985-988-1
eBook (PDF) ISBN 978-1-83881-934-7

For EU product safety concerns:
IN TECH d.o.o., Prolaz Marije Krucifikse Kozulić 3, 51000 Rijeka, Croatia, info@intechopen.com or visit our website at intechopen.com.

Meet the editors

Dr. Kırmusaoğlu, PhD, is an assistant professor of Microbiology at the Department of Molecular Biology and Genetics, T.C. Haliç University. She specialized in Microbiology at Abant Izzet Baysal University (Biology Department), Turkey. Her previous experience includes laboratory manager at microbiology laboratories in several research and private hospitals. Throughout her career, she collaborated with academicians/researchers from Abant Izzate Baysal University (AIBU), Middle East Technical University (METU), and Istanbul University Cerrahpaşa Faculty of Medicine, and has participated in various research projects. Dr. Kırmusaoğlu's research interests include medical microbiology, pathogenic bacteria, bacterial biofilms, antibiofilm and antibacterial activity, bacterial drug resistance, pathogen–host interactions, pathogenesis, molecular microbiology, and microbiota. She has published several international research articles, books, book chapters, and congress proceedings. She is also the editor of *Disinfection*, *Bacterial Pathogenesis and Antibacterial Control*, and *Antimicrobials, Antibiotic Resistance, Antibiofilm Strategies and Activity Methods* published by IntechOpen. In addition to these, she wrote the book *Genel Biyoloji Laboratuvar Kılavuzu* (*General Biology Laboratory Manual*) published by Hipokrat Publisher. She has contributed to a chapter translation of the book *Sherris Medical Microbiology* by Ryan et al. as one of the translation authors of *Sherris Tıbbi Mikrobiyoloji*, which is a Turkish translated book edited by Prof. Dr. Dürdal Us and Prof. Dr. Ahmet Başustaoğlu.

Dr. Sonia Bhonchal Bhardwaj is currently an assistant professor at the Department of Microbiology, Dr. Harvansh Singh Judge Institute of Dental Sciences and Hospital, Panjab University, Chandigarh. She graduated with honors in Microbiology and received her PhD from Postgraduate Institute of Medical Education and Research (PGIMER), Chandigarh. She has published nineteen international publications and three book chapters. She received grants from the Department of Science and Technology (India) and worked on biofilm formation by Enterococcus faecalis in periodontitis, Streptococcus mutans, and phage therapy.

Contents

Preface

Pathogenic bacteria reveal their pathogenicity and cause infection by their virulence factors such as surface proteins that lead bacterial adherence to host cells, cell surface carbohydrates, peptidoglycan, lipopolysaccharides of Gram-negative bacteria, biofilm, toxins that lead bacteria to invade and damage host cells, hydrolytic enzymes, antibiotic resistance genes, antiphagocytic factors, and so on. Virulence factors enable bacteria to survive and infect the host's cells. Bacterial pathogens are becoming the main problem in hospital- and community-acquired infections. Thus, antibacterial treatment strategies are needed to prevent bacterial adherence, invasion, and infection.

Due to widespread antimicrobial resistance, multidrug-resistant pathogens that are the main causes of morbidity and mortality have emerged worldwide. They have the ability to synthesize a number of enzymes such as extended spectrum β-lactamases, induced β-lactamases, carbapenamases, metallo-β-lactamases, New Delhi metallo-β-lactamases, and many others. It is hard to treat the strains that are resistant to antibiotics due to the causing recurrent and untreatable infections. This book presents examples of pathogenesis, virulence factors, and treatment strategies.

This book contains thirteen chapters from valued experts in Argentina, Brazil, China, India, Japan, Mexico, Nigeria, Pakistan, Qatar, Serbia, United Kingdom, and the United States. It is a useful resource for researchers interested in the study of bacteriology.

I would like to thank all the authors who contributed to this book. I would also like to thank Author Service Manager Sara Debeuc and IntechOpen for their concern and encouragement in publishing this book.

Dr. Sahra Kırmusaoğlu
Assistant Professor,
Faculty of Arts and Sciences,
Department of Molecular Biology and Genetics,
Haliç University,
Beyoğlu-Istanbul, Turkey

Dr. Sonia Bhonchal Bhardwaj
Panjab University,
India

Section 1

Microbial Mechanism of Pathogenicity

Chapter 1

Development of Biofilms for Antimicrobial Resistance

Asma Bashir, Neha Farid, Kashif Ali and Kiran Fatima

Abstract

Biofilms are a unit referred to as assemblage of microbial cells growing as surface-attached microbial communities within the natural surroundings. Their genetic and physiological aspects are widely studied. Biofilm development involves the assembly of extracellular compound substances that forms the most bailiwick network. Quorum sensing is one more crucial development specifically connected with biofilm formation in several microorganism species. In ecological purpose, the biofilm offers protection against unfavorable conditions and provides a platform for the genetic transfer. A biofilm-forming bacterium area unit is medically necessary, as they are resistant to several antibiotics and might spread resistant genes. This chapter provides the summary of microorganism biofilm formation and its significance in ecology.

Keywords: biofilms, resistance, microbes, disease, antimicrobial agents

1. Introduction

In the years which pursued the historical backdrop of microbiology, microscopic organisms have been for the most part contemplated as planktonic (free-floating) forms, the investigation of which contributed particularly to the comprehension of fundamental physiological procedures. It was just late 1960s and mid-1970s when the broad physical and chemical examinations of surface-attached microbes began coming up and the prevalence of surface-related microorganisms (biofilms) was perceived. A significant part of the prior work on biofilm characterization depended on the instruments, for example, scanning electron microscopy and standard microbiological culture procedures. The utilization of scanning electron microscopy by scientists uncovered that the biofilms are made out of a blend of various microorganisms; and the matrix material was predominantly made out of polysaccharide. The first genuine examination of biofilm was made by Costerton JW and KJ in 1978 when their examinations demonstrated that numerous microorganisms spend their most part of life inside surface-attached, sessile networks encased in a polymer network [1].

Initially the studies on biofilm were mostly focused on the structure of the polymer network or "glycocalyx" which was later portrayed by Costerton as an ion exchange network, thought to trap supplements from the surroundings [1]. Costerton found that the glycocalyx was a hydrated polyanionic polysaccharide network created by the polymerases inserted in the lipopolysaccharide part of bacterial cell wall [2]. In a watery situation (at the strong/fluid interface), biofilm generation assumes a noteworthy role in the assimilation and convergence of natural and

IntechOpen

inorganic supplements. In addition, the biofilm provides a physical barrier that ensures incomplete protection against antibacterial substances.

During the 1990s, researchers started to comprehend the complex association of bacterial biofilm network. With the quick advances in the molecular technologies and microscopic techniques and systems, empowering extensive investigations of the biofilm method of life, there has been a striking advancement of biofilm understanding in late years. The biofilm can be framed by a solitary bacterial species; be that as it may, in many biological systems, biofilm comprises of heterogeneous networks of microorganism including bacteria, fungi, algae, and protozoa. Biofilm arrangement usually happens when microorganisms attach to surfaces in fluid conditions and begin discharging extracellular fluid like slimy material that can anchor them to a variety of materials including metals, plastics, soil particles, medical implant materials, and tissue. Microbial biofilm arrangement is known to be a successive bacterial development process and is managed by a progression of hereditary and phenotypic determinants. Accurate screening strategies, for example, isolation of biofilm defective mutants, have contributed incredibly to understanding the hereditary qualities of biofilm formative procedure; furthermore, noteworthy data is included in the hereditary premise of biofilm development.

A biofilm is known to have the involvement of many associations of microorganisms which leads to the adherence of the cells to one another and also to the surface where they are growing [3]. These adherent cells become installed inside a slimy extracellular network that is made out of extracellular polymeric substances (EPS). The cells inside the biofilm produce the EPS components, which are ordinarily a polymeric aggregation of extracellular polysaccharides, proteins, lipids, and DNA [3].

Biofilms may form on living or nonliving surfaces and are common in natural, industrial, and hospital settings [4]. The microbial cells developing in a biofilm are physiologically distinct from planktonic cells of a similar life form, which, on the other hand, are unicellular which have the ability to buoy or swim in a liquid medium. Biofilms can also grow on the teeth structure of many creatures in the form of dental plaque. This dental plaque then leads to the oral diseases of tooth decay and gum illness.

Microbes form a biofilm by the contribution of many different factors which somehow help in the recognition of sites of attachments on a surface, help them to find the nutritional sources, or, in some cases, help to develop resistance to

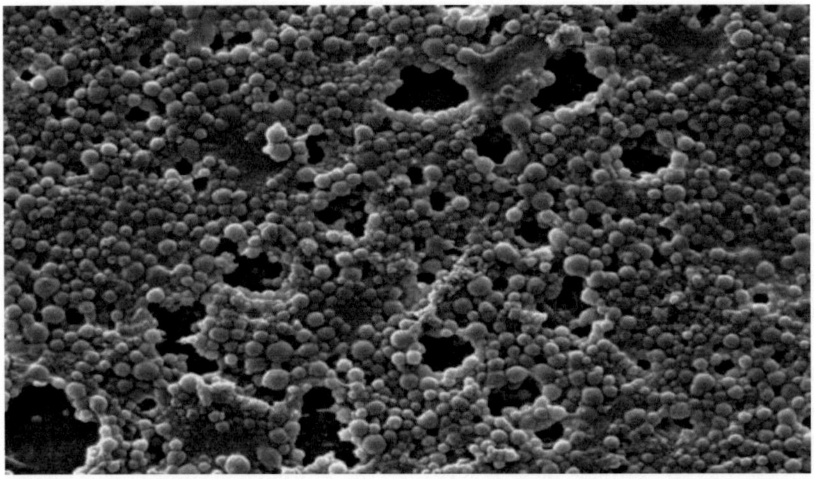

Figure 1.
Biofilm on the septum.

antibiotics. When a bacterial cell develops the property to form biofilm, it then undergoes phenotypic changes. These phenotypic changes also bring a change in the functioning of the genes.

A biofilm structure can be elucidated as hydrogel, made up of polymer which contains the dry mass enclosed in the water. Biofilms are layers formed of bacterial sludge along with the naturally occurring frameworks. This whole organization of network gives a look of well-structured meshwork of cells. Biofilms can connect to a surface, for example, a tooth, rock, or surface, and may incorporate a single microorganism category or various gatherings of microorganisms. The biofilm microscopic organisms can share nutrients and are shielded from harmful factors in the environment, for example, antitoxins, and a host body's insusceptible framework. A biofilm for the most part starts to frame whenever a free-swimming bacterium appends to a surface (**Figure 1**).

2. Origin and formation

2.1 Origin

Biofilms are known to have emerged on the primitive Earth for the purpose of defense for the prokaryotes at that time because the condition of the Earth in the early ages was very harsh and difficult for the survival of prokaryotic organism. Biofilms provide the prokaryotic cells with homeostatic conditions which empowers them with the advancement of complex interactions between the cells having biofilm.

2.2 Formation

The arrangement of a biofilm starts with the connection of free-skimming microorganisms to a surface [5]. Initially, the microbes of a biofilm may adhere tightly to the surface with the help of hydrophobic interactions and van der Waals forces. If the other colony-forming microbes are not isolated from the surface instantly, then they quickly attach themselves to the surface permanently by utilizing their cell griping structures such as pili.

Hydrophobicity has been observed to have effect on the ability of the microbes in the formation of biofilms. Microorganisms which have high amount of hydrophobicity are seen to have low amount of repulsive forces between the adherent surface and the attaching bacterium. In some cases, the microbes face difficulty in binding to the surface properly. This is because of their restricted motility, but however they can still adhere themselves to the matrix surface and to the other microbes which were initially present. The microbes having nil motility can neither attach to the surfaces nor have the ability to aggregate with each other effectively as that seen in the case of bacteria having motility.

In the process of surface colonization, the microbes have the ability to communicate by using the products of quorum sensing (QS). One of these products is N-acyl homoserine lactone (AHL). Once the cellular colonization starts, the development of biofilm also initiates by the combined effect of cell division and cell recruitment. The bacterial biofilms are mostly enclosed in the matrices made up of polysaccharides. Apart from the polysaccharides, these adherent matrices may also contain some other components such as different substances from the surrounding environment such as blood segments including fibrin and erythrocytes, minerals, particles of soil, and many other small substances. After all this comes the last phase of the arrangement of biofilm. This last stage is known as dispersion. Dispersion has been recognized as the stage in which the biofilm completely forms and may undergo some variations in shape and size.

3. Stages in the formation of biofilm

There are three stages in biofilm formation: initial attachment events, the development of complex biofilms, and separation events by clumps of microorganisms or by a "swarming" phenomenon within the interior of bacterial clusters, bringing about the so-called "seeding dispersal." Once a biofilm has fully formed, it frequently contains diverts in which supplements can flow. Cells in various locales of a biofilm additionally display diverse examples of gene expression. Since biofilms regularly build up their very own metabolism, they are in some cases contrasted with the tissues of higher creatures, in which firmly packed cells cooperate and make a system in which minerals can stream.

The biofilm life cycle is observed in three different stages: attachment, growth of colonies (advancement, and occasional detachment of planktonic cells: Free-drifting, or planktonic microorganisms experience an immersed surface and then within few minutes, they can become attached. They start producing slimy EPS and eventually begin to colonize the surface [1–4]. The formation of EPS allows the biofilm network to develop a three-dimensional and complex structure which is affected by various environmental factors. These complex networks of biofilm structures can be formed within few hours [5]. Biofilms have the sections of cluster of small or large portions of cells. It can also be observed by the process of "seeding dispersal" which helps to discharge the cells which are in singular property. Both the types of cellular separation allow the microbes to get connected either to a surface or to a unique network of biofilm [6, 7] (**Figure 2**).

3.1 Properties

Biofilms are mostly found on the solid substrates which are either submerged in or exposed in an aqueous environment. They are present in these environments apart from the fact that they can function as floating mats on liquid surfaces and also on the external surface of the leaves, which are present especially in the

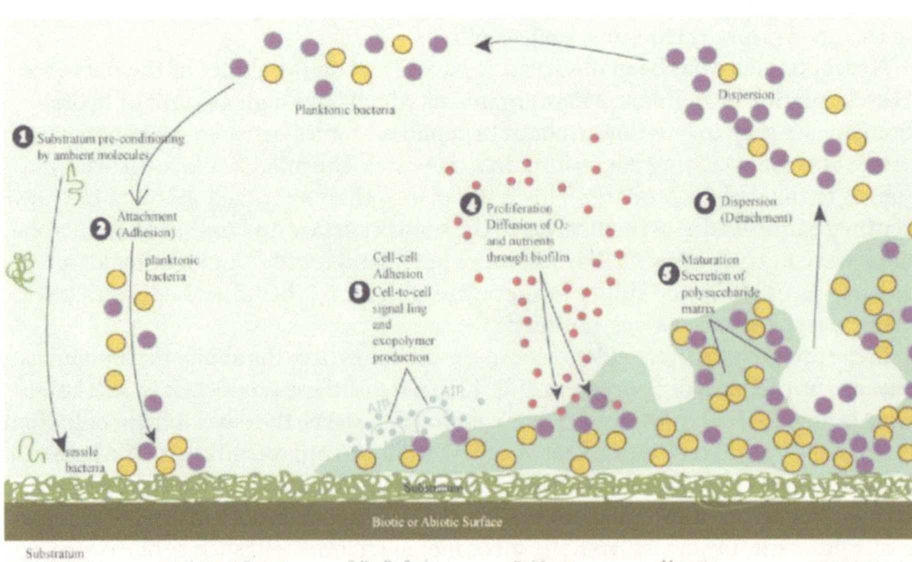

Figure 2.
Stages of biofilm development [8].

environment of high moisture. When the adequate resources for development are provided, there will be rapid development of a biofilm naturally in such a way that it will be visible clearly. Biofilms have the property to provide surface for the growth of a wide range of microorganisms which includes archaea, protozoa, bacteria, algae, and fungi with each organism having its own specific metabolic properties [9, 10].

3.2 Extracellular matrix

The EPS matrix is made up of exopolysaccharides, nucleic acids, and proteins. A major proportion of the EPS is somewhat hydrophilic along the hydrophobic portion. The example of such combination is cellulose which is made by many microbes. This matrix encloses the bacterial cells at intervals and also provides them the ability to communicate with each other through the biochemical signals and more importantly through gene exchange. The EPS matrix facilitates to trap the extracellular enzymes and then encloses them near the cells. This process shows that the EPS matrix has the ability of external digestion and it leads to the process of stable synergistic between various microbial species. There are some biofilms which have water channels. These water channels help them in the distribution of food and nutrients along with the signaling molecules [11].

Bacteria having the property of biofilm production are different from those which are free-floating bacterium of the same species. This is because of the dense and guarded setting of the biofilm which permits them to stick together [12]. The biofilm gives the microbe the advantage of resistance to different chemicals such as detergents and antibiotics. Thus, the dense matrix along with the external layer of cells provides a shield to the internal environment of the cells. In some instances, the biofilms increase the resistance several folds in the microbes [13]. It also helps in the lateral gene transfer in the normal microorganisms and the archaeal biofilms. This eventually makes a more stable biofilm structure [14]. But in some cases the biofilms have no contribution in the antimicrobial resistance. This can be seen in *Pseudomonas aeruginosa* which has no increased resistance to any antimicrobials as compared to the stationary-phase microbial cells which do not produce the biofilms. The biofilm production is seen in high rate in microbial cells present in the logarithmic phase of life cycle. This antimicrobial resistance seen in both the cells of the stationary phase and those of the biofilms may be contributed by the presence of persisted cells [15].

3.3 Quorum sensing

The role of quorum sensing in the regulation of biofilm has been first reported by Davies which initiated the dynamic research in the cell-to-cell signaling in biofilms [16, 17]. He demonstrated that lasI-mutant cells of *P. aeruginosa* that were unfit to blend the QS signaling molecule [3OC12-HSL (3-oxododecanoylhomoserine lactone)] created undifferentiated biofilm architecture and are additionally delicate to biocide SDS. Supplementation of lasI- mutant cells with 3OC12-HSL brought about a design similar to the wild sort biofilms. The procedure of cell-to-cell correspondence in bacterial populace is known to happen through small diffusible signaling molecules perceived as autoinducer. These signal molecules are created by the bacterial cells, and their concentration in the environment relies upon the density of the population. At the point when a limit focus is achieved, the signal can initiate other microbes leading to the induction or restraint of certain target genes [18].

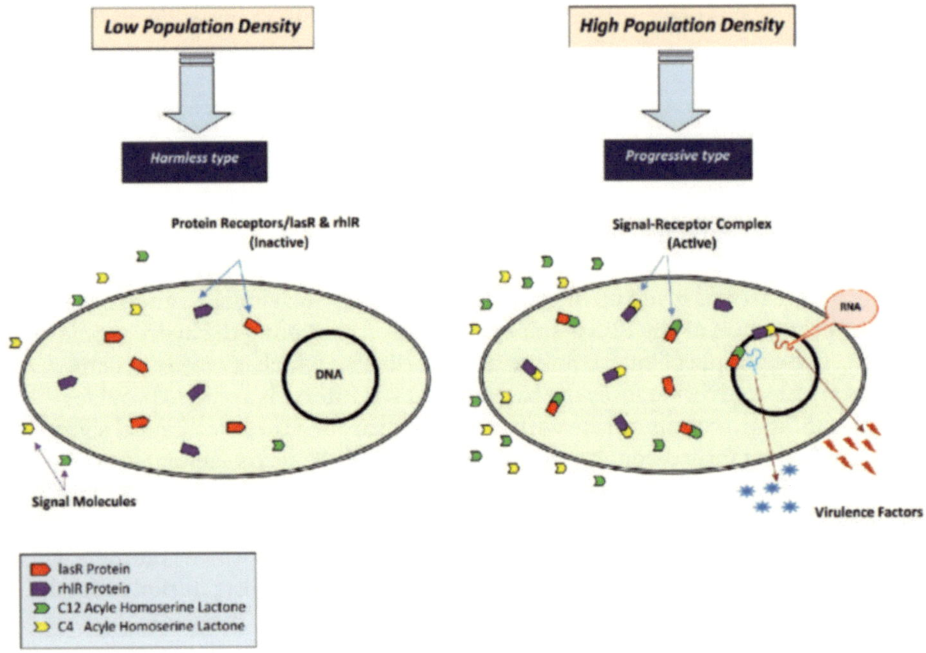

Figure 3.
Quorum sensing.

Cell density-dependent gene regulation phenomenon is otherwise called quorum sensing (QS). The chemical properties of signaling molecules associated with QS are differing; however gram-negative microbes most regularly utilize N-acylhomoserine lactones (AHLs). For instance, types of *Acidithiobacillus*, *Acinetobacter*, *Aeromonas*, *Agrobacterium*, *Brucella*, *Burkholderia*, *Erwinia*, *Enterobacter*, *Chromobacterium*, *Hafnia*, *Mesorhizobium*, *Methylobacter*, *Paracoccus*, *Pseudomonas*, *Ralstonia*, *Rhodobacter*, *Rhizobium*, *Rhanella*, *Serratia*, *Sinorhizobium*, *Vibrio*, and *Yersinia Williams* are referred to utilize AHLs as their major signaling molecules. In the biofilm arrangement as well as in the dispersal, QS assumes a noteworthy job. In *Rhodobactersphaeroides* (mutant cells), the addition of 7,8-cis-tetradecenoyl-HSL to the cell total brought about cell scattering prompting the development of free individual cells in suspension (**Figure 3**).

4. Taxonomic diversity

There are many different types of microorganisms which are known for their property to form biofilm. These include both the gram-positive and gram-negative species. The gram-positive bacteria include *Listeria monocytogenes*, *Bacillus species*, *Staphylococcus species*, and *lactic acid bacteria*, which includes *Lactobacillus plantarum* and *Lactococcus lactis*. And the gram-negative species include *Escherichia coli* and *Pseudomonas aeruginosa*. It is also been observed that other bacteria such as *Cyanobacteria* have the ability to form the biofilms in the aqueous environments. The production of biofilms is also the property of microbes which are known to colonize the plants. These microbes include *Pseudomonas putida*, *Pseudomonas fluorescens*, and connected *pseudomonads*. They are mostly the plant-associated microorganisms and are known to be present on roots, leaves, and within the soil. This is the reason which gives them the property of producing biofilms in botanical areas. Other than these microbes, there are many other nitrogen-fixing symbionts found in legumes such as the genus *Rhizobium leguminosarum,* and *Sinorhizobium meliloti* form biofilms on legume roots and different inert surfaces. Along with microorganisms, biofilms also are generated by archaea by a variety of eukaryotic organisms including fungi, e.g., *Cryptococcus laurentii* and *microalgae* [19].

5. Biological importance

5.1 Safety from the environment

The biofilm gives a safe house and homeostasis to the living beings living inside it, and the imperative segment of this safe house is the extracellular polymeric substance network. This network can possibly forestall the flood of certain antimicrobial operators in this way confining the dissemination of these mixes from the environment into the biofilm.

EPS has appeared to have metal binding property and consequently can sequester lethal metal particles and give defensive functions. In addition to metal binding capacity, the EPS can likewise sequester nutrients and minerals from the environment. This coupling property of EPS is basically because of the nearness of ionizable functional groups, for example, carboxyl, phosphoric, amine, and hydroxyl groups. Researchers found that the sanitized EPS from the container of a freshwater sediment bacterium is fit for restricting copper. Farag detailed the concentration of metals (Ar, Cd, Pb, Hg, and Zn) in various nourishment web segments [20]. Likewise, different authors have announced the stimulatory impact of metal particles on the biofilm development. Researchers in 1997 observed an enlistment of biofilm in the developing colony of *Archaeoglobus fulgidus* when exposed to high grouping of copper and nickel. Bereswill explained the creation of amylovoran: the fundamental polysaccharide of EPS in *Erwinia amylovora*, in the presence of copper [21]. Ordax demonstrated that the EPS removed from *E. amylovora* can bind copper cations and in this manner inferred that the EPS favors the survival of *E. amylovora* under copper pressure [22]. Comparable perceptions of increment in EPS generation within the sight of metal pressure have been accounted for other bacterial species. EPS is additionally known to give a certain level of assurance to the biofilm cells from different natural stresses, for example, UV radiation, pH shifts, osmotic shock, and desiccation.

5.2 Nutrient absorption

The developed biofilm regularly contains voids and water channels that give an expanded surface zone to nutrient trade. As the water channels are interconnected and dive deep into the biofilm, it guarantees supplement accessibility to microbial networks dwelling somewhere inside the biofilm. The biofilm traps the follow component and supplement from outside condition through physical trapping or electrostatic interaction. The complex biofilm design additionally gives the chance to metabolic cooperation, and specialties are framed inside these spatially composed structures. The microcolonies created in these specialties vary in their structure removal and redistribution of metabolic end product. As these microcolonies are orchestrated one next to the other, it gives a great chance to the trading of substrate, evacuation, and redistribution of metabolic finished result [23].

5.3 Gene transfer

Biofilm offer an appropriate niche during which bacterium of various microbial community will grow in shut proximity to every possible vicinity. This provides associate in nursing area for the exchange of extrachromosomal genetic parts like plasmid inclusion body. Indeed, the transfer of inclusion body deoxyribonucleic acid via conjugation occurs at higher frequency within the biofilm cells as compared to their planktonic counterparts. The horizontal transfer of conjugative plasmid adds to the event and stabilization of biofilm. Since inclusion body could have genes that provide resistance to several antimicrobials agents, biofilm formation also offers a mechanism for the unfolding of microorganism resistance to antimicrobial agents [24]. Conjugal transfer of deoxyribonucleic acid (plasmid) is not the sole mechanism of factor transfer in a very microbial biofilm; another mechanism like transformation also can be expected, as an amount of deoxyribonucleic acid is additionally found in the biofilm structure. This deoxyribonucleic acid is assumed to be discharged within the biofilm matrix by the lysis of microorganism cells as found within the case of *Streptococcus pneumonia* and *Acinetobacter calcoaceticus*. The dense population within the microcolonies of biofilm conjointly provides a wonderful chance for the uptake of this extracellular matrix deoxyribonucleic acid [25]. Researchers observed a high frequency transformation within the young and actively growing biofilms of *Acinetobacter sp.* BD413 and correlative enlarged transformation frequency with the deoxyribonucleic acid concentration and located no saturation [26].

5.4 Disease

The role of biofilm forming microorganism in mediating numerous infectious diseases is changing into rather more necessary with an increasing numbers of infections in humans. Biofilm infection in human includes microorganism endocarditis (infection of heart valves), otitis (infection of the middle ear), chronic microorganism inflammation (infection of the prostate gland), cystic fibrosis (infection of lower metabolic process system), dentistry diseases, and most medical device-connected infections [27]. These diseases are well reviewed by researchers. *Vibrion* infectious disease which is the causative agent of infectious disease has been famous to endure transition to conditionally viable environmental cells, once discharged into the environment. Recently, researchers showed that this process involves assemblage sensing dependent biofilm formation, the factors that enhances the waterborne unfold of infectious disease epidemic [28]. In *Acinetobacter baumannii*, a medical building pathogen, biofilm formation on abiotic and biological surfaces is understood to influence its virulence. Biofilm microorganisms are consistently resistant to the antimicrobial stress, and so their demolition with antibiotic treatment could be a prime concern of medical analysis [29, 30].

6. Conclusion

The nature of biofilm structure and therefore the physiological attributes of biofilm organisms have inherent resistance to antimicrobial agents, no matter these antimicrobial agents are antibiotics or disinfectants. From the results obtained from the study, it can be concluded that the microbial strains that have the ability to produce biofilms become methicillin resistant. This supports the argument that biofilms play major role in providing the antibiotic resistance to bacteria.

Conflict of interest

There is no conflict of interest.

Author details

Asma Bashir, Neha Farid*, Kashif Ali and Kiran Fatima
Department of Biosciences, Shaheed Zulfikar Ali Bhutto Institute of Science and Technology (SZABIST), Karachi, Pakistan

*Address all correspondence to: neha_farid@hotmail.com

References

[1] Costerton JW, Geesey GG, Cheng KJ. How bacteria stick. Scientific American. 1978;**238**(1):86-95

[2] Costerton JW, Irvin RT, Cheng KJ. The bacterial glycocalyx in nature and disease. Annual Reviews in Microbiology. 1981;**35**(1):299-324

[3] Percival SL, Malic S, Cruz H, Williams DW. Introduction to biofilms. In: Biofilms and Veterinary Medicine. Berlin, Heidelberg: Springer; 2011. pp. 41-68

[4] López D, Vlamakis H, Kolter R. Biofilms. Cold Spring Harbor Perspectives in Biology. 2010;**2**(7):a000398

[5] Christensen GD, Simpson WA, Younger JJ, Baddour LM, Barrett FF, Melton DM, et al. Adherence of coagulase negative *Staphylococci* to plastic tissue cultures: A quantitative model for the adherence of *Staphylococci* to medical devices. Journal of Clinical Microbiology. 1985;**22**:996-1006

[6] Stepanovic S, Cirkovic I, Ranin L, Svabic-Vlahovic M. Biofilm formation by *Salmonella* spp and listeria monocytogenes on plastic surface. Letters in Applied Microbiology. 2004;**38**(5):428-432

[7] Donlan RM, Costerton JW. Biofilms: Survival mechanisms of clinically relevant microorganisms. Clinical Microbiology Reviews. 2002;**15**(2):167-193

[8] Kırmusaoğlu S. Staphylococcal biofilms: Pathogenicity, mechanism and regulation of biofilm formation by quorum sensing system and antibiotic resistance mechanisms of biofilm embedded microorganisms. In: Dhanasekaran D, Thajuddin N, editors. Microbial Biofilms-Importance and Applications. Croatia: IntechOpen; 2016. pp. 189-209

[9] Dunne WM. Bacterial adhesion: Seen any good biofilms lately? Clinical Microbiology Reviews. 2002;**15**(2):155-166

[10] Carpentier BCO. Biofilms and their consequences with particular references to hygiene in the food industry. Journal of Applied Microbiology. 1993;**75**:499-511

[11] Stoodley P, Sauer K, Davies DG, Costerton JW. Biofilms as complex differentiated communities. Annual Reviews in Microbiology. 2002;**56**(1):187-209

[12] Kolari M. Attachment mechanisms and properties of bacterial biofilms on non-living surfaces. 2003. Electronic publication available from: http:// ethesis.helsinki.f

[13] Jefferson KK. What drives bacteria to produce a biofilm? FEMS Microbiology Letters. 2004;**236**(2):163-173

[14] Molin S, Tolker-Nielsen T. Gene transfer occurs with enhanced efficiency in biofilms and induces enhanced stabilisation of the biofilm structure. Current Opinion in Biotechnology. 2003;**14**(3):255-261

[15] Davies D. Understanding biofilm resistance to antibacterial agents. Nature Reviews Drug Discovery. 2003;**2**:114-122

[16] Annous BA, Fratamico Pina M, Smith James L. Quorum sensing in biofilms: Why bacteria behave the way they do. Journal of Food Science. n.d.;**74**:24-37

[17] O'Toole GA, Kolter R. Initiation of biofilm formation in *Pseudomonas*

fluorescens WCS365 proceeds via multiple, convergent signalling pathways: A genetic analysis. Molecular Microbiology. 1998;**28**(3):449-461

[18] Davies DG, Parsek MR, Pearson JP, Iglewski BH, Costerton JW, Greenberg EP. The involvement of cell-to-cell signals in the development of a bacterial biofilm. Science. 1998;**280**(5361):295-298

[19] Hall-Stoodley L, Costerton JW, Stoodley P. Bacterial biofilms: From the natural environment to infectious diseases. Nature Reviews Microbiology. 2004;**2**(2):95-108

[20] Farag AM, Woodward DF, Goldstein JN, Brumbaugh W, Meyer JS. Concentrations of metals associated with mining waste in sediments, biofilm, benthic macroinvertebrates, and fish from the Coeur d'Alene River basin, Idaho. Archives of Environment Contamination and Toxicology. 1998;**34**(2):119-127

[21] Bereswill S, Jock S, Bellemann P, Geider K. Identification of Erwinia amylovora by growth morphology on agar containing copper sulfate and by capsule staining with lectin. Plant Disease. 1998;**82**(2):158-164

[22] OrdaxM,Marco-NoalesE,LópezMM, Biosca EG. Exopolysaccharides favor the survival of Erwinia amylovora under copper stress through different strategies. Research in Microbiology. 2010;**161**(7):549-555

[23] Kamruzzaman M, Udden SN, Cameron DE, Calderwood SB, Nair GB, Mekalanos JJ, et al. Quorum-regulated biofilms enhance the development of conditionally viable, environmental *Vibrio cholerae*. Proceedings of the National Academy of Sciences of the United States of America. 2010;**107**(4):1588-1593

[24] Brown MR, Allison DG, Gilbert P. Resistance of bacterial biofilms to antibiotics a growth-rate related effect? The Journal of Antimicrobial Chemotherapy. 1988;**22**(6):777-780

[25] Fux CA, Stoodley P, Hall-Stoodley L, Costerton JW. Bacterial biofilms: A diagnostic and therapeutic challenge. Expert Review of Anti-Infective Therapy. 2003;**1**(4):667-683

[26] Molin S. Gene transfer occurs with enhanced efficiency in biofilms and induces enhanced stabilization of the biofilm structure. Current Opinion in Biotechnology. 2003;**14**:1-7

[27] Bekele ST, Abay GK, Gelaw B, Tessema B. Bacterial Biofilms; Links to Pathogenesis and Resistance Mechanism. Preprints.org; 2018

[28] Gordon RJ, Lowy FD. Pathogenesis of methicillin resistant Staphylococcus aureus infection. Clinical Infectious Diseases. 2008;**46**:350-359

[29] Baumann P. Isolation of Acinetobacter from soil and water. Journal of Bacteriology. 1968;**96**(1):39-42

[30] Maj Puneet Bhatt GS, Kundan Tandel PJ. Antimicrobial susceptibility profile of methicillin-resistant *Staphylococcus aureus* at a tertiary care centre. Archives of Clinical Microbiology. 2013

Chapter 2

Psuedomonas aeruginosa-Associated Acute and Chronic Pulmonary Infections

Nazish Mazhar Ali, Safia Rehman, Syed Abdullah Mazhar, Iram Liaqat and Bushra Mazhar

Abstract

Pseudomonas aeruginosa is highly successful in colonizing in all types of environments. *P. aeruginosa* colonizing in adverse environment due to the presence of its virulence factors include production of toxins, proteases hemolysins, and formation of biofilms. In man, the most common opportunist pathogen is *P. aeruginosa*. Metabolically *P. aeruginosa* is versatile. Most of the antibiotics targeted metabolically active cells and bacteria could contribute to decrease in biofilm susceptibility to the antimicrobial agents. Scientists suggested about *Pseudomonas* that it can be catabolized any hydrocarbon in specific time along with availability of oxygen and nitrite. If bacteria are not susceptible to one agent in three or more, it is called as multidrug-resistance strains. The antimicrobial treatments were not suitable when microorganism presented *in vitro* microorganism resistance to antimicrobials used for treatment of the patient which lack of treatment for 24 h after diagnosis of microbial infections. Bacteria have developed resistance against commonly used antibiotics. Treatment of *Pseudomonas* infections is coming harder day by day as its resistance against most of the antibiotics. Because of resistance of bacteria antibiotics, alternative methods are in consideration. These methods include use of lactic acid bacteria (LAB) and most recently nano-particles. That is why they are used as antibacterial agents.

Keywords: *P. aeruginosa*, pulmonary infections, acute lung infections, cystic fibrosis, quorum sensing system, virulent genes, antibacterial agents, LAB

1. Introduction

P. aeruginosa contributes its pathogenicity onwards respiratory infections in the hospitalized patients. Dasenbrook et al. has reported two types of airways infection acute and chronic spread by hospital community *P. aeruginosa* is a bacterium that lives in versatile environments [1]. It is Gram negative bacterium, metabolically able to regulate its systems and highly resistance to antibiotic causing it to spread in diverse habitats mainly in hospitals. *P. aeruginosa* is recognized as a human adaptable pathogen causing acute infections (bacteremia, pneumonia and urinary tract infections) in individuals with HIV infections, surgical wounds, cancer, carrying catheters or burns, or an organ transplantations. *P. aeruginosa* is persistence in

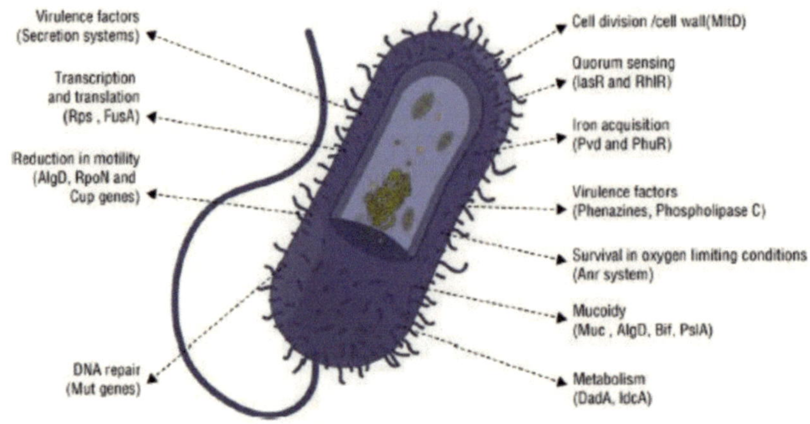

Figure 1.
P. aeruginosa *features relevant to pathogenicity and adaptation [2].*

chronic obstructive pulmonary diseases and chronic infections in individuals with cystic fibrosis (CF) [3, 4]. *P. aeruginosa* strains showed variation in their population as reported by characterization of the phenotypic clones. Four main clade of *P. aeruginosa* population are identified by the phylogenic analysis of single nucleotide polymorphisms (SNPs) which showed most different clade colonized by clonal outliers linked to the PA7 strain (**Figure 1**) [5].

2. Virulence factors

Acute infection usually observed in hospitalized patients having ventilator breathing. It is one of the main causative agents of hospital-acquired pneumonia and causing morbidity and mortality in the infected patients. Chastre and Fagon has reported 70–80% rate of mortality due to infection of *P. aeruginosa* in ventilator-associated pneumonia (Research has focused on type III secretion system (TTSS) secreting four exotoxins (ExoS, ExoT, ExoU and ExoY) [6]. ExoU is an effective virulence causing effector in TTSS. It is associated with morbidity and mortality in ventilator-associated pneumonia. Hogardt et al. reported that 40% of the isolates from such cases harbor the *exoU* gene [7].

3. Pathogenicity

The pathogenicity of *P. aeruginosa* makes its able to adhere and colonize in the presence of vast virulence factors and causing disease. Virulence factors used or synthesized by *P. aeruginosa* are enlisted in **Table 1**.

Virulence factor	**Action**
1. Colonization	
Flagella	Motility and invasion and adherence
Pili	Adhesion; transfer of secretions
Exopolysaccharides	Adherence and pathogen persistence
Lipopolysaccharide	Endotoxin; inflammatory agent; adherence and biofilm formation

Virulence factor	Action
2. Invasion	
Alkaline protease	Degrades immune system components
Elastase	Degrade elastin; disrupt membranes; impair
	Monocyte chemotaxis and degrade complement proteins
Lipase A and C	Involvement in degradation
Phospholipase C	Lung surfactant disruption
Protease	Degrades complement factors, plasmin, IgG, and fibrinogen
Pyocyanin	Inhibits lymphocyte proliferation; apoptosis of neutrophils
3. Pathogenesis	
Exotoxin A	Unknown role—possibly causes apoptosis of cells
Biofilm	Confers protection against biocides and immune system effectors as
	Impenetrable to antibodies (Ab), antibiotics, and biocides
Hydrogen cyanide	Unclear role, may be toxic agent.
Rhamnolipids	Dissolve phospholipids
Type III secretion	Exoenzyme (Exo) S, T and Y, and exotoxin U

Table 1.
Virulence factors produced and used by P. aeruginosa *[8–10].*

4. Biofilm

The pathogen colonized as planktonic form, and the cells convert to the sessile state to form biofilms. The hydrated structured matrices made up of exopolysaccharides and proteins, having 'slimy' characteristic can form on many surfaces from catheters to prokaryotic cells and eukaryotic. The main cause of persistent chronic infections is biofilm formation is essentially impenetrable inhabitants are protective for the bacterial strains from biocides [11]. The only one treatment to deal this situation is physical removal of the biofilm through surgery. Biofilms have heterogeneous populations of intra-species (phenotype and genotype, growth) and inter-species diversification. *P. aeruginosa* may be as dominant pathogen or with other pathogens such as Gram-negative *Burkholderia cenocepacia* and Gram-positive *S. aureus* [12]. The heterogeneous bacterial population of *P. aeruginosa* show distinct microenvironments for biofilms [13]. Metabolically active cells are at periphery and consume most of the oxygen, causing oxygen gradients in the biofilm [14]. The deeper layers of the biofilm have less metabolically active bacteria and are hypoxic. The actively growing peripheral bacterial cells of biofilms mostly susceptible to antibiotics or to the provided the drug which can penetrate slimy layer of the biofilm. Presence of a single polar flagellum made *P. aeruginosa* as motile. *P. aeruginosa is* exhibiting three distinctive types of motility and all of these types are required for development of biofilm which are;

- Swimming: It is aided by the flagellum
- Swarming: it requires both flagellum and type IV pili
- Twitching: it depends upon type IV pili

[15–20] (**Figure 2**).

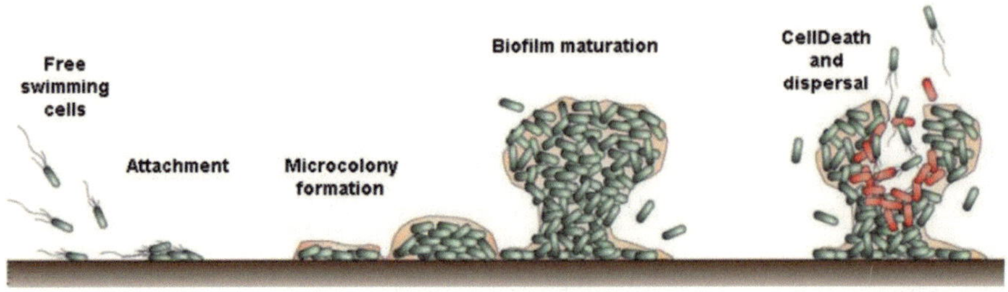

Figure 2.
Different stages of the biofilm development. Modified from [21].

5. Flagella

The bacterial flagellum is protrudes from the cell body in the form a long, thin filament that consists of the basal body, the hook and the filament. The basal body is rooted in the cytoplasmic membrane having three rings the outer membrane lipopolysaccharide (L) ring, the peptidoglycan (P) ring and the cytoplasmic membrane supra-membrane (MS) ring. The hook is exposed to the surface and is a flexible universal joint between the filament and basal body. The filament is made of polymerized flagellin monomers (up to 20,000 subunits) capped by the flagellar cap, FliD, which acts as mucin adhesion [22, 23].

The initial attachment of the bacteria needs flagella and have involvement in maturation of biofilm. Klausen et al. reported that the initial microcolony formation is occurred by clonal growth and flagella are not involved in biofilm development in *P. aeruginosa* during attachment [24].

6. Pili and type I fimbriae

The type IV pili are best characterized, which are composed of the Pil A subunit in a form of a helical polymer. Hahn and Solow reported that these IV pili are localized to the poles of the bacterial cells and facilitate the adhesive properties of *P. aeruginosa* [25]. Type IV pili appear to be required for biofilm formation and host colonization. Cell aggregation and formation of microcolonies are promoted by Type IV pili [26, 27]. *P. aeruginosa* having three sets of type I fimbriae (*CupA, CupB* and *CupC* fimbriae) which assembled by the chaperone usher pathway. CupA fimbriae demonstrated as important for adherence to abiotic surfaces causing biofilm formation and auto-aggregation of small colony variants (SCV) in *P. aeruginosa* [28].

7. Exopolysaccharides

Major components of the biofilm matrix are the exopolysachcarides produced by the *P. aeruginosa*. These exopolysaccharides include Alginate, and *Pel* polysaccharide and *Psl* polysaccharide. The *Pel* and *Psl* are associated with the non-mucoid strain [29] and Alginate is associated with the mucoid strains [30].

8. Alginate

P. aeruginosa produces alginate (an exopolysaccharide). This is a capsular polysaccharide and is overproduced in mucoid strains of *P. aeruginosa*. It is a high molecular weight polymer composed of monomers of D-mannuronic acids and β-1, 4 linked L-guluronic which are not repeating. Mutations in the negative regulator *MucA* is mainly caused to isolate the alginate-producing variants from chronically infected CF lungs [31]. Bacteria prevent itself from phagocytosis by this polymer acts as a physical barrier and an adherence factor, it gets oxygen free radicals resulting in enhancement of resistance of the biofilm against the host immune defense and antimicrobial agents. The mucoid strains to remain persistent and establish chronic infections in the CF lung by the influence of alginate [32, 33]. Wozniak et al. has demonstrated that in non-mucoid *P. aeruginosa* strains (PAO1 and PA14) alginate is not the main component of the biofilm matrix [34].

9. *Pel* and *Psl* polysaccharides

The *pel* and *psl* operon are encoded polysaccharide associated in biofilm formation in PAO1, ZK2870 and PA14, the non-mucoid *P. aeruginosa* strains. The main components of the extracellular polysaccharide matrix are constituted by these polysaccharides [35].

The *pel* is an operon having 7 genes (PA3058 to PA3064), which encoding *Pel* polysaccharide biosynthetic proteins. The structure of *Pel* is unknown and it is a glucose-rich matrix polysaccharide and found to be involved in maintenance of biofilms and pellicle formation in *P. aeruginosa* PA14 strain. Sozzi and Smiley reported that biofilm formation is inversely regulated by cytoplasmic protein SadB1 result in altering the expression of *Pel* polysaccharide [36].

The *psl*, is an operon consist of 15 genes (PA2231 to PA2245), which encoding the *Psl* biosynthetic machinery. It is composed of a galactose-rich and mannose-rich polysaccharide. The exact structure of *Psl* has not been clarified yet. During attachment, *Psl* is holds and anchored bacteria during biofilm formation on the surface. *Psl* was associated in differentiation and maturation of *P. aeruginosa* biofilms in non-mucoid strains [37]. In *P. aeruginosa psl* and *pel* operons expresions are controlled by intracellular level of signaling molecule c-di-GMP (bis-(3′,5′)-cyclic-dimeric-guanosine monophosphate), the GacS/GacA/RsmZ and the Wsp chemosensory system [38–40].

10. Rhamnolipids

P. aeruginosa produce rhamnolipids, the biosurfactants. Enzymes of the *rhlABC* operon synthesized the rhamnolipids. Rhamnolipids are required for biofilm formation by promoting the formation of microcolony at the initial phase of the biofilm formation. These are associated with to maintain channels and void spaces in mature biofilms and also involved in biofilm dispersions [41–43].

11. Infections caused by *P. aeruginosa*

The ability of P. aeruginosa, to survive in indifferent environments including aquatic or marshes or even in low O_2 or in very high temperatures (42°C) [44] resulting to withstand and survive on dry surfaces for more than 16 months by this pathogen [45]. It can colonize on dialysis machines, 'in-dwelling' appliances, sinks, floors, and toilet surfaces [46]. Immuno compromised host can be infected by *P. aeruginosa* by causing various clinical conditions, such as pneumonia, cystic fibrosis (CF), urinary tract infections, complications in clinical burns, and wounds [47, 48].

Sr. no.	Disease caused by *P. aeruginosa*
1	Respiratory tract infections (RTI)
2	Bacteremia; septicemia
3	Otitis externa
4	Skin infection; ecthyma gangrenosum, pyoderma, folliculitis, acne vulgaris
5	Eyes infections
6	Rare conditions like meningitis, perirectal infections and specific forms of osteomyelitis

Table 2.
Infectious diseases caused by P. aeruginosa *[55].*

P. aeruginosa is a common isolate from the patients who are hospitalized for more than a week. It is associated with high rate of mortality within 24 hours, infection can results in pneumonia, septicemia and urinary tract infection.

In sever chronic infection especially in patients with cystic fibrosis (CF), *P. aeruginosa* is involved. The main cause of mortality is *P. aeruginosa* lung infection in CF patients [49]. Murray et al. reported transmissible epidemic strains of *P. aeruginosa* emerged within the CF community. In earlier reports, CF patients considered to having their own strain of *P. aeruginosa* from their environment not from other infected individuals [50]. It is known as the Liverpool epidemic strain (LES) in recent research of UK due to the most common isolate recovered from CF patients [51, 52]. *P. aeruginosa* has a major focus in research as it is reporting to transmitted from a CF patient to non-CF parents, and causing significant morbidity in infected patients. *P. aeruginosa* is considered an opportunistic pathogen and this is an unusual characteristic to infect healthy individuals. Manchester and Midlands 1, Clone C are considered as predominant epidemic strains of *P. aeruginosa* [53].

P. aeruginosa causes two types of respiratory infections. Acute (if patient have extended periods of ventilation) and chronic (if patient suffer from cystic fibrosis). Patients with these two types of infections in hospitalized settings are likely to be infected by this pathogen. Acute murine respiratory models used to identify a number of virulence factors in mutants of *P. aeruginosa*. The detailed studied is the TTSS proteins including *ExoS*, *ExoT*, *ExoU* and *ExoY*. Main contributor towards morbidity and mortality are *ExoS*, *ExoT* and *ExoU* as observed in a murein acute respiratory model (**Table 2**) [54].

The presence or absence of components of the TTSS can be correlated in human clinical results [55]. *P. aeruginosa* excreted a blue pigment Pyocyanin having antibacterial properties against other bacterial strains. The pyocyanin production also cause significant damage to lungs in murine acute respiratory infection, which demonstrated by an intranasal infection of adult CD-1 mice [49]. Quorum-sensing systems that are *LasI, LasR* [56], and *RhlI* [57] in *P. aeruginosa* contributing to acute infections. Intranasal infection in adult female Balb/c mice, and they also analyzed bacterial loads in lungs, liver and spleen after 16–18 hours of infection.

12. Cystic fibrosis

Cystic fibrosis trans-membrane conductance regulator (CFTR) mutation caused reduced chloride ion transport result in Cystic fibrosis (CF). It is a recessive genetic disorder. CF affected the development and functioning of various organs including immune system, pancreas and intestine, resulting a low life expectancy. The tissue damage is promoted in acute or chronic infections by constant stimulation of immune system effectors. In young CF patient *P. aeruginosa* is an important pathogen which lasts later stages. CF lung is an enriched with the oxygen gradients and nutrient. The oxygen gradients contribute the uniqueness in development of mucus layer and excessive consumption of the epithelial cells in CF lungs. *P. aeruginosa is* adaptable to many phenotypes in these types of conditions. A single isolate genome showed 68 mutations over the period of 90 months. The pathology in the lung in acute infections is due to the presence of elastase, flagella, LPS with O-side chains proteases, and pyocyanin.

P. aeruginosa contributed biofilms formation, produce rough LPS (no O-side chains), lack flagella, and overproduction of alginate during chronic infections. In chronic infections typically mucoid phenotype with lesser production of pyocyanin, pyoverdine, and elastase was observed. *The antibiotic pressure* causes *P. aeruginosa* to mutate from non-mucoid form to mucoid form. Small colony variants after a continued antibiotic exposure resulted in production of mannose-rich (*psl*) or glucose- (*pel*) polysaccharides. These are hyperpiliated, which are characterized as persistent to this specific phenotype. Liverpool epidemic strain (LES) reported as *P. aeruginosa* strains by over-production of pyocyanin.

The isolates of *P. aeruginosa* form CF changed their genome to get rid form acute virulence factors [56]. The loss of *LasR* function is one of the main genetic changes observed in *P. aeruginosa* isolates from CF patients. Many other acute infection models shown that because of deficiency in quorum-sensing, *lasR* mutants are less virulent than wild-type. Bacteria used this selective pressure of genetic change in genome to escape from host immune system in acute virulence in CF airways. An effective chronic CF isolate would be one, that isolate must lose its ability to be an effective acute infection isolate. *lasR* mutants have been examined for its advantages. The *LasR* mutants can grow on selected carbon and nitrogen sources as compared to wild-type.

13. Innate immune system

Immune system is acting as natural defense mechanism to prevent the invasion of pathogens. PRRs (Porcine reproductive and respiratory syndrome) stimulus received by nonspecific innate system and respond the innate effectors responses;

- Phagocytosis by macrophages
- cell death or
- by the complement system with the membrane attack complex produced by the complement proteins or by natural killer (NK)

At the infection site the inflammatory response of effectors was induced by chemokines. Permanent tissue is damaged if continued stimulation of effectors by PAMPs (Pathogen associated molecular pattern) and virulence factors was induced [57].

14. Antibiotic resistance

P. aeruginosa is a free-living and aerobic bacillus that is isolated from soil and water in most of cases. Intrinsic resistance in *P. aeruginosa* causing high mortality due to a broad spectrum resistance of antibiotics and is able to quickly to acquire resistance genes by horizontal gene transfer. Fluoroquinolones, gentamicin and imipenem are restricted antibiotics as effective against *P. aeruginosa* and susceptibility to these antibiotics can vary between different strains. Bacterial infections are cured traditionally with the use of antibiotics and immune system is unable to have or eradicate this use of antibiotics.

Fluoroquinolones (ciprofloxacin) prevented DNA repair and replication [58]. Aminoglycosides, Beta-lactams (imipenam but not penicillins), 3rd and 4th generation cephalosporins, and fluoroquinolones are anti-pseudomonal drugs [59].

Colistin, is a drug having lesser side-effect profile, and mainly used against multi drug resistant strains (MDR) *P. aeruginosa* strains these days. The pattern or the use of antibiotic treatment now bettered towards the treatment of specific diseases including CF. The transmission of *P. aeruginosa* reduced by separation of infected and susceptible one and use of strict hygiene procedures [60].

The unavailability of the effective therapeutic option, the treatment of infections with pseudomonas is becoming difficult to deal a very few anti-pseudomonal drugs are being considered good for the treatment of emerging resistance strain, these include aminoglycosides beta-lactams, and fluoroquinolones [61–63].

The bactericidal MoA (mechanism of action) is significant for the survival by selection pressure for the fittest one. Antibiotic resistant bacteria are selected and propagated very well in the absence of the environmental resources competitions. Specific antibiotic resistance can be void of by the use of an alternative group of antibiotic. Bacteria have established active defense mechanisms which lead to MDR species such as methicillin-resistant *Staphylococcus aureus (S. aureus), Escherichia coli (E. coli), Acinetobacter baumannii (A. baumannii),* or *P. aeruginosa* are promoted as MDR strains and are difficult to eradicate this opportunistic pathogens [64].

15. *P. aeruginosa* and mechanisms of resistance

The mechanisms of resistance in *P. aeruginosa* against antibiotics can be intrinsic, adaptive, or acquired. Innately *P. aeruginosa* is resistant to many antibiotics. Intrinsically it has impermeable cell wall, outer membrane protein (Opr) channels, and multi-drug efflux pumps which give the bacteria resistance to certain antibiotics. Extended treatment and continuous use of higher therapeutic doses resulting in complete resistance [65].

16. Genomics of *P. aeruginosa*

The genome of a bacterium has two components first is the core genome and second the mobile genome. In a same species the core genome is common for all bacteria which include genes for the bacteria essential for development and the mobile genome can propagate within the whole genome [66, 67]. The mobile genome is varies between strains within a species. The mobile genome consists of a range of genetic elements such as insertion sequences, transposons, prophages, plasmids, and genomic islands. Horizontal gene transfer (HGT) processes such as conjugation, transformation, and transduction acquired the mobile genome. Genomic islands are the Clusters of genes in the mobile genome. Genomic islands encode gene products which enhance the fitness of a bacterium by survivability in new environment, increasing host range, and utilization of new nutrients. Genomic Island can be defined by various features [68].

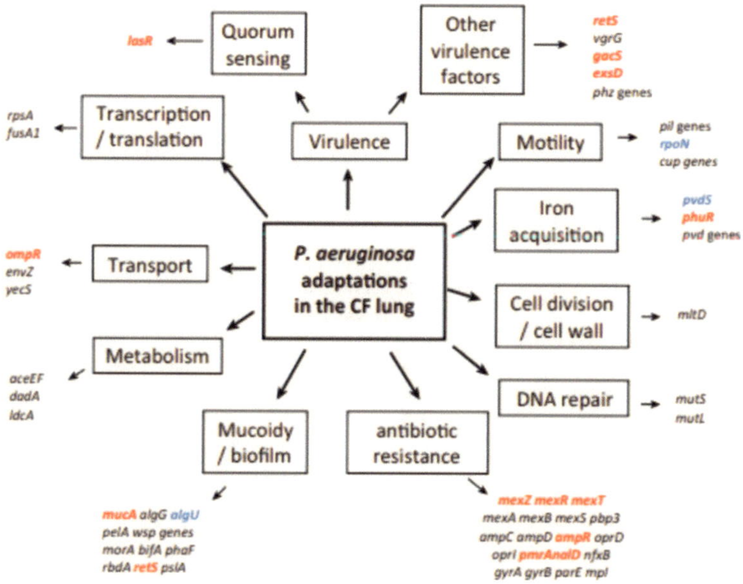

Figure 3.
The novel hypervirulent ExlA-ExlB system [69].

In *P. aeruginosa* classical clades, the most of homologous *exlB-exlA* loci are existing in *P. fluorescens, P. putida*, and *P. entomophila*, signifying this locus acquired by horizontal transfer in *Pseudomonas* spp. Strain toxicity data related to neutrophils and macrophages and ability of inflammatory cells to phagocytic *exlA*+ strains are not accessible for recognition the behavior of these strains *in vivo* in detail. Different mechanism for expression are used by bacteria for virulence effect to infect mammals and plants and that *ExlA* is not required for bacterial plants toxicity. It shows unequal level of virulence that could not be aid in one toxin, *Exl A* [70]. All strains have *lasB* gene which have *lasB* PAO1 sequence at same location and with the sequence identity up to 91–98%. Quorum sensing is regulatory for the expression of *LasB* and lacking of Quorum sensing genes affected the*lasB* expression. Quorum sensing genes are *lasR, lasI, vqsM, vqsR, rhlR, and rhlI* all existed at same locations in genomes of all the strains of *P. aeruginosa*. The sequences displayed identity of 98%, except for the strain PA39 (92%). Internally a frameshift mutation of *mvfR* gene, coding for an important regulator of the Quorum sensing, observed in PA7 genome [71]. Same mutation is also present in *mvfR* genes and absent in most of strains hence lacking *LasB* activity.It is reported that in the absence of *LasB* activity *lasB* sequences and Quorum sensing cannot explained.

PAO1 **Figure 3** (2001), PA14 (2005), PAS7 (2007) and LES (2008) are four complete sequenced genome available and *P. aeruginosa* is recent strain sequenced. Falagas et al. reported PAO1 had 44.2% of predicted ORFs in the class 4 with unknown function [72]. The labeling of pathogenicity islands in the genomic island must have virulence genes with known function other was this would be difficult to locate it (**Figure 4**).

P. aeruginosa PAO1 complete genome was available in 2001 and for remaining three strain genome sequences available in 2005–2008. The recent research data are based primarily on subtractive hybridization and microarray, for comparing the strains to reference strain that is PAO1 [73].

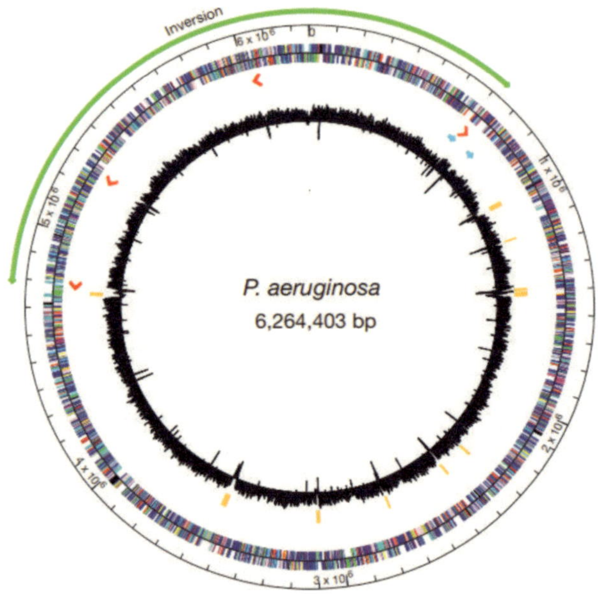

Figure 4.
Circular representation of the P. aeruginosa *genome.*

The differences in the genome targeted for further screening by using mutagenesis and virulence assays in models of *in vivo* infections. The current in vivo models commonly used for *P. aeruginosa* infection are:

- The nematode worm *Caenorhabditis elegans*
- The wax moth *Galleria mellonella* the plant *Arabidopsis thaliana*,
- The fruit fly *Drosophila melanogaster.*

Murine used for infection models that imitate the human mood of infection are not common [74].

The results of current studies of pathogenicity island mutant have explained the effectiveness in finding their influence towards virulence. PAI I_{536}–PAI V_{536} are five pathogenic island of *E. coli* and analyzed for their involvement in infection [75].

17. Quorum-sensing

Quorum-sensing systems controlled many virulent factors in *P. aeruginosa* such as the production of biofilm and the secretion of toxins [76]. Previously studies showed that insufficiency of the quorum-sensing systems can decrease the virulence in acute infection murine models. The use of insufficient quorum-sensing systems in a mutants murine burns injury model cause a decrease in mortality of the murine model. The ability of the bacteria to spread from the site of infection is also reduced. Lodise Jr. et al. reported the same results for a *LasR* mutant of PAO1 [77].

Quorum-sensing is a control system for coordination of gene expression in bacteria. As the level of an auto inducer goes to a threshold level, they caused the binding to specific bacterial receptors, in result gene transcription is initiated. In the result majority of the bacteria in a population expressing the same phenotype. *P. aeruginosa* is having two quorum-sensing systems, the *Las* system and the *Rhl* system.

18. Recent antibacterial agents against respiratory infections

18.1 Lactic acid bacteria

Food borne pathogenic and spoilage microorganisms are affected by the lactic acid bacteria (LAB) [78, 79], for example the growth of *B. subtilis* inhibited as it spoils bread [80]. Studies showed that *Lactobacillus* strains reported an inhibitory activity against *E. coli* [81]. Proteolytic activities and lipolytic activities of psychrophilic *Pseudomonas* causing food spoilage [82]. *Lactobacillus* species produced hydrogen peroxide which inhibits the growth of *Pseudomonas* species [42].

Lactic acid functions as the natural compound having antimicrobial activity and generally recognized as safe. Lactic acid has ability to inhibit the growth of Gram-negative species of *Pseudomonadaceae* and *Enterobacteriaceae* [83]. Lactic acid is used as a bio-preservative in fermented products.

Ribosome synthesized the bacteriocin extracellularly and secreted peptide complexes or bioactive peptides havingbacteriostatic orbactericidal effect [84]. Smart et al. reported *Lactococcus lactis produced a bacteriocin called* Nisin, which is studied in detailed, and applied as stabilizer to certain foods worldwide [85]. Bacteriocins are harmless due to quick proteolytic degradation by the gastrointestinal tract enzymes [86, 87].

Four major classes of bacteriocins are

i. Lantibiotics: which are smaller and are heat stable peptides acting on membrane structures of the pathogens

ii. Non-lantibiotics: which are small are also heat stable peptides,

iii. Larger heat-labile proteins

iv. Complex bacteriocins [52, 88].

Most of bacteriocin are related to classes I and II. Proteinaceous compounds which are synthesized by ribosomes have bactericidal effect towards Gram-positive bacteria as compared to Gram-negative which have an additional layer composed of proteins, lipopolysaccharides and phospholipids [89–91].

Bacteriocins considered as potentially food-grade to increase food safety these can decrease the occurrence of foodborne diseases. These helped to lessen the use of chemical based preservatives and intensity of heat treatments for food preservations, resulting more naturally preserved food that is richer in nutritional and organoleptic properties [92].

Schillinger and Lucke has observed that the fact that most of bacteriocins have a narrower host range, which made them effective against closely related bacteria having same nutritive demands [93]. Lactic acid bacteria produced lactic acid functioning as a natural antimicrobial agent, having a generally known as safe to use. The growth of many Gram-negative species *Pseudomonadaceae* and *Enterobacteriaceae was* inhibited by the lactic acid bacteria. Forbio-preservation of the naturally fermented products lactic acid is used instead of other organic salts. Lactic acid is penetrates to cytoplasmic membrane of the organisms, which result in disruption of trans-membrane proton motive force and decrease in intracellular pH [94].

19. Antimicrobial property by hydrogen peroxide production

Bacterial enzymatic activity is destroyed by hydrogen peroxide which is a thermodynamically unstable and produced by Lactobacillus [95]. Other lactic-acid-producing bacteria and lactobacillus both are lacking heme and thus not utilizing the cytochrome systems for terminal oxidation. Flavoproteins are used by Lactobacilli, which convert oxygen to hydrogen peroxide and this mechanism, results in the formation of hydrogen peroxide in amounts which are degraded by the organism [96].

20. Antibacterial activity of nanoparticles (NPs)

The nanotechnology applications used in the food industry for food safety, disease treatment, for molecular and cellular biology as new tools, and for pathogen detection and protection [94]. NPs reported as applied in the nano tracer and nano-sensor fields in food industries [97, 98]. Nanotechnology used in food packaging to prevent contamination and to improve the shelf life of food [99, 100].

There are many types of NPs, and a variety of others are expected to introduce by the future researchers. Antibacterial agents are important and used in many industries, mainly in food industry. Cintas currently used antibacterial agents in the food industry are classified into two groups: inorganic agents and organic [101]. Inorganic antibacterial agents including NPs are used in food industry as they are stable under high pressures and temperatures conditions are required in food-processing, and regarded as safe to use for human and animals, as compared to organic materials [102, 103]. Studies showed that few NPs have selective toxicity to bacteria having lesser effects on human cells [104]. Foodborne outbreaks all over the world are increasing day by day and is important to control the causes. NPs are useful antibacterial agents that applied in the food industry. Silver (Ag) NPs are used in the medical and pharmaceutical industries. Ag NPs are very significant for the potential use in wide range of biological applications, as an antibacterial and antifungal agent for antibiotic resistant organisms to prevent infections. The concentration of NPs is linked with antibacterial activity. However, studies disagree with one another, indicating the mechanisms of NPs which causing antibacterial activity and toxicity to bacterial cells are very complex one [105–107]. Thus, it is challenging to classify the NPs as or adverse NPs or beneficial NPs towards bacteria. The tolerance property of bacteria having lesser growth rate is associated with the expression of genes related to stress-response [108].

After exposure to zinc oxide (ZnO) NP minimal inhibition concentration *Staphylococcus aureus (S. aureus) and Salmonella typhimurium* (S. *typhimurium*) were reduced to 0 within 4 and 8 h, respectively. Scanning electron micrographs of the targeted cell showed the completely lysis of the cells. *Pseudomonas* spp. were the most resistant and *Bacillus cereus* was the most sensitive among all of the studied strains against ZnO NPs.

Higher concentrations of Ag NPs showed the stronger antimicrobial activity. Ag NPs are used as antibacterial agents against *Escherichia coli (E. coli)* (Gram negative bacteria) [109].

The studied showed same results of Ag NPs between *S. aureus*, *E. coli* [110]. It was reported that smaller Ag NPs had effective antibacterial activity but having higher cytotoxicity. The antibacterial activity of Ag NPs is not only against the *S. aureus* and *E. coli* and, but also, *P. aeruginosa* [111]. ZnO NP inform of powders are widely used in coating electronic devices, cosmetics, catalysts and pigments. Instead of the extensive use of ZnO NPs, the safety of ZnO NPs for humans is not clear yet. In many studies the toxicity of metal oxide NPs and ZnO NPs towards mammalian cell and organs reported [112, 113]. Concentrations of ZnO less than 100 g/mL caused a substantial decrease of mitochondrial function decreased up to substantial level by concentration of ZnO (less than 100 g/mL). Weiss and Takhistov reported that Ag NPs decreased the cell viability of epithelial cells in lung [114]. For extracellular biosynthesis of gold nanoparticles *P. aeruginosa* used.

Moraru (2003) observed the AuNPs were prepared by reduction of gold ion in bacterial cell supernatant solutions [115]. Silver nanoparticles showed excellent antibacterial effect against pathogenic bacteria, *Klebsiella pneumoniae and S. aureus* [116, 117].

21. Conclusion

From the total *P. aeruginosa* isolates 66 to >90% were from the Lahore region and showed *in vitro* resistance to many of the commercially available antibiotics tested. Meropenem, Piperacillin, and Amoxicillin were the drugs for which there was the greatest susceptibility and represent recommended treatments for infections due to *P. aeruginosa* in our region. A significant killing of these *resistant P. aeruginosa* strains by factors present in supernates of *Lactobacilli* spp. was observed, suggesting that the use of *Lactobacilli* spp. as probiotics may be of value for the treatment or prevention of *P. aeruginosa* colonization. We also found strong *in vitro* anti-bacterial efficacy of Ag, Zn and Fe3 oxide NPS against the local *P. aeruginosa* isolates, suggestive of additional research into their practical application in a healthcare department [117]. The differences in pathogenicity due to between the *P. aeruginosa* isolates, which could be due to genes involved in the quorum sensing and biofilm formation which having the ability to develop infections. The research could be important in future studies as the already reported isolates used are have same environmental conditions which having the multidrug resistant *P. aeruginosa* strains. The genomic variations between the isolates of *P. aeruginosa* are also observed for detection of virulence genes in strains of *P. aeruginosa* could highlight the link between acute and chronic respiratory infections. The collected and provided data could conclude that the virulence genes are important in severity of acute and chronic respiratory infections in human beings.

Author details

Nazish Mazhar Ali*, Safia Rehman, Syed Abdullah Mazhar, Iram Liaqat and Bushra Mazhar
Microbiology Laboratory, Department of Zoology, GCU, Lahore, Pakistan

*Address all correspondence to: nazipak@hotmail.com

References

[1] Dasenbrook EC, Checkley W, Merlo CA, Konstan MW, Lechtzin N, Boyle MP. Association between respiratory tract methicillin-resistant *Staphylococcus aureus* and survival in cystic fibrosis. JAMA. 2010;**303**(23):2386-2392

[2] Wong K, Roberts MC, Owens L, Fife M, Smith AL. Selective media for the quantitation of bacteria in cystic fibrosis sputum. J of medical microbiology. 1984;**17**(2):113-119

[3] Gellatly SL, Hancock RE. P. aeruginosa: New insights into pathogenesis and host defenses. Pathogens and disease. 2013;**67**(3):159-173

[4] Thrane S, Pedersen AM, Thomsen MBH, Kirkegaard T, Rasmussen BB, Duun-Henriksen AK, et al. A kinase inhibitor screen identifies Mcl-1 and Aurora kinase a as novel treatment targets in antiestrogen-resistant breast cancer cells. Oncogene. 2015;**34**(32):4199

[5] Wolf P, Elsässer-Beile U. *Pseudomonas* exotoxin a: From virulence factor to anti-cancer agent. International J of Medical Microbiology. 2009;**299**(3):161-176

[6] Chastre J, Fagon JY. Ventilator-associated pneumonia. American J of respiratory and critical care medicine. 2002;**165**(7):867-903

[7] Hogardt M, Hoboth C, Schmoldt S, Henke C, Bader L, Heesemann J. Stage-specific adaptation of hypermutable *P. aeruginosa* isolates during chronic pulmonary infection in patients with cystic fibrosis. The J of infectious diseases. 2007;**195**(1):70-80

[8] Jiang H, Jiang D, Shao J, Sun X. Magnetic molecularly imprinted polymer nanoparticles based electrochemical sensor for the measurement of gram-negative bacterial quorum signaling molecules (N-acyl-homoserine-lactones). Biosensors and Bioelectronics. 2016;**75**:411-419

[9] Jimenez PN, Koch G, Thompson JA, Xavier KB, Cool RH, Quax WJ. The multiple signaling systems regulating virulence in *P. aeruginosa*. Microbiology and Molecular Biology Reviews. 2012;**76**(1):46-65

[10] Lyczak JB, Cannon CL, Pier GB. Establishment of *P. aeruginosa* infection: Lessons from a versatile opportunist1. Microbes and Infection. 2000;**2**(9):1051-1060

[11] Estela CRL, Alejandro PR. Biofilms: A survival and resistance mechanism of microorganisms. In: Antibiotic Resistant Bacteria-A Continuous Challenge in the New Millennium. Rijeka: InTech; 2012

[12] Jageethadevi A, Saranraj P, Ramya N. Inhibitory effect of chemical preservatives and organic acids on the growth and organic acids on the growth of bacterial pathogens in poultry chicken. Asian J of Biochemical and Pharmaceutical Research. 2012;**1**(2):1-9

[13] Ren S, Cai M, Shi Y, Xu W, Zhang XD. Influence of bronchial diameter change on the airflow dynamics based on a pressure-controlled ventilation system. International J for numerical methods in biomedical engineering. 2018;**34**(3):e2929

[14] Kaye KS, Pogue JM. Infections caused by resistant gram-negative Bacteria: Epidemiology and management. Pharmacotherapy: The J of Human Pharmacology and Drug Therapy. 2015;**35**(10):949-962

[15] Bradley JV. Nonrobustness in classical tests on means and variances:

A large-scale sampling study. Bulletin of the Psychonomic Society. 1980;**15**(4):275-278

[16] Whitchurch CB, Hobbs M, Livingsston SP, Krishnapillai V, Mattick JS. Characterisation of a *P. aeruginosa* twitching motility gene and evidence for a specialised protein export system widespread in eubacteria. Gene. 1991;**101**(1):33-44

[17] Darzins A. Characterization of a *P. aeruginosa* gene cluster involved in pilus biosynthesis and twitching motility: Sequence similarity to the chemotaxis proteins of enterics and the gliding bacterium *Myxococcus xanthus*. Molecular Microbiology. 1994;**11**(1):137-153

[18] Feldman GJ, Cousins RD. Unified approach to the classical statistical analysis of small signals. Physical review D. 1998;**57**(7):3873

[19] Kohler KJ. Investigating unscripted speech: Implications for phonetics and phonology. Phonetica. 2000;**57**(2-4):85-94

[20] Rashid MH, Kornberg A. Inorganic polyphosphate is needed for swimming, swarming, and twitching motilities of *P. aeruginosa*. Proceedings of the National Academy of Sciences. 2000;**97**(9):4885-4890

[21] McDougald LR. Histomoniasis (Blackhead) and other protozoan diseases of the intestinal tract. In: Saif YM, Fadly AM, Glisson JR, LR MD, Nolan LK, et al., editors. Diseases of Poultry. Ames: Blackwell Academic Publishing Professional; 2008, 2008. pp. 1095-1105

[22] Arora S, Lund C, Motwani R, Sudan M, Szegedy M. Proof verification and the hardness of approximation problems. J of the ACM (JACM). 1998;**45**(3):501-555

[23] Chevance FF, Hughes KT. Coordinating assembly of a bacterial macromolecular machine. Nature reviews microbiology. 2008;**6**(6):455

[24] Klausen M, Aaes-Jørgensen A, Molin S, Tolker-Nielsen T. Involvement of bacterial migration in the development of complex multicellular structures in *P. aeruginosa* biofilms. Molecular Microbiology. 2003;**50**(1):61-68

[25] Hahn F, Solow RM. A Critical Essay on Modern Macroeconomic Theory. USA: MIT Press, AWS; 1997

[26] Khalil R, Gomaa M. Evaluation of the microbiological quality of conventional and organic leafy greens at the time of purchase from retail markets in Alexandria, Egypt. Polish Journal of Microbiology. 2014;**63**:237-243

[27] Vallet-Regi M, Ramila A, Del Real RP, Pérez-Pariente J. A new property of MCM-41: Drug delivery system. Chemistry of Materials. 2001;**13**(2):308-311

[28] Lewis K. Riddle of biofilm resistance. Antimicrobial Agents and Chemotherapy. 2001;**45**(4):999-1007

[29] Friedman L, Kolter R. Two genetic loci produce distinct carbohydrate-rich structural components of the *P. aeruginosa* biofilm matrix. J of bacteriology. 2004;**186**(14):4457-4465

[30] Govan JR, Deretic V. Microbial pathogenesis in cystic fibrosis: Mucoid P. *aeruginosa* and *Burkholderia cepacia*. Microbiological Reviews. 1996;**60**(3):539-574

[31] Ramsey BW, Davies J, McElvaney NG, Tullis E, Bell SC, Dřevínek P, et al. A CFTR potentiator in patients with cystic fibrosis and the G551D mutation. New England J of Medicine. 2011;**365**(18):1663-1672

[32] Krieg M, Aramendia PF, Braslavsky SE, Schaffner K. 124-kDa Phytochrome in model membrane systems: Studies of the ii700 intermediates with the protein covalently bound to preformed liposomes. Photochemistry and Photobiology. 1988;**47**(2):305-310

[33] Worlitzsch D, Tarran R, Ulrich M, Schwab U, Cekici A, Meyer KC, et al. Effects of reduced mucus oxygen concentration in airway *Pseudomonas infections* of cystic fibrosis patients. The J of clinical investigation. 2002;**109**(3):317-325

[34] Wozniak MA, Desai R, Solski PA, Der CJ, Keely PJ. ROCK-generated contractility regulates breast epithelial cell differentiation in response to the physical properties of a three-dimensional collagen matrix. The Journal of Cell Biology. 2003;**163**(3):583-595

[35] Matsuo Y, Ishizuka M. Keyword extraction from a single document using word co-occurrence statistical information. International J on Artificial Intelligence Tools. 2004;**13**(01):157-169

[36] Sozzi T, Smiley MB. Antibiotic resistances of yogurt starter cultures *Streptococcus thermophilus* and *Lactobacillus bulgaricus*. Applied Environmental Microbiology. 1980;**40**:862-865

[37] Caiazza NC, Merritt JH, Brothers KM, O'Toole GA. Inverse regulation of biofilm formation and swarming motility by P.aeruginosa PA14. J of bacteriology. 2007;**189**(9):3603-3612

[38] Sadikot RT, Blackwell TS, Christman JW, Prince AS. Pathogen–host interactions in *P. aeruginosa* pneumonia. American J of respiratory and critical care medicine. 2005;**171**(11):1209-1223

[39] Rickard AH, Gilbert P, High NJ, Kolenbrander PE, Handley PS. Bacterial coaggregation: An integral process in the development of multi-species biofilms. Trends in Microbiology. 2003;**11**(2):94-100

[40] Hickman JW, Tifrea DF, Harwood CS. A chemosensory system that regulates biofilm formation through modulation of cyclic diguanylate levels. Proceedings of the National Academy of Sciences of the United States of America. 2005;**102**(40):14422-14427

[41] Ventre I, Goodman AL, Vallet-Gely I, Vasseur P, Soscia C, Molin S, et al. Multiple sensors control reciprocal expression of *P. aeruginosa* regulatory RNA and virulence genes. Proceedings of the National Academy of Sciences of the United States of America. 2006;**103**(1):171-176

[42] Lee H, Dellatore SM, Miller WM, Messersmith PB. Mussel-inspired surface chemistry for multifunctional coatings. Science. 2007;**318**(5849):426-430

[43] Davey Smith G, Ebrahim S. 'Mendelian randomization': Can genetic epidemiology contribute to understanding environmental determinants of disease? International J of epidemiology. 2003;**32**(1):1-22

[44] Boles BR, Thoendel M, Singh PK. Rhamnolipids mediate detachment of *P. aeruginosa* from biofilms. Molecular Microbiology. 2005;**57**(5):1210-1223

[45] Pamp SJ, Tolker-Nielsen T. Multiple roles of biosurfactants in structural biofilm development by *P. aeruginosa*. J of bacteriology. 2007;**189**(6):2531-2539

[46] Bjarnsholt T, Givskov M. Quorum-sensing blockade as a strategy for enhancing host defences against bacterial pathogens. Philosophical Transactions of the

Royal Society B: Biological Sciences. 2007;**362**(1483):1213-1222

[47] Kramer A, Schwebke I, Kampf G. How long do nosocomial pathogens persist on inanimate surfaces? A systematic review. BMC infectious diseases. 2006;**6**(1):130

[48] De Kievit TR, Gillis R, Marx S, Brown C, Iglewski BH. Quorum-sensing genes in *P. aeruginosa* biofilms: Their role and expression patterns. Applied and Environmental Microbiology. 2001;**67**(4):1865-1873

[49] Mesaros N, Nordmann P, Plésiat P, Roussel-Delvallez M, Van Eldere J, Glupczynski Y, et al. *P. aeruginosa*: Resistance and therapeutic options at the turn of the new millennium. Clinical Microbiology and Infection. 2007;**13**(6):560-578

[50] Murray TS, Egan M, Kazmierczak BI. *P. aeruginosa* chronic colonization in cystic fibrosis patients. Current Opinion in Pediatrics. 2007;**19**(1):83-88

[51] Scott FW, Pitt TL. Identification and characterization of transmissible *P. aeruginosa* strains in cystic fibrosis patients in England and Wales. J of medical microbiology. 2004;**53**(7):609-615

[52] Salunkhe P, Smart CH, Morgan JAW, Panagea S, Walshaw MJ, Hart CA, et al. A cystic fibrosis epidemic strain of *P. aeruginosa* displays enhanced virulence and antimicrobial resistance. J of bacteriology. 2005;**187**(14):4908-4920

[53] Shorr AF, Tabak YP, Gupta V, Johannes RS, Liu LZ, Kollef MH. Morbidity and cost burden of methicillin-resistant *Staphylococcus aureus* in early onset ventilator-associated pneumonia. Critical care. 2006;**10**(3):R97

[54] McCallum AK. Mallet: A Machine Learning for Language Toolkit. USA: AWS; 2002

[55] Shaver CM, Hauser AR. Relative contributions of *P. aeruginosa* ExoU, ExoS, and ExoT to virulence in the lung. Infection and Immunity. 2004;**72**(12):6969-6977

[56] Hauser MD, Chomsky N, Fitch WT. The faculty of language: What is it, who has it, and how did it evolve? Science. 2002;**298**(5598):1569-1579

[57] Lau GK, Piratvisuth T, Luo KX, Marcellin P, Thongsawat S, Cooksley G, et al. Peginterferon Alfa-2a, lamivudine, and the combination for HBeAg-positive chronic hepatitis B. New England J of Medicine. 2005;**352**(26):2682-2695

[58] Pearson SF. Behavioral asymmetries in a moving hybrid zone. Behavioral Ecology. 2000;**11**(1):84-92

[59] Allewelt M, Coleman FT, Grout M, Priebe GP, Pier GB. Acquisition of expression of the *P. aeruginosa* ExoU cytotoxin leads to increased bacterial virulence in a murine model of acute pneumonia and systemic spread. Infection and Immunity. 2000;**68**(7):3998-4004

[60] Laskowski MA, Kazmierczak BI. Mutational analysis of RetS, an unusual sensor kinase-response regulator hybrid required for *P. aeruginosa* virulence. Infection and Immunity. 2006;**74**(8):4462-4473

[61] Siddique S, Shah ZH, Shahid S, Yasmin F. Prepara-tion, characterization and antibacterial activity of ZnO nanoparticles on broad spectrum of microorganisms. Acta Chimica Slovenica. 2013;**60**:660-665

[62] Smith SC, Allen J, Blair SN, Bonow RO, Brass LM, Fonarow GC, et al. AHA/ACC guidelines for

secondary prevention for patients with coronary and other atherosclerotic vascular disease: 2006 update: Endorsed by the National Heart, Lung, and Blood Institute. J of the American College of Cardiology. 2006;**47**(10):2130-2139

[63] D'argenio DA, Wu M, Hoffman LR, Kulasekara HD, Déziel E, Smith EE, et al. Growth phenotypes of *P. aeruginosa* lasR mutants adapted to the airways of cystic fibrosis patients. Molecular Microbiology. 2007;**64**(2):512-533

[64] Tang P, Hung MC, Klostergaard J. Length of the linking domain of human pro-tumor necrosis factor determines the cleavage processing. Biochemistry. 1996;**35**(25):8226-8233

[65] Rumbaugh KP, Griswold JA, Iglewski BH, Hamood AN. Contribution of quorum sensing to the virulence of *P. aeruginosa* in burn wound infections. Infection and Immunity. 1999;**67**(11):5854-5862

[66] Christensen JH, Christensen OB. A summary of the PRUDENCE model projections of changes in European climate by the end of this century. Climatic Change. 2007;**81**(1):7-30

[67] Nowroozi J, Mirzaii M, Norouzi M. Study of *Lactobacillus* as probiotic bacteria. Iranian J of Public Health. 2004;**33**:1-7

[68] Lau SK, Woo PC, Fung AM, Chan KM, Woo GK, Yuen KY. Anaerobic, non-sporulating, gram-positive bacilli bacteraemia characterized by 16S rRNA gene sequencing. J of medical microbiology. 2004;**53**(12):1247-1253

[69] Sonnleitner E, Haas D. Small RNAs as regulators of primary and secondary metabolism in *Pseudomonas* species. Applied Microbiology and Biotechnology. 2011;**91**(1):63-79

[70] Mikkelsen H, Sivaneson M, Filloux A. Key two-component regulatory systems that control biofilm formation in *P. aeruginosa*. Environmental Microbiology. 2011;**13**(7):1666-1681

[71] Breidenstein EB, de la Fuente-Núñez C, Hancock RE. *P. aeruginosa*: All roads lead to resistance. Trends in Microbiology. 2011;**19**(8):419-426

[72] Falagas ME, Kasiakou SK, Saravolatz LD. Colistin: The revival of polymyxins for the management of multidrug-resistant gram-negative bacterial infections. Clinical Infectious Diseases. 2005;**40**(9):1333-1341

[73] Arancibia F, Bauer TT, Ewig S, Mensa J, Gonzalez J, Niederman MS, et al. Community-acquired pneumonia due to gram-negative bacteria and *P. aeruginosa*: Incidence, risk, and prognosis. Archives of Internal Medicine. 2002;**162**(16):1849-1858

[74] Mah TFC, O'toole GA. Mechanisms of biofilm resistance to antimicrobial agents. Trends in Microbiology. 2001;**9**(1):34-39

[75] Moore NM, Flaws ML. Epidemiology and pathogenesis of *P. aeruginosa* infections. Clinical laboratory science. 2011;**24**(1):43

[76] Alekshun MN, Levy SB. Molecular mechanisms of antibacterial multidrug resistance. Cell. 2007;**128**(6):1037-1050

[77] Lodise TP Jr, Lomaestro B, Drusano GL. Piperacillin-tazobactam for *P. aeruginosa* infection: Clinical implications of an extended-infusion dosing strategy. Clinical Infectious Diseases. 2007;**44**(3):357-363

[78] Hart CA, Winstanley C. Persistent and aggressive bacteria in the lungs of cystic fibrosis children. British Medical Bulletin. 2002;**61**(1):81-96

[79] Ou HH, Lo SL. Review of titania nanotubes synthesized via the hydrothermal treatment: Fabrication, modification, and application. Separation and Purification Technology. 2007;**58**(1):179-191

[80] Winstanley C, Langille MG, Fothergill JL, Kukavica-Ibrulj I, Paradis-Bleau C, Sanschagrin F, et al. Newly introduced genomic prophage islands are critical determinants of in vivo competitiveness in the liverpool epidemic strain of *P. aeruginosa*. Genome Research. 2009;**19**(1):12-23

[81] Hilker M, Fatouros NE. Plant responses to insect egg deposition. Annual Review of Entomology. 2015;**60**:493-515

[82] Roy A, Kucukural A, Zhang Y. I-TASSER: A unified platform for automated protein structure and function prediction. Nature protocols. 2010;**5**(4):725

[83] Stover CK, Pham XQ, Erwin AL, Mizoguchi SD, Warrener P, Hickey MJ, et al. Complete genome sequence of *P. aeruginosa* PAO1, an opportunistic pathogen. Nature. 2000;**406**(6799):959

[84] Choi CW, Vanhatalo A, Ahvenjärvi S, Huhtanen P. Effects of several protein supplements on flow of soluble non-ammonia nitrogen from the forestomach and milk production in dairy cows. Animal Feed Science and Technology. 2002;**102**(1-4):15-33

[85] Smart SK, Cassady AI, Lu GQ, Martin DJ. The biocompatibility of carbon nanotubes. Carbon. 2006;**44**(6):1034-1047

[86] Mahajan-Miklos S, Tan MW, Rahme LG, Ausubel FM. Molecular mechanisms of bacterial virulence elucidated using a *P. aeruginosa–Caenorhabditis elegans* pathogenesis model. Cell. 1999;**96**(1):47-56

[87] Rahme LG, Ausubel FM, Cao H, Drenkard E, Goumnerov BC, Lau GW, et al. Plants and animals share functionally common bacterial virulence factors. Proceedings of the National Academy of Sciences. 2000;**97**(16):8815-8821

[88] Duan K, Surette MG. Environmental regulation of *P. aeruginosa* PAO1 Las and Rhl quorum-sensing systems. J of bacteriology. 2007;**189**(13):4827-4836

[89] Tan H, Feng Y, Chen J, Tang M, Li C, Xu L, et al. Preparation and characterization of Nano-ZnO antibacterial cardboard. Nanoscience and Nanotechnology Letters. 2017;**9**(3):241-246

[90] Sartor RB. Therapeutic manipulation of the enteric microflora in inflammatory bowel diseases: Antibiotics, probiotics, and prebiotics. Gastroenterology. 2004;**126**(6):1620-1633

[91] Schillinger U, Lücke FK. Antibacterial activity of *Lactobacillus* sake isolated from meat. Applied and Environmental Microbiology. 1989;**55**(8):1901-1906

[92] Sawai J. Quantitative evaluation of antibacterial activities of metallic oxide powders (ZnO, MgO and CaO) by conductimetric assay. J of microbiological methods. 2003;**54**(2):177-182

[93] Vogel P, Beacom JF. Angular distribution of neutron inverse beta decay, $\bar{\nu}$ e+ p→ e++. Physical review D. 1999;**60**(5):053003

[94] Urga G. The Econometrics of Panel Data: A Selective Introduction (No. 99151). UK: EconPapers; 1992

[95] Rodríguez LF, Reipurth B. Detection of radio continuum emission from the

Herbig-Haro objects 80 and 81 and their suspected energy source. Revista Mexicana de Astronomia y Astrofisica. 1989;**17**:59-63

[96] Ünlütürk A, Turantaş F. Gıda Mikrobiyolojisi. İzmir: Mengi Tan Basımevi; 1998

[97] Daeschel MA. Antimicrobial substances from lactic acid bacteria for use as food preservatives. J of FoodPreservatives. 2011;**44**:164-167

[98] DOORE S. Organic acids. In: Antimicrobials in foods. Aspen, NY. 1993. p. 95

[99] Garneau S, Martin NI, Vederas JC. Two-peptide bacteriocins produced by lactic acid bacteria. Biochimie. 2002;**84**(5-6):577-592

[100] Dalié DKD, Deschamps AM, Richard-Forget F. Lactic acid bacteria–potential for control of mould growth and mycotoxins: A review. Food Control. 2010;**21**(4):370-380

[101] Cintas P. Synthetic organoindium chemistry: What makes indium so appealing? Synlett. 1995;**1995**(11):1087-1096

[102] De Vuyst L, Vandamme EJ. Antimicrobial potential of lactic acid bacteria. In: Bacteriocins of lactic acid bacteria. Boston, MA: Springer; 1994. pp. 91-142

[103] Klaenhammer TR. Bacteriocins of lactic acid bacteria. Biochimie. 1988;**70**(3):337-349

[104] González-Martínez BE, Gómez-Treviño M, Jiménez-Salas Z. Bacteriocinas de probióticos. Revista Salud Pública y Nutrición. 2003;**4**(2):64-70

[105] Deegan LH, Cotter PD, Hill C, Ross P. Bacteriocins: Biological tools for bio-preservation and shelf-life extension. International dairy J. 2006;**16**(9):1058-1071

[106] Abee T, Krockel L, Hill C. Bacteriocins: Modes of action and potentials in food preservation and control of food poisoning. Inter J of food Microb. 1995;**28**(2):169-185

[107] Dortu C, Thonart P. Bacteriocins from lactic acid bacteria: Interest for food products biopreservation. Biotechnologie, Agronomie, Société et Environnement. 2009;**13**(1):143-154

[108] Bromberg L, Temchenko M, Alakhov V, Hatton TA. Bioadhesive properties and rheology of polyether-modified poly (acrylic acid) hydrogels. International Journal of Pharmaceutics. 2004;**282**(1-2):45-60

[109] Gálvez A, Abriouel H, López RL, Omar NB. Bacteriocin-based strategies for food biopreservation. International J of food microbiology. 2007;**120**(1-2):51-70

[110] Davoodabadi A, Dallal MMS, Foroushani AR, Douraghi M, Harati FA. Antibacterial activity of *Lactobacillus* spp. isolated from the feces of healthy infants against enteropathogenic bacteria. Anaerobe. 2015;**34**:53-58

[111] Rangasamy S, Tak YK, Kim S, Paul A, Song JM. Bifunctional therapeutic high-valence silver-pyridoxine nanoparticles with proliferative and antibacterial wound-healing activities. J of biomedical nanotechnology. 2016;**12**(1):182-196

[112] Collins EB, Aramaki K. Production of hydrogen peroxide by *Lactobacillus acidophilus*. J of dairy science. 1980;**63**(3):353-357

[113] Dahiya RS, Speck ML. Hydrogen peroxide formation by lactobacilli and its effect on *Staphylococcus aureus*. J of Dairy Science. 1968;**51**(10):1568-1572

[114] Weiss J, Takhistov P. J Mc Clements. Functional materials in food nanotechnology. Journal of Food Science. 2006;**71**(9):R107-R116

[115] Moraru CI, Panchapakesan CP, Huang Q, Takhistov P, Liu S, Kokini JL. Nanotechnology: A new frontier in food science. Food Technology. 2003;**57**(12):24-29

[116] Jin T, Sun D, Su JY, Zhang H, Sue HJ. Antimicrobial efficacy of zinc oxide quantum dots against *Listeria monocytogenes*, *Salmonella enteritidis*, and *Escherichia coli* O157: H7. J of food science. 2009;**74**(1):46-52

[117] Rehman S, Ali NM, Pier BG, Liaqat I. In vitro susceptibility of *P. aeruginosa* isolated from acute and chronic pulmonary infection to antibiotics, *Lactobacillus* competition and metal nanoparticles. PJZ. 2018;**50**(6)

Chapter 3

Exploitation of Phosphoinositides by the Intracellular Pathogen, *Legionella pneumophila*

Colleen M. Pike, Rebecca R. Noll and M. Ramona Neunuebel

Abstract

Manipulation of host phosphoinositide lipids has emerged as a key survival strategy utilized by pathogenic bacteria to establish and maintain a replication-permissive compartment within eukaryotic host cells. The human pathogen, *Legionella pneumophila*, infects and proliferates within the lung's innate immune cells causing severe pneumonia termed Legionnaires' disease. This pathogen has evolved strategies to manipulate specific host components to construct its intracellular niche termed the *Legionella*-containing vacuole (LCV). Paramount to LCV biogenesis and maintenance is the spatiotemporal regulation of phosphoinositides, important eukaryotic lipids involved in cell signaling and membrane trafficking. Through a specialized secretion system, *L. pneumophila* translocates multiple proteins that target phosphoinositides in order to escape endolysosomal degradation. By specifically binding phosphoinositides, these proteins can anchor to the cytosolic surface of the LCV or onto specific host membrane compartments, to ultimately stimulate or inhibit encounters with host organelles. Here, we describe the bacterial proteins involved in binding and/or altering host phosphoinositide dynamics to support intracellular survival of *L. pneumophila*.

Keywords: bacteria, infection, effector proteins, pneumonia, *Legionella pneumophila*, phosphoinositides, host-pathogen interactions, membrane traffic

1. Introduction

Bacterial pathogens have evolved diverse and effective strategies to promote their survival in human cells. Some bacteria can circumvent the innate immune response, managing to replicate within macrophages, which are the first line of defense against microbial pathogens and genetically programmed to eradicate foreign particles. Mechanisms that bacteria employ to survive in macrophages include (i) acclimating to the acidic environment within the host lysosome, (ii) escaping the phagosome to persist inside the host cell cytoplasm, and (iii) eluding the endolysosomal pathway by establishing a replication permissive vacuole within the host [1]. The Gram-negative facultative intracellular bacterium, *Legionella pneumophila*, has adopted a survival strategy that relies on the establishment of a protective vacuole that avoids encounters with the endolysosomal pathway. By phagocytosis, macrophages internalize *L. pneumophila* into a membrane-bound compartment termed as phagosome. Upon uptake, *L. pneumophila* directs

IntechOpen

membrane remodeling of the phagosomal compartment, employing a sizeable artillery of bacterial proteins that subvert multiple host cellular processes without compromising survival of the host cell throughout infection [2–4]. A specialized secretion system is responsible for translocating these proteins, known as effector proteins, from the bacterial milieu into the host cytosol [5–7]. Effector proteins do not share extensive homology with each other and are often composed of multiple domains that are functionally distinct [8, 9]. An emerging feature among effector proteins is their ability to recognize and bind host phosphoinositides (PIPs) [10], which are a series of phospholipids that play critical roles in coordinating cell signaling and membrane trafficking events in eukaryotic cells [11]. *L. pneumophila* effector proteins exploit the spatiotemporal regulation of host PIPs to facilitate the formation of the *Legionella*-containing vacuole (LCV) and to avoid the endolysosomal pathway. Disruption of the PIP distribution on the LCV membranes leads to bacterial degradation, illustrating that controlling PIP dynamics on and around the LCV is crucial for intracellular survival of *L. pneumophila* [12]. Here we will discuss the *L. pneumophila* effector proteins that contribute to vacuole biogenesis and maintenance through the exploitation of host phosphoinositides.

2. *Legionella pneumophila* replicates in protozoan and innate immune cells

L. pneumophila is ubiquitously found in aquatic environments forming close associations with protozoans and often found as an intracellular parasite of free-living amoeba [13]. In the human lung, *L. pneumophila* infects resident alveolar macrophages leading to severe pneumonia, known as Legionnaires' disease, which can be fatal in immunocompromised individuals [14]. Outbreaks stem from contaminated water systems such as those supplying water towers, cooling systems, and decorative fountains [15]. In 2017, a study by the Centers for Disease Control and Prevention (CDC) found that *L. pneumophila* was the leading bacterial agent responsible for public drinking water-associated outbreaks within the United States [14]. The number of reported Legionnaires' disease cases has been escalating since 2000, presumably due to an increase in urbanization, reliance on industrial water systems, as well as improved diagnostic methods [16]. *Legionella* spp. can exist within biofilms or amoebal hosts in freshwater systems, transitioning between a replicative and a transmissive/virulent phase life cycle [17, 18]. Nutrient deprivation within a biofilm or host triggers the upregulation of genes encoding virulence traits such as motility, osmotic stress resistance, pigmentation production, and multiple virulence factors [17]. This change in gene expression primes the bacterium to be engulfed by a new host cell and tap into their nutrient resources.

Inter-kingdom horizontal gene transfer events and circulating mobile genetic elements over long-term coevolution with multiple hosts have extensively reshaped the plasticity of the *Legionella* spp. genomes [19]. All *Legionella* spp. contain a highly conserved type IV secretion system (T4SS), yet there are differences in the combination of effectors present in each species. An analysis of 38 *Legionella* spp. genomes revealed that DNA exchange between species is rare and only seven core effectors are shared by all sequenced species [8]. *Legionella* effectors share more similarity with eukaryotic proteins than prokaryotic proteins, suggesting *Legionella* spp. have acquired their effector arrays from their hosts [20]. A striking number of effectors across the genus (>18,000) contain eukaryotic-derived domains [9]. This extensive combination of effectors likely stems from intimate coevolution between *Legionella* spp. with diverse protozoan hosts, such as *Acanthamoeba castellanii* [13], *Hartmannella vermiformis* [21], *Dictyostelium discoideum* [22, 23], *Tetrahymena pyriformis* [24], and *Naegleria fowleri* [25]. Only 20 of the 65 known species have been associated with human disease, suggesting that perhaps *Legionella* species are better adapted for infection within their amoebal hosts [9]. A clear set of effectors that render *Legionella* better suited for human infection is not apparent, although conservation of ankyrin motifs, F-box, or Set18 domains was predominantly found in more virulent strains [9].

The prevailing thought is that the mechanisms that enable *L. pneumophila* to infect and proliferate within protozoa have equipped this bacterium with the ability to survive within innate immune cells. This ability could be due to the high conservation of the pathways involved in uptake and microbial degradation between protozoa and human macrophages. In the lung, resident macrophages and neutrophils engulf *L. pneumophila* by phagocytosis but are often unable to degrade it through phagosome maturation [26–28], a process that entails sequential fusion of the phagosome with endocytic compartments and ultimately the lysosome [29]. *L. pneumophila* is initially encased within a phagosome after macrophage engulfment, but within minutes, the membrane of this phagosome is drastically remodeled into a compartment resembling the endoplasmic reticulum (ER) [2, 4]. Tubular ER and secretory vesicles are rapidly routed toward the phagosome where some eventually fuse with the phagosomal membrane, allowing the phagosome to adopt the identity of the recruited host membrane [30]. While promoting LCV membrane fusion with the ER and Golgi-derived vesicles, *L. pneumophila* prevents fusion with endosomal compartments. Studies have found that *L. pneumophila* effector proteins can target specific host membrane compartments, including early endosomes, recycling endosomes, and autophagosomes. Collectively, these effectors help *L. pneumophila* evade the macrophage's pre-programmed lysosomal degradation pathway [10], although precisely how these events are choreographed is not well understood.

The extensive remodeling of the vacuolar membrane is entirely dependent on a specialized Dot/Icm T4SS that delivers a staggering number of bacterial effector proteins (over 350) [8] into the host cytosol, many of which target membrane transport pathways [31, 32]. Disruption of the T4SS results in lysosomal degradation of the bacterium, indicating that the actions of effector proteins are paramount to bacterial survival [33]. However, it is often a challenge to identify an observable phenotype that can be attributed to a single effector because of functional redundancy among bacterial effectors [34]. Many advances have been made to dissect the molecular contribution of individual effectors toward bacterial infection (reviewed in [35]). A number of these effectors have been reported to hijack host vesicular trafficking pathways. An emerging feature among some of the effectors that target membrane trafficking is the ability to bind key host regulatory lipids, phosphoinositides (PIPs).

3. Phosphoinositides as crucial regulators of vesicular trafficking

Membrane compartments within eukaryotic cells are highly abundant, dynamic, and functionally distinct structures. Their movement must be tightly regulated to ensure that cargo carried by these structures is delivered to the proper destination. The cellular machinery recognizes and distinguishes these compartments based on the unique protein and lipid composition on the cytosolic leaflet of the membrane lipid bilayer [11]. Phosphoinositides are glycerophospholipids that amount to less than 15% of phospholipids within membranes but are essential for coordinating the spatiotemporal regulation of membrane trafficking events [11]. Phosphatidylinositol (PI), the precursor of phosphoinositides, can be reversibly phosphorylated at positions 3, 4, and 5 of its *myo*-inositol ring resulting in the generation of seven PIP species [11]. Membrane compartments are characterized in part by the presence of distinct PIP species that essentially act as molecular anchors to facilitate protein recruitment and attachment to specific compartments [11]. PI is synthesized in the endoplasmic reticulum and delivered to membrane-bound compartments via vesicular transport or cytosolic PI transfer proteins [11]. The Golgi and plasma membrane are highly enriched with PI(4)P, while lower levels of PI(4)P are also found within membranes of the ER and late endosomes [11, 36, 37]. PI(3)P is mainly found on phagosomes, early endosomes, late endosomes, and multivesicular bodies (MVBs). MVBs and late endosomes also contain PI(3,5)P_2, which is the dominant PIP on lysosomes. Phagocytosis and phagosome maturation are entirely dependent on phosphoinositide dynamics [38]. PI(4,5)P_2 and PI(3,4,5)P_3 are present on the plasma membrane and are critical for recruiting the cellular machinery for initiating phagocytosis. Once phagosomes have been formed, PI(3)P is the predominant PIP on the organelle [29]. PI(3)P then triggers the recruitment of proteins to the phagosome, such as EEA1 and its subsequent effector Rab5, to facilitate docking and fusion with early endosomes and progression down the phagolysosomal maturation pathway [39]. Blocking the formation of these PIP species results in robust inhibition of phagocytosis [40]. Given the crucial importance of PIPs for particle uptake and degradation, it is not surprising that intracellular bacteria have evolved molecular mechanisms to take command of these eukaryotic lipids.

4. Phosphoinositide dynamics on the LCV

The PIP composition on the LCV membrane has profound effects on the fate of the bacteria-bearing vacuole. PI conversion that accompanies LCV maturation was deciphered by tracking the localization of fluorescent PI probes produced in the soil amoeba, *Dictyostelium discoideum*, which serves as a model organism for the study of host-pathogen interactions [41]. As *L. pneumophila* enters *D. discoideum*, the phagocytic cup is coated with PI(3,4,5)P_3. On the membrane of the newly formed phagosome, PI(3,4,5)P_3, PI(3,4)P_2, and PI(4)P persist for less than 60 s on average. By 60 s, the phagosome begins to accumulate PI(3)P. Over the next 2 h, PI(4)P levels increase, the LCV lumen expands, and PI(3)P is slowly lost and excluded from the maturing LCV. The mature LCV maintains a discrete pool of PI(4)P separate from the surrounding ER, in which it acquires 30 to 60 min after uptake. As the bacterium continues to replicate, PI(4)P levels are steadily maintained on the LCV but are present in pools distinct from the surrounding ER network. The conversion from a PI(3)P to a PI(4)P-positive compartment is secretion system-dependent: a mutant strain lacking a functional T4SS accumulates PI(3)P on the LCV, PI(4)P is never acquired, and the LCV is destined for lysosomal degradation [12]. Thus, translocated effectors control the PIP composition of the LCV and potentially other host membranes.

In a recent study, Weber and colleagues [42] pursued the source of the PI(4)P on the LCV membrane. Real-time high-resolution confocal laser scanning microscopy (CLSM) revealed that LCVs of infected *D. discoideum* capture PI(4)P from trans-Golgi-derived vesicles. PI(4)P-enriched vesicles accumulate close to the LCV, even in the absence of the T4SS, but retention of these vesicles relies on the T4SS. This observation indicates that while PI(4)P-positive compartments localize to phagosomes regardless of the internalized cargo, effector proteins are needed to prolong this interaction. The removal of PI(3)P from the phagosome membrane was thought to occur through the actions of PIP-modifying enzymes; however, CSLM imaging of infected *D. discoideum* revealed shedding of PI(3)P-positive vesicles from the LCV. Moreover, the timing of PI(3)P shedding coincided with the gradual accumulation of PI(4)P-compartments around the LCV [42]. Together, these observations support the notion that *L. pneumophila* adopts a combined strategy to convert the LCV from a PI(3)P- to PI(4)P-enriched compartment, employing both direct modification of PIPs on the LCV membrane and selective association with host vesicles.

5. *L. pneumophila* effector proteins alter the PIP composition of the LCV membrane

To manipulate the PIP composition on the LCV, *L. pneumophila* uses both genetically encoded and host-derived PI kinases and phosphatases (**Figure 1**). Converting the PI(3)P-enriched phagosome to a predominantly PI(4)P-positive compartment requires a concerted effort between enzymes that add and remove

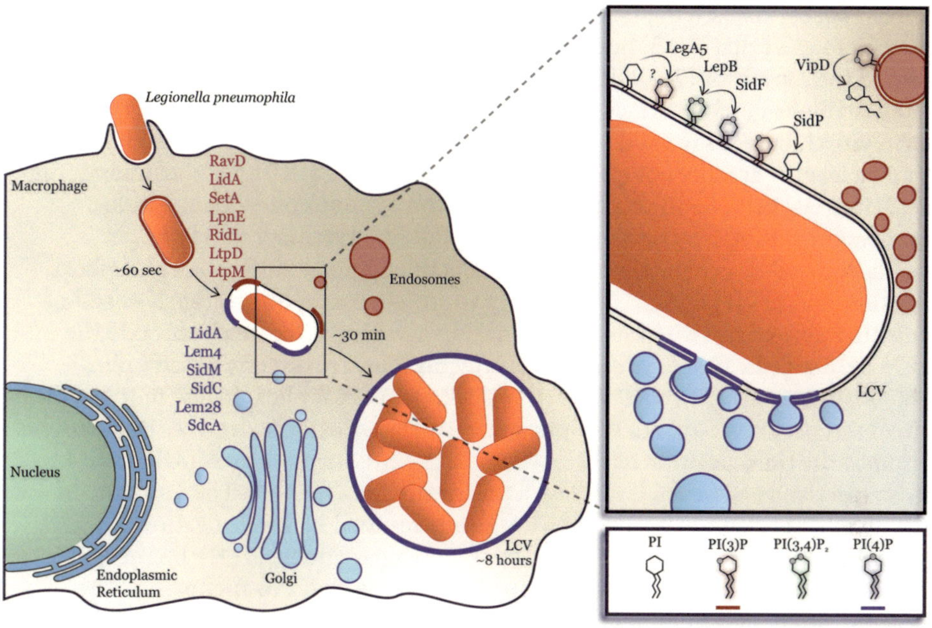

Figure 1.
L. pneumophila converts the phagosome to a PI(4)P-rich vacuole. Within a minute of uptake into the host cell, the LCV acquires the endosomal phosphoinositide, PI(3)P. Within an hour of infection, the LCV starts to accumulate PI(4)P until the bacteria are completely encapsulated in a PI(4)P-rich membrane. To avoid progression down the phagosome maturation pathway, L. pneumophila translocates effectors that alter the phosphoinositide composition on the LCV membrane to a PI(4)P-positive compartment (inset). This process is a result of close association and fusion with host vesicles as well as the direct conversion of existing phosphoinositides by kinases and phosphatases. Golgi-derived PI(4)P-positive vesicles accumulate around the LCV and later fuse with the vacuolar membrane. In contrast, PI(3)P-containing vesicles traffic toward the LCV but do not fuse with it. Additionally, the Legionella effector LepB is a PI kinase that phosphorylates PI(3)P and generates PI(3,4)P2 on the LCV membrane. This PI is a substrate for SidF which dephosphorylates PI(3,4)P2 to PI(4)P. While the origin of PI(3)P that LepB utilizes as a substrate is undetermined, LegA5 is a PI 3-kinase produced by Legionella that phosphorylates PI and could lead to additional PI(3)P on the LCV for conversion to PI(4)P. In combination, LegA5, LepB, and SidF may provide a cascade of enzymatic events for converting the LCV into a PI(4)P-positive compartment. SidP, another direct modifier of phosphoinositides produced by Legionella, may also contribute to the avoidance of the endocytic pathway by removing the phosphate from PI(3)P to hinder vesicle fusion. VipD localizes to endosomes and hydrolyzes a lipid tail from PI(3)P to potentially limit their interaction with the LCV. During this phosphoinositide conversion, Legionella effectors associate with the LCV through phosphoinositide binding domains. Some effectors localize by binding PI(3)P (RavD, LidA, SetA, LpnE, RidL, LtpD, LtpM), and some can associate via PI(4)P-binding (LidA, Lem4, SidM, SidC, Lem28, SdcA). During the later stages of infection, PI(3)P is undetectable and PI(4)P has become enriched on the expanding vacuole.

a phosphate group of the *myo*-inositol head group. The direct PI 4-kinase activity of the effector LepB could potentially initiate the conversion process to a PI(4)P-positive membrane by converting PI(3)P to PI(3,4)P_2. LepB was initially identified as an effector that is involved in bacterial egress [43]. Since then, LepB was found to localize to the LCV, where it contributes to the dynamics of Rab1 by acting as a GTPase-activating protein (GAP) [44, 45]. Found between amino acids 313 and 618, the structure and mechanism of the GAP domain is now well understood [46–48]. The N-terminal domain consisting of the first 311 amino acids garnered interest as this domain alone could disrupt the structure and function of the Golgi. The crystal structure of $LepB_{1–618}$ revealed homology to atypical kinases such as CtkA from *Helicobacter pylori* and actin-fragmin kinase from *Physarum polycephalum*. When mutating residues capable of performing phosphorylation, the yeast toxicity phenotype was found to be suppressed. While this suggested a kinase functionality, the pocket for a substrate was too small to accommodate proteins any larger than Rab GTPases. However, LepB did not phosphorylate any of the tested Rab GTPases. Instead, the LepB substrates were revealed to be phosphoinositides. Studies showed that LepB, but not the catalytically inactive mutant $LepB_{H154A}$, caused the sensor for PI(3,4)P_2 and PI(3,4,5)P_3 to relocate from a cytosolic to punctate distribution, while the signal for PI(3)P diminished dramatically. Ultimately, an in vitro kinase assay validated that LepB is a PI 4-kinase with specificity for PI(3)P and a level of activity comparable to the host kinase PI4KIII. By phosphorylating PI(3)P on the LCV, LepB could be initiating the vacuole's phosphoinositide conversion to PI(4)P by providing the PI(3,4)P_2 intermediate step [49].

As PI(3,4)P_2 is generated on the LCV, it is thought that SidF can dephosphorylate this lipid to PI(4)P. SidF is a membrane protein containing a large N-terminal domain followed by two transmembrane domains triggering localization to the LCV. SidF was the first *L. pneumophila* effector found to directly modify phosphoinositides through a screen for the well-known CX_5R phosphatase motif in the effector repertoire. SidF is a PI 3-phosphatase: this effector hydrolyzes the phosphate group at the third position of PI(3,4)P_2 or PI(3,4,5)P_3 to produce PI(4)P or PI(4,5)P_2, respectively; however, it displays a preference for PI(3,4)P_2. The mutation of the catalytic cysteine at residue 645 to a serine resulted in the abrogation of this phosphatase activity. As described in the following sections, PI(4)P on the LCV membrane can serve as a means for effectors, such as SidC, to anchor onto the LCV. Infection with a mutant lacking *sidF* shows significantly fewer vacuoles positive for SidC, suggesting that SidF contributes to the generation of PI(4)P on the LCV.

Ultimately, the functions of LepB and SidF suggest that PI(3)P can be converted to PI(4)P through the sequential efforts of these enzymes. The deletion of *lepB* and *sidF* individually shows a significant deficiency of SidC on the LCV membrane at similar levels [49, 50]. The deletion of both effectors simultaneously causes a decrease in SidC acquisition on the membrane no greater than the single-mutant strains, suggesting that these effectors are functioning in a linear pathway [49]. However, the complete loss of SidC was not seen in the infection with a *lepB sidF* double-deletion mutant. Additionally, both *lepB* and *sidF* are not always found in other *Legionella* species. Together, this suggests that there are other *Legionella* effectors or host proteins manipulating the LCV PIP landscape.

The screen that identified SidF as a PI phosphatase also yielded SidP as another direct modifier of phosphoinositides. SidP was identified as a candidate due to its CX_5R motif. It was found to have PI 3-phosphatase activity, cleaving PI(3)P and PI(3,5)P_2 in vitro. The *L. longbeachae* orthologue of SidP was only found to hydrolyze PI(3)P, suggesting this lipid may be the true target. SidP was also found to act as a PI 3-phosphatase *in vivo* when it suppressed yeast toxicity in a PI 3-phosphatase-deficient mutant but not a mutant lacking PI 4-phosphatases. This

DOI: http://dx.doi.org/10.5772/intechopen.89158

activity was confirmed when the levels of PI(3)P, but not PI(4)P or PI(4,5)P_2, were decreased in the presence of SidP [51]. Nonetheless, the purpose of SidP's phosphatase activity for successful infection has not yet been determined. We can speculate that SidP may work alongside LepB to quickly eliminate PI(3)P from the vacuole. As LepB converts PI(3)P to PI(3,4)P_2, SidP may be dephosphorylating PI(3)P to PI to completely deplete the membrane of this phospholipid that would otherwise trigger the phagocytic maturation.

As part of an effort to determine the function of a *Francisella* effector, OpiA, LegA5 was found to possess PI 3-kinase activity. LegA5 contains two motifs, DXHXXN and IDH, separated by 14 amino acids that are characteristic of the catalytic and activation loops of PI 3-kinases (and PI 4-kinases) [52]. PI(3)P has been shown to accumulate on the LCV early during infection in a manner independent of effector protein translocation [12]. This lipid is speculated to be the substrate LepB that acts on to initiate the PI(3)P to PI(4)P conversion on the phagosome membrane. However, *Legionella* may encode an effector that also contributes to the PI(3) P pool. These proteins may be delivered to the LCV in a complex so that PI is efficiently converted to PI(4)P. Alternatively, or perhaps in addition, PI(3)P-positive vesicles that accumulate around the nascent *Legionella*-containing phagosome may serve as a source for the initial wave of PI(3)P.

Aside from kinases and phosphatases that change the phosphorylation state of PIPs, *Legionella* also encodes 19 phospholipases. Phospholipases differ from PI phosphatases by cleaving the phospholipid backbone instead of hydrolyzing a phosphate on the *myo*-inositol head group. While these proteins can enter the host through different systems such as the Sec, Tat, T2SS, T4SS, and outer membrane vesicles (OMVs), only phospholipases translocated via T4SS will be discussed here [53]. The best characterized T4SS-secreted phospholipase is VipD. VipD has three paralogs, VpdA, VpdB, and VpdC, which are also T4SS substrates but have yet to be studied in detail [54]. The structure of VipD shows two distinct domains: the N-terminal domain has phospholipase activity A, indicating cleavage of the ester bond releasing a fatty acid chain, and the C-terminal domain causes localization to early endosomes and interacts directly with Rab5 and Rab22 [55]. The phospholipase activity is activated when VipD is bound to Rab5 due to a conformational change that exposes the active site [56, 57]. The activation of VipD causes cleavage of PI(3)P on endosomal membranes that prevents normal localization of membrane trafficking regulators, contributing to endosomal avoidance by *Legionella* [56].

While VipD has phospholipase A activity, *Legionella* also translocates two T4SS effectors with phospholipase C and D activity. A phospholipase C hydrolyzes the phosphorus-oxygen bond, releasing the phosphate of the phospholipid and the attached head group, and a phospholipase D solely cleaves off the attached head group. The phospholipase C effector protein, PlcC, is able to cleave phospholipids such as phosphatidylglycerol, phosphatidylcholine, and phosphatidylinositol [58]. Alone or in combination with two other phospholipase C effectors, PlcA and PlcB, translocated by the T2SS, these effectors were dispensable for growth in amoeba and macrophages. However, a triple mutant of these phospholipases displayed inefficient killing of larvae in the *G. mellonella* infection model compared to the wild type [58]. It is not yet known how this function may contribute to intracellular survival. We speculate that perhaps removing the head group on these phospholipids, specifically PI, would render them incapable of being modified by PI kinases and phosphatase and prevent the vacuole from being quickly converted to an endosome-like membrane. It would be interesting to determine if this phospholipase activity alters the PI composition of the LCV.

Lastly, the phospholipase D effector, LpdA, was first identified due to its homology with known phospholipase D enzymes [59]. LpdA specifically cleaves the head group from PI, PI(3)P, PI(4)P, and phosphatidylglycerol *in vitro* [60]. While LpdA localizes to the LCV [59], it is not known if or how this effector contributes to phosphoinositide dynamics. Nonetheless, deleting this gene results in the attenuation of growth in a mouse model [60].

LppA is a phytase enzyme that dephosphorylates the compound *myo*-inositol hexakisphosphate, known as phytate. While LppA's phosphatase activity on phytate may play various roles during infection, of interest to this review are its effects on phosphoinositides. The inositol phosphate head group of PIPs is similar in structure to phytate. LppA was shown to dephosphorylate PI(3,4)P_2 and PI(4,5)P_2 as well as, but less efficiently, PI(3,4,5)P_3 to PI(4)P in vitro. However, infection with an *lppA* deletion strain did not impact the presence of PI(4)P on the LCV [61]. It is possible that lack of LppA generates a more subtle phenotype that requires more sensitive detection methods.

In addition to directly manipulating the phosphoinositide composition of the vacuolar membrane, *Legionella* may change the PIP landscape by enlisting host enzymes. For instance, the host PI 5-phosphatase OCRL1 is recruited to the LCV in a T4SS-dependent manner. OCRL1 preferentially removes a phosphate from PI(4,5)P_2 to generate PI(4)P [62]. The homolog of OCRL1 in *Dictyostelium*, Dd5P4, was found to localize to LCVs where it is catalytically active and therefore able to dephosphorylate PIPs [63]. How OCRL1 is recruited to the LCV is not yet clear, but it is thought that *Legionella* protein LpnE may contribute to this process. LpnE is a Sel1-like repeat protein translocated into host cells in a T4SS-independent manner, and it seems to be exported extracellularly through an unknown mechanism [64]. LpnE is important for entry into amoebae and macrophages as well as intracellular replication. *In vitro* LpnE binds PI(3)P and interacts with OCRL1, but it does not seem to be essential for recruitment of OCRL1 to the LCV. It may be that LpnE synergizes with other effectors to stably recruit OCRL1, but this idea remains to be tested [63].

6. *L. pneumophila* effector proteins specifically bind phosphoinositides

Central to the ability of *L. pneumophila* to grow within both mammalian and protozoan cells is the remodeling of the phagosomal membrane through the manipulation of host secretory and endosomal trafficking. The loss of PI(3)P and the acquisition of PI(4)P on the phagosome membrane are achieved through a concerted mechanism carried out by the actions of multiple effector proteins. The acquisition of PI(4)P on the phagosome membrane is imperative for the subsequent recruitment of membranes to promote vacuole expansion [12]. PI(4)P on the LCV can serve as a docking site for effector attachment to ensure effectors are directed to the correct compartment within the cell [65]. Many effectors that bind to PI(4)P on the LCV are involved in the recruitment and fusion of secretory vesicles and ER. In addition to directly producing PI(4)P on the LCV via effector-driven phosphorylation and dephosphorylation of PIPs, it was recently reported that the phagosome also derives PI(4)P from the membrane material of secretory vesicles, demonstrating *L. pneumophila* employs multiple tactics to acquire PI(4)P. A number of PI(3)P-binding effectors have also been identified [66]. The few whose functions have been characterized interfere with phagosomal maturation, retrograde trafficking, and autophagy [67–70]. An overview of *L. pneumophila* effectors that target PIPs is in **Table 1**.

6.1 *L. pneumophila* T4SS effectors that bind PI(4)P

Bacterial effectors translocated early during infection have been shown to facilitate the recruitment and fusion of ER/secretory vesicles with the LCV. SidM

Name	PIP target	PIP-binding domain	Function	Citation
RavB	PI(3)P	LED035	Not Determined	66
CegC2	PI(3)P?	LED006	Not Determined	66
RavD	PI(3)P, PI(4)P		Prevents accumulation of linear ubiquitin chains on the LCV through deubiquitinase activity and prevents endolysosomal maturation of the LCV	67, 87
LtpM	PI(3)P		Glucosyltransferase activity stimulated by PI(3)P-binding	98
AnkX	PI(3)P, PI(4)P		Phosphocholinates Rab1 & Rab35; prevents lysosome-LCV fusion and endocytic recycling	70, 83, 86
LidA	PI(3)P, PI(4)P		RabGTPase interacting protein; contributes to retention of activated Rab1 on LCV and recruitment of secretory vesicles	48, 73, 74, 76, 80
Lem4	PI(4)P	P4M	Phosphotyrosine phosphatase	77, 78
Ceg19	PI(3)P	LED027	Causes secretory trafficking defects in yeast	66
LegK1	PI(3)P	LED006	Activates NF-κB by phosphorylating regulatory proteins	66
Ceg22	PI(3)P	LED006	Not Determined	66
LegC5/Lgt3	PI(3)P	LED006	Glucosylates eEF1A to inhibit translation	66
Lem9	PI(3)P	LED006	Not Determined	66
LegC6	PI(3)P	LED006	Not Determined	66
RavZ	PI(3)P	LED027	Inhibits autophagy through irreversible deconjucation of LC3 from autophagosome membranes	66, 69, 94
Lpg1961	PI(3)P?	LED027	Not Determined	66
SetA	PI(3)P		Glucosyltransferase activity stimulated by PI(3)P-binding	95, 98
LpnE	PI(3)P		Interacts with OCRL1 on the LCV, promotes intracellular uptake	63, 64
Lem21/LotA	PI(3)P	LED035	Prevents accumulation of ubiquitin chains on the LCV through deubiquitinase activity	66, 88
RidL/Ceg28	PI(3)P		Binds the retromer complex to inhibit retrograde trafficking	68, 90
Lpg2327	PI(3)P	LED006	Not Determined	66
MavH	PI(3)P	LED035	Not Determined	66
SidM/DrrA	PI(4)P	P4M	Promotes the recruitment and fusion of secretory vesicles with the LCV, AMPylates Rab1, interacts with exocyst complex	73-75, 85
SidC, SdcA	PI(4)P		Involved in ER recruitment to the LCV and ubiquitination through E3 ligase activity	50, 81, 82
Lem28	PI(4)P	P4M	Not Determined	77

L. pneumophila **effectors with PIP-modifying activity**

Name	Substrate	Product	Enzymatic activity	Citation
SidP	PI(3)P, PI(3,5)P2	PI, PI(5)P	PI 3-phosphatase	51
LepB	PI(3)P	PI(3,4)P2	Rab1 GAP; PI 4-kinase that generates PI(4)P on the LCV membrane	44, 48, 49, 66
LegA5/AnkK	PI	PI(3)P	PI 3-kinase	52
SidF	PI(3,4)P2, PI(3,4,5)P3	PI(4)P, PI(4,5)P2	PI 3-phosphatase that acts on the LCV	50
VipD	PI(3)P	PI	Rab5-activated phospholipase activity cleaves PI(3)P on endosomal membranes	54-57

Table 1.
Legionella pneumophila *effectors targeting PI(3)P and PI(4)P.*

(DrrA), an effector protein translocated immediately upon infection, localizes to the LCV and plays a crucial role in ER recruitment by exploiting the activity of Rab1, a small GTPase responsible for the transport of vesicles between the ER and Golgi [71–74]. SidM is a modular protein consisting of an N-terminal adenylyltransferase domain, a C-terminal PI(4)P-binding domain, and a central guanine nucleotide exchange factor (GEF) domain that activates the small GTPase Rab1 by facilitating the exchange of GDP with GTP [73]. SidM's adenylyltransferase activity covalently adds an adenosine monophosphate moiety onto Tyr 77 of Rab1, locking this small GTPase in its active conformation. Activated Rab1 is required for the recruitment of secretory vesicles to the LCV [73, 74]. SidM then promotes the tethering and fusion of these compartments with the phagosome membrane by interacting with an exocyst complex comprised of Sec5 and Sec15 [75].

A high-resolution crystal structure of SidM revealed a novel fold within the protein structure, termed P4M, that was responsible for binding PI(4)P with an unprecedented high affinity in the nanomolar range [76]. Two additional PI(4)P-binding effectors, Lem4 and Lem28, contain C-terminal domains similar to the P4M domain [77]. While Lem4 and Lem28 localize to the LCV through their PI(4)P-binding domains, they do not act on Rab1. Lem4 was recently demonstrated to be a phosphotyrosine phosphatase [78], although how this enzymatic function contributes to infection has yet to be determined.

Multiple effectors manipulate Rab1 to exploit secretory trafficking [44, 79]. While SidM is required for activating this small GTPase on the LCV, the PI(3)P and PI(4)P binder, LidA, protects Rab1 from being inactivated [73, 74, 80]. LidA also localizes to the early LCV as well as other uncharacterized membrane compartments [73, 74, 80]. Unlike P4M-containing effectors, LidA interacts with PIPs through a central coiled-coil region. LidA interacts with AMPylated Rab1 through the same coiled-coil domain, preventing GAPs from accessing Rab1 to deactivate it. It is unknown whether the PIP interaction contributes to LidA's function.

In addition to SidM, the PI(4)P binders SidC and its paralogue, SdcA, are also required for the recruitment of ER proteins to the LCV. In the absence of *sidC*, only 20% of LCVs acquire the ER marker calnexin, indicating that the interaction of LCVs with the ER is severely impaired upon deletion of this gene [81]. SidC and SdcA interact with PI(4)P using a 20 kDa C-terminal-binding domain (P4C) that does not share similarities with P4M or other eukaryotic PIP-binding motifs. Mutations that abolish P4C-PI(4)P interactions reduced ER recruitment to the LCV, indicating that SidC's PI(4)P-binding activity is critical for remodeling the LCV membrane [82].

6.2 *L. pneumophila* T4SS effectors that bind PI(3)P

Multiple PI(3)P-binding effectors have been identified, and several were shown to be involved in preventing the LCV from entering the phagosomal maturation pathway. AnkX binds both PI(3)P and PI(4)P *in vitro*, and in macrophages infected with a mutant strain lacking AnkX, the lysosomal marker, LAMP1, accumulates around the LCV indicating it is being routed for endolysosomal degradation [70]. AnkX's N-terminal FIC domain harbors phosphocholine transferase activity catalyzing, the covalent attachment of a phosphocholine moiety onto a serine or threonine residue of Rab1 and Rab35 [83, 84]. It is unknown whether AnkX localizes to the LCV, and despite its ability to covalently modify Rab1, it does not enhance retention of Rab1 on the LCV as observed for SidM-catalyzed adenylylation of Rab1. Phosphocholination locks Rab35 in an inactive conformation by preventing interaction with its cognate GEF, connecdenn; however, phosphocholinated Rab1 was still able to interact with SidM, which also acts as a GEF [85]. AnkX disrupts endocytic recycling in infected macrophages in a phosphocholination-dependent manner, suggesting that phosphocholination of Rab35, a key regulator of endocytic recycling, may be responsible for this phenotype [86].

The PI(3)P-binding effector, RavD, also contributes to preventing encounters between lysosomes and the LCV. Transmission electron microscopy and structured-illumination microscopy revealed RavD is present on the LCV membrane and vesicles adjacent to the LCV; however the identity of these vesicles has not yet been revealed. RavD binds PI(3)P via a C-terminal region [67]. A recent study reported that RavD's N-terminal region harbors deubiquitinase activity (DUB) that specifically cleaves linear ubiquitin chains from the LCV using a Cys-His-Ser triad [87]. Deletion of *ravD* causes the LCV to become decorated with linear ubiquitin and triggers subsequent activation of the NF-κB pathway [87]. Since *Legionella* species have not coevolved with macrophages, it is possible that RavD's DUB activity would be functional in both macrophages and protozoan hosts. It would be interesting to determine RavD's substrates in the context of a macrophage versus amoebae infection. Understanding the functional link between RavD's DUB activity and its contribution to the prevention of LCV-endolysosomal fusion could provide novel insight into why pathogens exploit ubiquitin during infection.

L. pneumophila's cohort of effectors includes multiple deubiquitinases that have evolved to act on different ubiquitin chains. Effector LotA localizes to the LCV through interaction with PI(3)P and harbors dual DUB activity to remove ubiquitin from the LCV [88]. LotA uses a C13 residue that acts against K6 linkages and a C303 residue that acts against K48 and K63 linkages, although C303 has a more considerable contribution to removing ubiquitin from the LCV. A *Legionella* strain lacking LotA and the ubiquitin-associated SidE family of effectors resulted in impaired bacterial growth within murine bone marrow-derived macrophages, indicating LotA has coordinated activity with other *L. pneumophila* ubiquitin-modifying enzymes [88]. While it has not been reported whether the SidE effector family interacts with PIPs, it cannot be ruled out that these ubiquitin-modifying enzymes may also rely on PIPs to correctly direct them to the sites where their enzymatic activity is required.

The effector RidL binds PI(3)P and inhibits retrograde transport through molecular mimicry. Retrograde trafficking serves as a conduit that connects endosomes, the trans-Golgi network, and the ER [89]. Cargo that is cycled from endosomes to the Golgi is recognized and sorted by a retromer complex. Ectopically expressed RidL blocks retrograde trafficking at endosome exit sites through interactions with the retromer complex protein, Vps29 [68]. RidL is present on the LCV membrane and endosomes but does not localize to endosomes through interactions with PI(3)P. Instead, RidL inserts itself into the endosomal retromer complex through interactions with Vps29, displacing Vps29 from binding to the Rab7 GAP, TBC1D5. RidL interacts with Vps29 using a hairpin loop that mimics the same manner in which TBC1D5 interacts with Vps29 [90]. This displacement blocks the movement of retrograde vesicles through an unknown mechanism. In the absence of *ridL*, LCVs accumulate lysosomal markers and retrograde cargo such as CI-MPR, which delivers acidic hydrolases to endocytic compartments [90]. This suggests the LCV may accept cargo or membranes from a subset of endosomal pathways and that RidL could intercept these incoming vesicles.

PI(3)P is also present on autophagosomes [91], and studies found that indeed *L. pneumophila* effectors also interfere with the dynamics of these compartments [69]. Autophagy is a conserved process across eukaryotic species that is triggered by cellular stress and serves as an additional defense mechanism against intracellular pathogens. Autophagy progression relies on a series of membrane reconstruction events, starting with phagophore membrane nucleation, to phagophore elongation and fusion to form the PI(3)P-rich autophagosome and ultimately fusion with lysosomes to degrade the internal cargo [91]. Early phagophore formation events are dependent on the presence of PI(3)P, which stimulates the recruitment of PI(3)P-binding proteins on ER-derived omegasomes [91]. Phagophore closure is completed through conjugation of LC3 to phosphoethanolamine (PE) on the phagophore membrane [92, 93]. Effector RavZ inhibits autophagy by extracting lipidated LC3 from autophagosome membranes and generating a modified LC3 product that lacks the essential C-terminal glycine required for reconjugation back onto autophagosome membranes.

RavZ localizes to autophagosome membranes through a C-terminal domain that recognizes PI(3)P. $RavZ_{1-331}$ contains catalytic activity yet displays reduced LC3-PE extraction, indicating proper localization to phagosomes is needed to inhibit autophagy [94]. This high-affinity PI(3)P-binding domain, termed LED027, contains two conserved tyrosine and lysine residues that are key for PI(3)P binding. LED027 is found in two other effectors, Lpg1121 (Ceg19) and Lpg1961, although Lpg1961 did not display lipid-binding activity when tested *in vitro* [66]. It would be interesting to determine if these LED027-containing effectors also preferentially localize to PI(3)P on autophagosomes, possibly unveiling a novel conserved domain that confers autophagy-related activity in bacterial effectors.

While effectors rely on PIPs for proper localization, binding to PIPs can also induce the enzymatic activity of some effectors. Effector protein SetA possesses an N-terminal region with glucosyltransferase activity and a C-terminal PI(3) P-binding region responsible for LCV localization [95]. Notably, PI(3)P binding enhances SetA's glucosyltransferase activity [96]. *In vitro* SetA has multiple substrates including actin, vimentin, and the chaperonin CCT5 [96], although it is unclear if these substrates are modified during infection.

The cohort of T4SS substrates is not conserved across all *L. pneumophila* strains. Strains harbor variations in their combinations of effectors that have been presumably acquired during the course of coevolution with a variety of protozoan hosts [97]. Despite these variations, PIP binding is emerging as a common feature among effectors of *L. pneumophila* strains. The *L. pneumophila* Paris strain encodes the glucosyltransferase LtpM that resembles the Philadelphia strain effector, SetA, in domain structure and the ability to cause a growth defect in yeast [96, 98]. Unlike SetA which uses a typical DxD motif for catalysis, LtpM harbors a noncanonical DxN motif. The glucosyltransferase activity LtpM is also stimulated by PI(3)P, indicating multiple effectors have evolved to exploit PI(3)P for purposes other than directing proper localization.

7. Eukaryotic and bacterial phosphoinositide-binding domains

In eukaryotes, proteins bind PIPs via domains that are highly conserved. Protein-lipid binding typically occurs through electrostatic interactions between positively charged amino acid residues and the negative phosphate(s) on the *myo*-inositol ring. These protein domains vary in their binding affinity and specificity for the seven PIP species [99]. The well-characterized pleckstrin homology (PH) domain is the eleventh most common domain in humans, found in 275 proteins [100]. Proteins harboring the PH domain are recruited to membranes through interactions with either $PI(3,4)P_2$, $PI(4,5)P_2$, or $PI(3,4,5)P_3$. The FYVE domain confers high specificity for PI(3)P and is present in many proteins that localize to endosomes [101, 102]. The phox domain (PX) is commonly found in sorting nexins and preferentially binds PI(3)P and in some cases $PI(3,4)P_2$ [103]. Intriguingly, bacterial proteins that specifically bind host PIPs do not use eukaryotic-like domains.

Bacteria can acquire protein domains by horizontal gene transfer from the hosts they infect [97]. A number of *L. pneumophila* effectors harbor eukaryotic-like domains such as ankyrin repeats, U-Box, F-box, and Sel1 repeats [9]. Interestingly, prokaryotic PIP-binding domains were not derived from their eukaryotic hosts. Global bioinformatic analysis of 38 *Legionella* genomes revealed a conserved PI(4)P-binding domain found in 36 putative effectors, while a domain termed LED006 is found in 136 effectors from 30 species [8]. The PI(4) P-binding domain was experimentally validated to be functional in SidM, Lpg1101, and Lpg2603 [73, 77].

A recent study identified three conserved PI(3)P-binding domains present in 14 *Legionella* effectors across 41 *Legionella* species: LED006, LED027, and LED025 [66]. All three domains rely on positively charged or aromatic residues confined to the C-terminus and are accompanied by an adjacent enzymatic or protein-binding domain. LED006 displayed the weakest affinity for PI(3)P yet is the most conserved, found in eight *L. pneumophila* effectors: CegC2, LegK1, Ceg22, LegC5, Lem9, LegC6, LepB, and Lpg2327. Only LegK1, LegC5, and LepB have been studied and shown to possess catalytic activity. While the C-terminal region of these proteins is conserved, the catalytic activity harbored by their N-terminal region varies. LegK1 is a serine/threonine kinase that targets the NF-κB pathway, LegC5 is a glucosyltransferase that modifies eEF1A, and LepB has dual PI 4-kinase activity and a Rab GAP domain. LED027 binds PI(3)P with high affinity and is found in RavZ, Lpg1121 (Ceg19), and Lpg1961, although Lpg1961 did not display lipid-binding activity when tested in vitro. LED035 is present in RavB, Lem21, and MavH, although none have been functionally characterized.

Biochemical analysis of a *Vibrio parahaemolyticus* effector revealed a conserved type III secreted bacterial phosphoinositide-binding domain (BPD) domain that mediates membrane localization in eukaryotic cells. The BPD domain is the first instance of a domain found in both plant and animal pathogens yet shares no homology to eukaryotes suggesting this domain is the result of convergent evolution [104]. Despite the recent discoveries of novel PIP-binding domains, the PI(3) P-binding regions in effectors SetA, RavD, LotA, and AnkX have not been linked to any conserved domains. We could speculate that perhaps this is because phosphoinositide binding is mediated by small, variable motifs or that lipid-binding domains may be quite diverse, as is the case for eukaryotic proteins. A clear perspective on this issue requires further identification, domain mapping, and computational analysis of known and novel phosphoinositide-binding effectors. Therefore, there is much to be learned about the molecular details underlying interactions between bacterial proteins and host phosphoinositides.

8. Conclusions and perspectives

What enables *L. pneumophila* effectors to target multiple membrane trafficking pathways stems in part from their modular structures consisting of various combinations of protein domains. Many of the PIP binding effectors are characterized by the presence of a C-terminal PIP-binding region and an N-terminal region that harbors enzymatic activity or interacts with host proteins.

The presence of PI(3)P on phagosomal membranes serves as a signpost for the recruitment of endocytic proteins that promote fusion with subsequent endocytic compartments and ultimately the lysosome. PI(3)P is therefore an attractive target for intracellular pathogens to eliminate entry into the phagosomal maturation pathway. It is well-established that after phagocytosis, PI(3)P on the nascent phagosome is rapidly depleted in conjunction with PI(4)P acquisition [12, 42]. Multiple studies have supported that this lipid rearrangement is accomplished through the actions of PIP-modifying effectors and effectors that promote the recruitment and fusion of PI(4)P-rich compartments with the LCV (reviewed in [105]). The recent evidence demonstrated that this lipid can also be removed from on or around the LCV in the form of PI(3)P-positive vesicles that are shed from the LCV. This would indicate that somehow microdomains of PI(3)P within membranes are being recognized, sequestered, and sorted into vesicles for removal or that perhaps PI(3)P-positive vesicles do not stably interact with the LCV. How the LCV can distinguish the simultaneous shedding of PI(3)P-compartments with the fusion of PI(4)P-compartments has yet to be determined. We can speculate that *L. pneumophila* has evolved cohorts of effectors that can independently regulate the acceptance of PI(4)P-rich membrane or the egress of PI(3)P-rich membrane from the LCV.

PI(3)P is completely lost from the LCV membrane after 2 hours; however, it is unclear why there is a strong presence of PI(3)P-binding effectors that are on the LCV membrane after this time point (LpnE, SetA, LotA, RidL, LtpM, LtpD, RavD). At later stages of infection, an accumulation of stagnant PI(3)P-positive vesicles can be seen surrounding the LCV. It is possible that effectors anchored to the LCV could be interacting with these vesicles by recognizing multiple membrane compartments. Most LCV localization studies are assessed using light microscopy, in which the resolution may not be high enough to visualize smaller distinct structures around the LCV. Light microscopy showed RavD is present on the LCV membrane; however higher-resolution imaging techniques like structured illumination and transmission electron microscopy revealed RavD is also present on a subset of unidentified vesicles adjacent to the LCV. It is most likely these vesicles are PI(3)P-rich, as RavD does not localize to PI(4)P-positive compartments. Moreover, RavD does not rely on PI(3)P binding to anchor to the LCV, supporting that effectors may exhibit dual localization patterns and that RavD may interact with the LCV and vesicles through different domains.

L. pneumophila has developed intricate strategies to facilitate intracellular growth by circumventing essential host cellular processes. The arsenal of effectors secreted by the type IV secretion system has evolved to target specific eukaryotic components such as proteins and lipids. Localization to the correct compartments within this host cell is imperative for protein function. A number of *Legionella* effectors rely on phosphoinositides to confer this directionality during infection. Not only are phosphoinositides needed to govern organelle identity, but they also dictate the path the phagosome embarks on once engulfed into the host cell. Thus, some effectors are ingeniously equipped to directly modify the lipid content on the phagosome membrane to avoid being routed toward degradation. Only a small percentage of effectors have been reported to interact with or modify phosphoinositides. Future studies that continue to expand on the repertoire of PIP-binding effectors will undoubtedly enhance our understanding of how intracellular pathogens survive within membrane-bound compartments within eukaryotic hosts.

DOI: http://dx.doi.org/10.5772/intechopen.89158

Author details

Colleen M. Pike, Rebecca R. Noll and M. Ramona Neunuebel*
Department of Biological Sciences, University of Delaware, Newark, DE, USA

*Address all correspondence to: neunr@udel.edu

References

[1] Di Russo CE, Samuel JE. Contrasting lifestyles within the host cell. Microbiology Spectrum. 2016;**4**(1)

[2] Swanson MS, Isberg RR. Association of *Legionella pneumophila* with the macrophage endoplasmic reticulum. Infection and Immunity. 1995;**63**(9):3609-3620

[3] Hubber A, Roy CR. Modulation of host cell function by *Legionella pneumophila* type IV effectors. Annual Review of Cell and Developmental Biology. 2010;**26**:261-283

[4] Robinson CG, Roy CR. Attachment and fusion of endoplasmic reticulum with vacuoles containing *Legionella pneumophila*. Cellular Microbiology. 2006;**8**(5):793-805

[5] Segal G, Shuman HA. Intracellular multiplication and human macrophage killing by *Legionella pneumophila* are inhibited by conjugal components of IncQ plasmid RSF1010. Molecular Microbiology. 1998;**30**(1):197-208

[6] Berger KH, Isberg RR. Two distinct defects in intracellular growth complemented by a single genetic locus in *Legionella pneumophila*. Molecular Microbiology. 1993;**7**(1):7-19

[7] Vogel JP, Andrews HL, Wong SK, Isberg RR. Conjugative transfer by the virulence system of *Legionella pneumophila*. Science. 1998;**279**(5352):873-876

[8] Burstein D, Amaro F, Zusman T, Lifshitz Z, Cohen O, Gilbert JA, et al. Genomic analysis of 38 *Legionella* species identifies large and diverse effector repertoires. Nature Genetics. 2016;**48**(2):167-175

[9] Gomez-Valero L, Rusniok C, Carson D, Mondino S, Perez-Cobas AE, Rolando M, et al. More than 18,000 effectors in the *Legionella* genus genome provide multiple, independent combinations for replication in human cells. Proceedings of the National Academy of Sciences of the United States of America. 2019;**116**(6):2265-2273

[10] Steiner B, Weber S, Hilbi H. Formation of the *Legionella*-containing vacuole: Phosphoinositide conversion, GTPase modulation and ER dynamics. International Journal of Medical Microbiology. 2018;**308**(1):49-57

[11] Di Paolo G, De Camilli P. Phosphoinositides in cell regulation and membrane dynamics. Nature. 2006;**443**(7112):651-657

[12] Weber S, Wagner M, Hilbi H. Live-cell imaging of phosphoinositide dynamics and membrane architecture during *Legionella* infection. mBio. 2014;**5**(1):e00839-e00813

[13] Rowbotham TJ. Preliminary report on the pathogenicity of *Legionella pneumophila* for freshwater and soil amoebae. Journal of Clinical Pathology. 1980;**33**(12):1179-1183

[14] (CDC) CfDCaP. Legionellosis–United States, 2000-2009. MMWR. Morbidity and Mortality Weekly Report. 2011;**60**:1083-1086

[15] Llewellyn AC, Lucas CE, Roberts SE, Brown EW, Nayak BS, Raphael BH, et al. Distribution of *Legionella* and bacterial community composition among regionally diverse US cooling towers. PLoS One. 2017;**12**(12):e0189937

[16] Nguyen TM, Ilef D, Jarraud S, Rouil L, Campese C, Che D, et al. A community-wide outbreak of legionnaires disease linked to industrial cooling towers–how far can contaminated aerosols spread?

The Journal of Infectious Diseases. 2006;**193**(1):102-111

[17] Bruggemann H, Hagman A, Jules M, Sismeiro O, Dillies MA, Gouyette C, et al. Virulence strategies for infecting phagocytes deduced from the in vivo transcriptional program of *Legionella pneumophila*. Cellular Microbiology. 2006;**8**(8):1228-1240

[18] Declerck P. Biofilms: The environmental playground of *Legionella pneumophila*. Environmental Microbiology. 2010;**12**(3):557-566

[19] Cazalet C, Rusniok C, Bruggemann H, Zidane N, Magnier A, Ma L, et al. Evidence in the *Legionella pneumophila* genome for exploitation of host cell functions and high genome plasticity. Nature Genetics. 2004;**36**(11):1165-1173

[20] Gomez-Valero L, Buchrieser C. Genome dynamics in *Legionella*: The basis of versatility and adaptation to intracellular replication. Cold Spring Harbor Perspectives in Medicine. 2013;**3**(6):a009993

[21] Fields BS, Fields SR, Loy JN, White EH, Steffens WL, Shotts EB. Attachment and entry of *Legionella pneumophila* in *Hartmannella vermiformis*. The Journal of Infectious Diseases. 1993;**167**(5):1146-1150

[22] Hagele S, Kohler R, Merkert H, Schleicher M, Hacker J, Steinert M. *Dictyostelium discoideum*: A new host model system for intracellular pathogens of the genus *Legionella*. Cellular Microbiology. 2000;**2**(2):165-171

[23] Solomon JM, Isberg RR. Growth of *Legionella pneumophila* in *Dictyostelium discoideum*: A novel system for genetic analysis of host-pathogen interactions. Trends in Microbiology. 2000;**8**(10):478-480

[24] Fields BS, Shotts EB Jr, Feeley JC, Gorman GW, Martin WT. Proliferation of *Legionella pneumophila* as an intracellular parasite of the ciliated protozoan *Tetrahymena pyriformis*. Applied and Environmental Microbiology. 1984;**47**(3):467-471

[25] Newsome AL, Baker RL, Miller RD, Arnold RR. Interactions between *Naegleria fowleri* and *Legionella pneumophila*. Infection and Immunity. 1985;**50**(2):449-452

[26] Horwitz MA, Silverstein SC. Legionnaires' disease bacterium (*Legionella pneumophila*) multiples intracellularly in human monocytes. The Journal of Clinical Investigation. 1980;**66**(3):441-450

[27] Horwitz MA. Formation of a novel phagosome by the Legionnaires' disease bacterium (*Legionella pneumophila*) in human monocytes. The Journal of Experimental Medicine. 1983;**158**(4):1319-1331

[28] Nash TW, Libby DM, Horwitz MA. Interaction between the legionnaires' disease bacterium (*Legionella pneumophila*) and human alveolar macrophages. Influence of antibody, lymphokines, and hydrocortisone. The Journal of Clinical Investigation. 1984;**74**(3):771-782

[29] Kinchen JM, Ravichandran KS. Phagosome maturation: Going through the acid test. Nature Reviews. Molecular Cell Biology. 2008;**9**(10):781-795

[30] Haenssler E, Ramabhadran V, Murphy CS, Heidtman MI, Isberg RR. Endoplasmic reticulum tubule protein reticulon 4 associates with the *Legionella pneumophila* vacuole and with translocated substrate Ceg9. Infection and Immunity. 2015;**83**(9):3479-3489

[31] Segal G, Shuman HA. *Legionella pneumophila* utilizes the same genes to multiply within *Acanthamoeba castellanii* and human macrophages. Infection and Immunity. 1999;**67**(5):2117-2124

[32] Luo ZQ, Isberg RR. Multiple substrates of the *Legionella pneumophila* Dot/Icm system identified by interbacterial protein transfer. Proceedings of the National Academy of Sciences of the United States of America. 2004;**101**(3):841-846

[33] Roy CR, Berger KH, Isberg RR. *Legionella pneumophila* DotA protein is required for early phagosome trafficking decisions that occur within minutes of bacterial uptake. Molecular Microbiology. 1998;**28**(3):663-674

[34] Ghosh S, O'Connor TJ. Beyond paralogs: The multiple layers of redundancy in bacterial pathogenesis. Frontiers in Cellular and Infection Microbiology. 2017;**7**:467

[35] Schroeder GN. The toolbox for uncovering the functions of *Legionella* Dot/Icm type IVb secretion system effectors: Current state and future directions. Frontiers in Cellular and Infection Microbiology. 2017;**7**:528

[36] Behnia R, Munro S. Organelle identity and the signposts for membrane traffic. Nature. 2005;**438**(7068):597-604

[37] Blumental-Perry A, Haney CJ, Weixel KM, Watkins SC, Weisz OA, Aridor M. Phosphatidylinositol 4-phosphate formation at ER exit sites regulates ER export. Developmental Cell. 2006;**11**(5):671-682

[38] Kale SD, Gu B, Capelluto DG, Dou D, Feldman E, Rumore A, et al. External lipid PI3P mediates entry of eukaryotic pathogen effectors into plant and animal host cells. Cell. 2010;**142**(2):284-295

[39] Duclos S, Diez R, Garin J, Papadopoulou B, Descoteaux A, Stenmark H, et al. Rab5 regulates the kiss and run fusion between phagosomes and endosomes and the acquisition of phagosome leishmanicidal properties in RAW 264.7 macrophages. Journal of Cell Science. 2000;**113**(Pt 19):3531-3541

[40] Araki N, Johnson MT, Swanson JA. A role for phosphoinositide 3-kinase in the completion of macropinocytosis and phagocytosis by macrophages. The Journal of Cell Biology. 1996;**135**(5):1249-1260

[41] Swart AL, Harrison CF, Eichinger L, Steinert M, Hilbi H. *Acanthamoeba* and *Dictyostelium* as cellular models for *Legionella* infection. Frontiers in Cellular and Infection Microbiology. 2018;**8**:61

[42] Weber S, Steiner B, Welin A, Hilbi H. *Legionella*-containing vacuoles capture PtdIns(4)P-rich vesicles derived from the golgi apparatus. mBio. 2018;**9**(6):e02420-e02418

[43] Chen J, de Felipe KS, Clarke M, Lu H, Anderson OR, Segal G, et al. *Legionella* effectors that promote nonlytic release from protozoa. Science. 2004;**303**(5662):1358-1361

[44] Ingmundson A, Delprato A, Lambright DG, Roy CR. *Legionella pneumophila* proteins that regulate Rab1 membrane cycling. Nature. 2007;**450**(7168):365-369

[45] Hardiman CA, Roy CR. AMPylation is critical for Rab1 localization to vacuoles containing *Legionella pneumophila*. mBio. 2014;**5**(1):e01035-e01013

[46] Mihai Gazdag E, Streller A, Haneburger I, Hilbi H, Vetter IR, Goody RS, et al. Mechanism of Rab1b deactivation by the *Legionella pneumophila* GAP LepB. EMBO Reports. 2013;**14**(2):199-205

[47] Mishra AK, Del Campo CM, Collins RE, Roy CR, Lambright DG. The *Legionella pneumophila* GTPase activating protein LepB accelerates Rab1 deactivation by a non-canonical

hydrolytic mechanism. The Journal of Biological Chemistry. 2013;**288**(33):24000-24011

[48] Yu Q, Hu L, Yao Q, Zhu Y, Dong N, Wang DC, et al. Structural analyses of *Legionella* LepB reveal a new GAP fold that catalytically mimics eukaryotic RasGAP. Cell Research. 2013;**23**(6):775-787

[49] Dong N, Niu M, Hu L, Yao Q, Zhou R, Shao F. Modulation of membrane phosphoinositide dynamics by the phosphatidylinositide 4-kinase activity of the *Legionella* LepB effector. Nature Microbiology. 2016;**2**:16236

[50] Hsu F, Zhu W, Brennan L, Tao L, Luo ZQ, Mao Y. Structural basis for substrate recognition by a unique *Legionella* phosphoinositide phosphatase. Proceedings of the National Academy of Sciences of the United States of America. 2012;**109**(34):13567-13572

[51] Toulabi L, Wu X, Cheng Y, Mao Y. Identification and structural characterization of a *Legionella* phosphoinositide phosphatase. The Journal of Biological Chemistry. 2013;**288**(34):24518-24527

[52] Ledvina HE, Kelly KA, Eshraghi A, Plemel RL, Peterson SB, Lee B, et al. A phosphatidylinositol 3-kinase effector alters phagosomal maturation to promote intracellular growth of *Francisella*. Cell Host & Microbe. 2018;**24**(2):285-295 e8

[53] Hiller M, Lang C, Michel W, Flieger A. Secreted phospholipases of the lung pathogen *Legionella pneumophila*. International Journal of Medical Microbiology. 2018;**308**(1):168-175

[54] VanRheenen SM, Luo ZQ, O'Connor T, Isberg RR. Members of a *Legionella pneumophila* family of proteins with ExoU (phospholipase a) active sites are translocated to target cells. Infection and Immunity. 2006;**74**(6):3597-3606

[55] Ku B, Lee KH, Park WS, Yang CS, Ge J, Lee SG, et al. VipD of *Legionella pneumophila* targets activated Rab5 and Rab22 to interfere with endosomal trafficking in macrophages. PLoS Pathogens. 2012;**8**(12):e1003082

[56] Gaspar AH, Machner MP. VipD is a Rab5-activated phospholipase A1 that protects *Legionella pneumophila* from endosomal fusion. Proceedings of the National Academy of Sciences of the United States of America. 2014;**111**(12):4560-4565

[57] Lucas M, Gaspar AH, Pallara C, Rojas AL, Fernandez-Recio J, Machner MP, et al. Structural basis for the recruitment and activation of the *Legionella phospholipase* VipD by the host GTPase Rab5. Proceedings of the National Academy of Sciences of the United States of America. 2014;**111**(34):E3514-E3523

[58] Aurass P, Schlegel M, Metwally O, Harding CR, Schroeder GN, Frankel G, et al. The *Legionella pneumophila* Dot/Icm-secreted effector PlcC/CegC1 together with PlcA and PlcB promotes virulence and belongs to a novel zinc metallophospholipase C family present in bacteria and fungi. The Journal of Biological Chemistry. 2013;**288**(16):11080-11092

[59] Viner R, Chetrit D, Ehrlich M, Segal G. Identification of two *Legionella pneumophila* effectors that manipulate host phospholipids biosynthesis. PLoS Pathogens. 2012;**8**(11):e1002988

[60] Schroeder GN, Aurass P, Oates CV, Tate EW, Hartland EL, Flieger A, et al. *Legionella pneumophila* effector LpdA is a palmitoylated phospholipase D virulence factor. Infection and Immunity. 2015;**83**(10):3989-4002

[61] Weber S, Stirnimann CU, Wieser M, Frey D, Meier R, Engelhardt S, et al. A type IV translocated *Legionella cysteine* phytase counteracts intracellular growth restriction by phytate. The Journal of Biological Chemistry. 2014;**289**(49):34175-34188

[62] Zhang X, Jefferson AB, Auethavekiat V, Majerus PW. The protein deficient in Lowe syndrome is a phosphatidylinositol-4,5-bisphosphate 5-phosphatase. Proceedings of the National Academy of Sciences of the United States of America. 1995;**92**(11):4853-4856

[63] Weber SS, Ragaz C, Hilbi H. The inositol polyphosphate 5-phosphatase OCRL1 restricts intracellular growth of *Legionella*, localizes to the replicative vacuole and binds to the bacterial effector LpnE. Cellular Microbiology. 2009;**11**(3):442-460

[64] Newton HJ, Sansom FM, Dao J, McAlister AD, Sloan J, Cianciotto NP, et al. Sel1 repeat protein LpnE is a *Legionella pneumophila* virulence determinant that influences vacuolar trafficking. Infection and Immunity. 2007;**75**(12):5575-5585

[65] Hilbi H, Weber S, Finsel I. Anchors for effectors: Subversion of phosphoinositide lipids by *Legionella*. Frontiers in Microbiology. 2011;**2**:91

[66] Nachmias N, Zusman T, Segal G. Study of *Legionella* effector domains revealed novel and prevalent phosphatidylinositol 3-phosphate binding domains. Infection and Immunity. 2019;**87**(6):e00153-e00119

[67] Pike CM, Boyer-Andersen R, Kinch LN, Caplan JL, Neunuebel MR. The *Legionella* effector RavD binds phosphatidylinositol-3-phosphate and helps suppress endolysosomal maturation of the *Legionella*-containing vacuole. The Journal of Biological Chemistry. 2019;**294**(16):6405-6415

[68] Finsel I, Ragaz C, Hoffmann C, Harrison CF, Weber S, van Rahden VA, et al. The *Legionella* effector RidL inhibits retrograde trafficking to promote intracellular replication. Cell Host & Microbe. 2013;**14**(1):38-50

[69] Choy A, Dancourt J, Mugo B, O'Connor TJ, Isberg RR, Melia TJ, et al. The *Legionella* effector RavZ inhibits host autophagy through irreversible Atg8 deconjugation. Science. 2012;**338**(6110):1072-1076

[70] Pan X, Luhrmann A, Satoh A, Laskowski-Arce MA, Roy CR. Ankyrin repeat proteins comprise a diverse family of bacterial type IV effectors. Science. 2008;**320**(5883):1651-1654

[71] Zhu Y, Hu L, Zhou Y, Yao Q, Liu L, Shao F. Structural mechanism of host Rab1 activation by the bifunctional *Legionella* type IV effector SidM/DrrA. Proceedings of the National Academy of Sciences of the United States of America. 2010;**107**(10):4699-4704

[72] Suh HY, Lee DW, Lee KH, Ku B, Choi SJ, Woo JS, et al. Structural insights into the dual nucleotide exchange and GDI displacement activity of SidM/DrrA. The EMBO Journal. 2010;**29**(2):496-504

[73] Brombacher E, Urwyler S, Ragaz C, Weber SS, Kami K, Overduin M, et al. Rab1 guanine nucleotide exchange factor SidM is a major phosphatidylinositol 4-phosphate-binding effector protein of *Legionella pneumophila*. The Journal of Biological Chemistry. 2009;**284**(8):4846-4856

[74] Machner MP, Isberg RR. Targeting of host Rab GTPase function by the intravacuolar pathogen *Legionella pneumophila*. Developmental Cell. 2006;**11**(1):47-56

[75] Arasaki K, Kimura H, Tagaya M, Roy CR. *Legionella* remodels the plasma membrane-derived vacuole by

utilizing exocyst components as tethers. The Journal of Cell Biology. 2018;**217**(11):3863-3872

[76] Schoebel S, Blankenfeldt W, Goody RS, Itzen A. High-affinity binding of phosphatidylinositol 4-phosphate by *Legionella pneumophila* DrrA. EMBO Reports. 2010;**11**(8):598-604

[77] Hubber A, Arasaki K, Nakatsu F, Hardiman C, Lambright D, De Camilli P, et al. The machinery at endoplasmic reticulum-plasma membrane contact sites contributes to spatial regulation of multiple *Legionella* effector proteins. PLoS Pathogens. 2014;**10**(7):e1004222

[78] Beyrakhova K, Li L, Xu C, Gagarinova A, Cygler M. *Legionella pneumophila* effector Lem4 is a membrane-associated protein tyrosine phosphatase. The Journal of Biological Chemistry. 2018;**293**(34):13044-13058

[79] Neunuebel MR, Chen Y, Gaspar AH, Backlund PS Jr, Yergey A, Machner MP. De-AMPylation of the small GTPase Rab1 by the pathogen *Legionella pneumophila*. Science. 2011;**333**(6041):453-456

[80] Neunuebel MR, Mohammadi S, Jarnik M, Machner MP. *Legionella pneumophila* LidA affects nucleotide binding and activity of the host GTPase Rab1. Journal of Bacteriology. 2012;**194**(6):1389-1400

[81] Ragaz C, Pietsch H, Urwyler S, Tiaden A, Weber SS, Hilbi H. The *Legionella pneumophila* phosphatidylinositol-4 phosphate-binding type IV substrate SidC recruits endoplasmic reticulum vesicles to a replication-permissive vacuole. Cellular Microbiology. 2008;**10**(12):2416-2433

[82] Luo X, Wasilko DJ, Liu Y, Sun J, Wu X, Luo ZQ, et al. Structure of the *Legionella* virulence factor, SidC reveals a unique PI(4)P-specific binding domain essential for its targeting to the bacterial phagosome. PLoS Pathogens. 2015;**11**(6):e1004965

[83] Mukherjee S, Liu X, Arasaki K, McDonough J, Galan JE, Roy CR. Modulation of Rab GTPase function by a protein phosphocholine transferase. Nature. 2011;**477**(7362):103

[84] Tan Y, Arnold RJ, Luo ZQ. *Legionella pneumophila* regulates the small GTPase Rab1 activity by reversible phosphorylcholination. Proceedings of the National Academy of Sciences of the United States of America. 2011;**108**(52):21212-21217

[85] Murata T, Delprato A, Ingmundson A, Toomre DK, Lambright DG, Roy CR. The *Legionella pneumophila* effector protein DrrA is a Rab1 guanine nucleotide-exchange factor. Nature Cell Biology. 2006;**8**(9):971-977

[86] Allgood SC, Romero Dueñas BP, Noll RR, Pike C, Lein S, Neunuebel MR. *Legionella* effector AnkX disrupts host cell endocytic recycling in a phosphocholination-dependent manner. Frontiers in Cellular and Infection Microbiology. 2017;**7**(397)

[87] Wan M, Wang X, Huang C, Xu D, Wang Z, Zhou Y, et al. A bacterial effector deubiquitinase specifically hydrolyses linear ubiquitin chains to inhibit host inflammatory signalling. Nature Microbiology. 2019;**4**(8):1282-1293

[88] Kubori T, Kitao T, Ando H, Nagai H. LotA, a *Legionella deubiquitinase*, has dual catalytic activity and contributes to intracellular growth. Cellular Microbiology. 2018;**20**(7):e12840

[89] Barlocher K, Welin A, Hilbi H. Formation of the *Legionella* replicative compartment at the crossroads of retrograde trafficking. Frontiers in Cellular and Infection Microbiology. 2017;**7**:482

[90] Barlocher K, Hutter CAJ, Swart AL, Steiner B, Welin A, Hohl M, et al. Structural insights into *Legionella* RidL-Vps29 retromer subunit interaction reveal displacement of the regulator TBC1D5. Nature Communications. 2017;**8**(1):1543

[91] Axe EL, Walker SA, Manifava M, Chandra P, Roderick HL, Habermann A, et al. Autophagosome formation from membrane compartments enriched in phosphatidylinositol 3-phosphate and dynamically connected to the endoplasmic reticulum. The Journal of Cell Biology. 2008;**182**(4):685-701

[92] Tanida I, Ueno T, Kominami E. LC3 conjugation system in mammalian autophagy. The International Journal of Biochemistry & Cell Biology. 2004;**36**(12):2503-2518

[93] Wild P, McEwan DG, Dikic I. The LC3 interactome at a glance. Journal of Cell Science. 2014;**127**(Pt 1):3-9

[94] Yang A, Pantoom S, Wu YW. Elucidation of the anti-autophagy mechanism of the *Legionella* effector RavZ using semisynthetic LC3 proteins. eLife. 2017;**6**:e23905

[95] Jank T, Bohmer KE, Tzivelekidis T, Schwan C, Belyi Y, Aktories K. Domain organization of *Legionella* effector SetA. Cellular Microbiology. 2012;**14**(6):852-868

[96] Levanova N, Steinemann M, Bohmer KE, Schneider S, Belyi Y, Schlosser A, et al. Characterization of the glucosyltransferase activity of *Legionella pneumophila* effector SetA. Naunyn-Schmiedeberg's Archives of Pharmacology. 2019;**392**(1):69-79

[97] Best A, Kwaik YA. Evolution of the arsenal of *Legionella pneumophila* effectors to modulate protist hosts. mBio. 2018;**9**(5):e01313-e01318

[98] Levanova N, Mattheis C, Carson D, To KN, Jank T, Frankel G, et al. The *Legionella* effector LtpM is a new type of phosphoinositide-activated glucosyltransferase. The Journal of Biological Chemistry. 2019;**294**(8):2862-2879

[99] Itoh T, Takenawa T. Phosphoinositide-binding domains: Functional units for temporal and spatial regulation of intracellular signalling. Cellular Signalling. 2002;**14**(9):733-743

[100] Lemmon MA, Ferguson KM. Signal-dependent membrane targeting by pleckstrin homology (PH) domains. Biochemical Journal. 2000;**350**(1):1-8

[101] Gaullier JM, Simonsen A, D'Arrigo A, Bremnes B, Stenmark H, Aasland R. FYVE fingers bind PtdIns (3) P. Nature. 1998;**394**(6692):432

[102] Lawe DC, Patki V, Heller-Harrison R, Lambright D, Corvera S. The FYVE domain of early endosome antigen 1 is required for both phosphatidylinositol 3-phosphate and Rab5 binding critical role of this dual interaction for endosomal localization. Journal of Biological Chemistry. 2000;**275**(5):3699-3705

[103] Kanai F, Liu H, Field SJ, Akbary H, Matsuo T, Brown GE, et al. The PX domains of p47phox and p40phox bind to lipid products of PI (3) K. Nature Cell Biology. 2001;**3**(7):675

[104] Salomon D, Guo Y, Kinch LN, Grishin NV, Gardner KH, Orth K. Effectors of animal and plant pathogens use a common domain to bind host phosphoinositides. Nature Communications. 2013;**4**:2973

[105] Haneburger I, Hilbi H. Phosphoinositide lipids and the *Legionella* pathogen vacuole. Current Topics in Microbiology and Immunology. 2013;**376**:155-173

Section 2

Antimicrobial Resistance

Chapter 4

Carbapenem Resistance: Mechanisms and Drivers of Global Menace

Bilal Aslam, Maria Rasool, Saima Muzammil, Abu Baker Siddique, Zeeshan Nawaz, Muhammad Shafique, Muhammad Asif Zahoor, Rana Binyamin, Muhammad Waseem, Mohsin Khurshid, Muhammad Imran Arshad, Muhammad Aamir Aslam, Naveed Shahzad, Muhammad Hidayat Rasool and Zulqarnain Baloch

Abstract

The emergence of carbapenem-resistant bacterial pathogens is a significant and mounting health concern across the globe. At present, carbapenem resistance (CR) is considered as one of the most concerning resistance mechanisms and mainly found in gram-negative bacteria of the *Enterobacteriaceae* family. Although carbapenem resistance has been recognized in *Enterobacteriaceae* from last 20 years or so, recently it emerged as a global health issue as CR clonal dissemination of various *Enterobacteriaceae* members especially *E. coli*, and *Klebsiella pneumoniae* are reported from across the globe at an alarming rate. Phenotypically, carbapenems resistance is in due to the two key mechanisms, like structural mutation coupled with β-lactamase production and the ability of the pathogen to produce carbapenemases which ultimately hydrolyze the carbapenem. Additionally, penicillin-binding protein modification and efflux pumps are also responsible for the development of carbapenem resistance. Carbapenemases are classified into different classes which include Ambler classes A, B, and D. Several mobile genetic elements (MGEs) have their potential role in carbapenem resistance like Tn4401, Class I integrons, IncFIIK2, IncF1A, and IncI2. Taking together, resistance against carbapenems is continuously evolving and posing a significant health threat to the community. Variable mechanisms that are associated with carbapenem resistance, different MGEs, and supplementary mechanisms of antibiotic resistance in association with virulence factors are expanding day by day. Timely demonstration of this global health concern by using molecular tools, epidemiological investigations, and screening may permit the suitable measures to control this public health menace.

Keywords: carbapenems, antibiotic resistance, public health, global health concern, *Enterobacteriaceae*

IntechOpen

1. Antibiotic resistance as a global threat

The global burden of antibiotic resistance is mounting continuously; preferably it piles up the pressure on veterinary medicine and on human. The WHO made a landmark by promoting and declaring AMR as a global health concern. The agenda of global health concerns are at the developmental stages, for example, a book named as *The Evolving Threat of Antimicrobial Resistance: Options for Action* is a precious addition to the archive [1]. Currently, the world is experiencing dramatic pre-antibiotic era, and many of the untreated infection emerge on a large-scale; clinicians often encounter many patients with such infections that normally reported as PDR or MDR bacteria by many laboratories and not responding to already available therapeutics. It has been estimated that yearly about two million people acquire vulnerable infections just because of these antibiotic-resistant pathogens, and as a result of this, about 23,000 people die according to Centers for Disease Control and Prevention (CDC) [2].

In a historical perspective, antibiotic resistance is a mounting and compelling concern. New types of antibiotic-resistant bacteria are taking control of ancient drugs. We may be entering the post-antibiotic era, because of increased persistence, spread, and the emergence of superbugs. It has been reported that annually, in the USA, about 99,000 deaths are caused by antibiotic-resistant pathogen-related hospital-acquired infections [3]. While in America, the annual death rate is about 50,000 caused by two usual HAIs, known as sepsis and pneumonia, which cost around $8 billion to the economy of the USA. The patients infected with bacterial strains that are resistant to antibiotics must stay in the hospital minimum for 13 days, which adds to 8 million days annually. An annual report of the cost of economy loss with regard to a productivity loss of around $35 billion has been demonstrated within healthcare settings [3].

2. Causes of antibiotic resistance

Currently, the multifarious causes of resistance constitute many factors including improper use and regulations, lack of awareness, aberrant antibiotic usage, the use of antibiotics as a growth promoter in livestock as well as in poultry for infection control, and online marketing [4]. Fundamentally, the reason behind the resistance evolution is the improper and excessive use of antimicrobials. The powerful drivers of antibiotic resistance include infection control standards, sanitation system, drug quality, water hygiene systems, diagnostics and therapeutics, and migration or travel quarantine. Genetic mutations and exchange of genetic material between organisms play a key role in the distribution of antibiotic resistance [5]. MDR organisms in hospital wastes are associated with public health illnesses because they are ultimately disseminated to humans. In this regard, recently a study has been conducted in Pakistan to find the occurrence of ESBL producing *K. pneumoniae* in hospital wastes including hospital sludge and wastewater, operation theater waste. They found the significant percentage of extended-spectrum β-lactamases (ESBL) producing MDR *K. pneumoniae* in these wastes [6]. Similarly another study conducted by [7] reported the patterns of antibiotic-resistant *K. pneumoniae* in clinical isolates with special reference to fluoroquinolones, depicting an alarming threat of antibiotic resistance among *K. pneumoniae*-related nosocomial infections.

3. Carbapenems

Carbapenems are effective β-lactam antimicrobials and have very potent efficacy against many ESBL-producing bacteria and are also administered intravenously.

In order to treat bacterial infections, carbapenems are considered as the most reliable and the last resort class of antimicrobials. Carbapenem agent has a very unique structure, usually defined by carbapenem coupled to B-lactam ring, which provide protection against the majority of b-lactamases as well as metallo-b-lactamases, and thus possess extended antibacterial activity [8]. Carbapenems work by penetrating the cell wall of bacteria, binding with penicillin-binding proteins (PBPs), and result in inactivation of intracellular autolytic inhibitor enzymes, ultimately killing the bacterial cell.

In addition, carbapenems mainly target "transpeptidase inhibition enzyme" during bacterial cell wall synthesis, preventing peptide cross-linking activity, leading to enhanced autolytic activity, and thus resulting in cell death. Therefore, carbapenems are considered as effective antimicrobials to treat life-threatening and invasive infections due to their "concentration-independent killing effect" on infecting bacteria [9, 10].

4. Carbapenemases

Carbapenemases are versatile b-lactamases, having the capability to hydrolyze carbapenems, cephalosporins, penicillins, and monobactams. Carbapenemases typically belong to two molecular families, namely, "metallo-carbapenemases" in which activity is inhibited by EDTA, used zinc molecule at their active sites, and "serine-based carbapenemases" in which activity is not inhibited by EDTA rather used serine residues at their active sites and inactivated through β-lactamase inhibitors like tazobactam and clavulanic acid [11].

β-Lactamases are classified based on two properties: functional and molecular ones. Functional classification was proposed by a scientist "Bush" in 1988, who classified β-lactamases into four functional groups namely, groups 1–4. Carbapenems fall under subgroup the 2f and group 3 [12]. Later on another scientist, Rasmussen, suggested that group 3 can be further divided into three functional subgroups on the basis of substrate specificity [13].

The molecular classification was proposed by scientist "Frere" and colleagues, who classified carbapenemases into class A, class B, and class D carbapenemases (**Table 1**).

Class A carbapenemases require a serine active site at position number 70 in Ambler numbering system, fall under the group 2f, and have the ability to hydrolyze carbapenems, penicillins, aztreonam, and cephalosporins [14].

Classification	Enzymes	Common bacteria
Class A	SME, NMC, KPC, IMI, GES	All *Enterobacteriaceae*, rarely *P. aeruginosa*
Class B	VIM, SPM, GIM, IMP	*Acinetobacter* species, *P. aeruginosa*, *Enterobacteriaceae*
Subclass B1	VIM-2, IMP-1, SPM-1, CcrA and BcII	
Subclass B2	Sfh-1, CphA	
Subclass B3	Gob-1, FEZ-1, CAU-1 & L1	
Class D	OXA	*Acinetobacter* species

Table 1.
Molecular classification scheme of carbapenemases [16].

Class B metallo-B-lactamases require a zinc ion at their active sites and have the ability to hydrolyze carbapenems, penicillins, and cephalosporins but do not hydrolyze aztreonam [15].

Class D carbapenemases were firstly described in 1993; among these class D OXA β-lactamases are the most important and were anciently named as penicillinases and have the ability to hydrolyze oxacillin, penicillin, cloxacillin, and ceftazidime but do not hydrolyze imipenem [11].

5. The emergence of carbapenem resistance

Carbapenem resistance is a leading and major public health concern around the globe. It mainly occurs among the *Enterobacteriaceae* family, particularly in healthcare settings. In the UK and the USA, carbapenem-resistant enteric bacterial strain has been identified and isolated from such patients who recently received medical care in Bangladesh, Pakistan, and India. Such strains possess a gene called New Delhi metallo-β-lactamases (NDM), responsible for producing metallo-β-lactamase enzyme that causes hydrolysis of carbapenems [17].

Factors that play a critical role in the emergence of carbapenem resistance are improper antibiotic prescription, uncontrolled public access to antimicrobials, poor sales regulation, lack of infection control measures within healthcare centers, the use of sub-therapeutic doses in agricultural settings [18].

In gram-negative bacteria, the development of carbapenem resistance (particularly in the presence of carbapenemases) is a leading factor associated with the emergence of MDR pathogens which may ultimately lead to the development of pandrug resistant (PDR) bacterial strains. Undoubtedly, among the carbapenemase-producing organisms, resistance to the last resort agents rapidly emerge and spread particularly when such agents are used in healthcare centers [18]. It has also been demonstrated that this carbapenem-resistant-nosocomial pathogens continually emerge, thus accruing more carbapenem resistance determinants, mechanisms, as well as carbapenem encoding genes that ultimately lead to increase carbapenem MICs ruling out yet the best therapeutic choice against such carbapenemase producers [18].

6. Mechanisms of carbapenem resistance

The emergence of resistance against these antibiotics reflects a growing health concern around the globe. Carbapenem resistance is mainly caused by two basic mechanisms including the production of carbapenemases (carbapenem-hydrolyzing enzymes) and B-lactamase activity coupled with structural mutations (ESBLs and AmpC cephalosporinases) [19, 20] (**Figure 1**).

Carbapenem resistance can be developed either due to acquired or intrinsic resistance mechanisms or sometimes both, since the bacteria have acquired numerous resistance mechanisms including mutations in the target site, efflux pumps, and enzymatic inactivation. Among these, enzymatic inactivation [acquired carbapenemases (plasmid-mediated)] is the most emerging and well-established mechanism. Acquired carbapenem resistance mechanisms include (1) destruction of carbapenems which are resistant to hydrolysis by plasmid AmpCs in conjunction with ESBL enzymes, contributing insusceptibility towards carbapenem agent [21]; (2) transfer of ESBL genes between the organisms; and (3) porin mutation with expression modulation. Loss of *OprD* porin and efflux pump overexpression is a usual mechanism of carbapenem resistance in the case of *Pseudomonas aeruginosa* [22]. Intrinsic carbapenem resistance mechanism includes reduced uptake (due to altered porin channels) and reduced outer membrane permeability of B-lactam drugs [16].

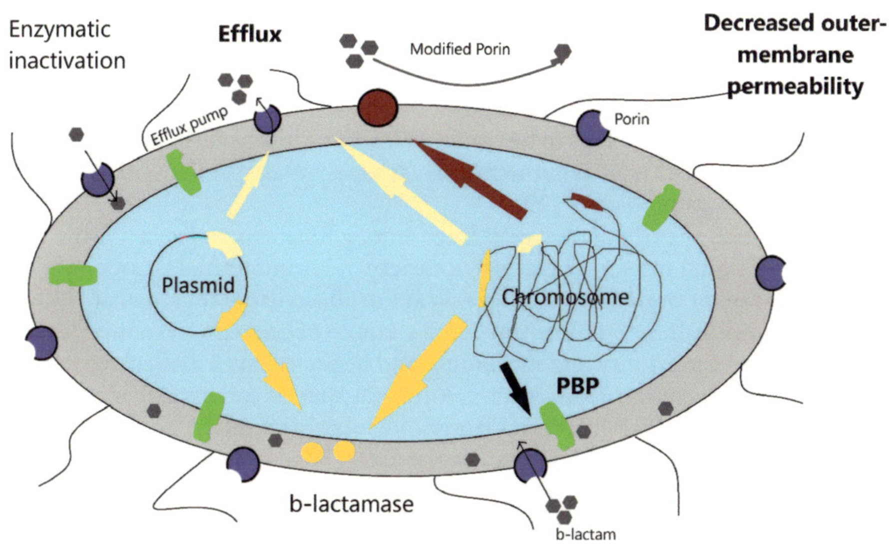

Figure 1.
Mechanism of carbapenem resistance in Enterobacteriaceae. (1) Reduced membrane permeability through modified porins, expression loss or shift in porin proteins in outer-membrane; (2) enzymatic inactivation by plasmid mediated or chromosomal enzymes (having hydrolytic activity); and (3) antibiotic efflux through efflux pump.

Several mobile genetic elements have their potential role in carbapenem resistance like Tn4401, Class I integrons, IncFIIK2, IncF1A, and IncI2 [17]. Transposon Tn4401 contains tnpR and tnpA genes, coding for "resolvase" and "transposase," respectively, and is mainly associated with $bla_{KPC\text{-}2}$ type [23]. Plasmids IncFIIK2, IncF1A, and IncI2 belong to ST101 *K. pneumoniae* type-2 found from bloodstream infections in the Asian region particularly in India.

7. Drivers of carbapenem resistance

To date, drivers for the acquisition of Carbapenem resistance among gram-negative bacteria have not been emphasized. But some of the known drivers for carbapenem resistance are prior long-term use of metronidazole and imipenem drugs in hospital settings, prior long-term hospital stays, and the presence of biliary drain catheters. It has been described that the disruption of normal flora by metronidazole increases the frequency of translocation, hence promoting carbapenem resistance among *Enterobacteriaceae* [24].

It has also been demonstrated that carbapenem resistance accelerated, once the gene for these enzymes became associated with acquired genetic elements like integrons and plasmids [25]. Thus the circulation of carbapenem-resistant genes among different strains isolated from clinic and hospital sewerage system coupled with the transfer of such genes by bacteriophage carrying B-lactamases genes coding for OXA-B-lactamases is now been considered as potential drivers for the increased spread and emergence of Carbapenem resistance [26].

8. Carbapenem-resistant *Enterobacteriaceae*: a mounting health concern

The *Enterobacteriaceae* is responsible for causing healthcare-related infections. Recent studies reported by the regulatory authority "Centre for Disease Control and Prevention" reveal that more than 21.3% of healthcare-related infections are due to *Enterobacteriaceae* [27]. Spread and the emergence of Carbapenem-resistant *Enterobacteriaceae* is a mounting health concern around the globe [28]. Regulatory authority "Centre for Disease Control and Prevention" defines CRE as "Enterobacteriaceae that seems to be tested as resistant to any carbapenem agent including ertapenem or may demonstrate as carbapenemase production through molecular or phenotypic assay" [29].

The emergence of carbapenem resistance among *Enterobacteriaceae* (CRE) possessing additional resistance genes to a variety of antimicrobial classes had led to the creation of organisms nearly resistant to all available therapeutics [30]. Carbapenem-resistant *Enterobacteriaceae* are a family of bacteria, responsible for causing significant mortality and morbidity, and hence are very difficult to treat. Among the *Enterobacteriaceae*, *E. coli* and *Klebsiella* species can easily become carbapenem resistant. CRE infections commonly occur in healthcare and hospital settings as well as in nursing homes, while the patients on-going long-term antibiotic treatment are also highly susceptible to these CRE infections [31].

Epidemiological data on carbapenemase-producing carbapenem-resistant *Enterobacteriaceae* (CP-CRE) varies country to country. An important carbapenemase-producing carbapenem resistance (KPC) was the first identified carbapenemase in the USA in 1996, and the prevalence is distributed unevenly among the US states [32]. Since epidemiology of CRE varies differently, so in this regard, KPC is endemic in Israel, while VIM, IPM, NDM, and OXA-48 carbapenemases are endemic in Greece, Japan, India, and Turkey, respectively, and are also disseminated successfully around the globe [33]. The continuous movement of subjects infected or colonized with CP-CRE in conjunction with the continuous exposure of these subjects to medical care is a significant contributor to the spread of CP-CRE [34]. Therefore, the decisive detection of CP-CRE may be the initial step to combat such a mounting health concern [29].

9. Treatment options

Since CRE infections are very difficult to treat, some of the treatment options for addressing the threat of "Carbapenem-resistant Enterobacteriaceae" include tigecycline, polymyxins, aminoglycosides, fosfomycin, meropenem/vaborbactam, and ceftazidime/avibactam. Combinations of B-lactamase are also available and are safer and more effective for treating CRE infections. It has been reported that polymyxin monotherapy can also lead to the emergence of resistance; therefore, polymyxin in combination with carbapenems must be administered in an appropriate dose [35]. Similar is the case with fosfomycin. The use of fosfomycin intravenously is recommended for urinary tract infections [36]. Clinicians should be vigilant in exploring new treatment options as well as for detection of CRE infections. Many of the new treatment options are in process, but none of them represent a magic bullet to address this concerned threat.

10. Conclusion

The rapid spread of carbapenem resistance as well as carbapenem-resistant *Enterobacteriaceae* into the community is a growing and emerging threat to public health. Despite of the large efforts being made to control this public menace, it is very essential to look for some definite solution which still seems to be far off. Until a potential alternative solution to overcome this problem is found, application of infection control measures whenever CR is detected, rationalization of antibiotic use as well as ensuring active surveillance system may be some steps to control this menace.

An interdisciplinary and global assess should be examined for the formulation of new diagnostic and screening tools. In this regard, alternative strategies to antibiotics like the use of phage therapy and probiotics can reduce this resistance burden. The spread of resistance can be minimized by immunization, application of infection control measures, rationalization of antibiotic usage, proper screening and treatment, and education and awareness programs. At global, national, and regional level, tracking and bio-surveillance system and preventive approaches of MDR and AMR pathogens can control this "global resistome."

Author details

Bilal Aslam[1*], Maria Rasool[1,3], Saima Muzammil[1], Abu Baker Siddique[1], Zeeshan Nawaz[1], Muhammad Shafique[1], Muhammad Asif Zahoor[1], Rana Binyamin[2], Muhammad Waseem[1], Mohsin Khurshid[1,3], Muhammad Imran Arshad[4], Muhammad Aamir Aslam[4], Naveed Shahzad[5], Muhammad Hidayat Rasool[1] and Zulqarnain Baloch[6]

1 Department of Microbiology, Government College University Faisalabad, Pakistan

2 University of Agriculture, Sub Campus, Burewala-Vehari, Faisalabad, Pakistan

3 College of Allied Health Professionals, Directorate of Medical Sciences, Government College University Faisalabad, Pakistan

4 Institute of Microbiology, University of Agriculture Faisalabad, Pakistan

5 School of Biological Sciences, The University of Punjab, Lahore, Pakistan

6 College of Veterinary Medicine, South China Agricultural University, Guangzhou, China

*Address all correspondence to: drbilalaslam@gcuf.edu.pk

References

[1] WHO. The Evolving Threat of Antimicrobial Resistance: Options for Action. Geneva: World Health Organization; 2012

[2] Lammie SL, Hughes JM. Antimicrobial resistance, food safety, and one health: The need for convergence. Annual Review of Food Science and Technology. 2016;**7**:287-312

[3] Robert JG. IDSA public policy: Combating antimicrobial resistance: Policy recommendations to save lives. Clinical Infectious Diseases: An Official Publication of the Infectious Diseases Society of America. 2011;**52**:S397

[4] Bartlett JG, Gilbert DN, Spellberg B. Seven ways to preserve the miracle of antibiotics. Clinical Infectious Diseases. 2013;**56**:1445-1450

[5] Aslam B, Wang W, Arshad MI, et al. Antibiotic resistance: A rundown of a global crisis. Infection and Drug Resistance. 2018;**11**:1645

[6] Chaudhry TH, Aslam B, Arshad MI, Nawaz Z, Waseem M. Occurrence of ESBL-producing Klebsiella pneumoniae in hospital settings and waste. Pakistan Journal of Pharmaceutical Sciences. 2019;**32**:773-778

[7] Alvi RF, Aslam B, Shahzad N, Rasool MH, Shafique M. Molecular basis of quinolone resistance in clinical isolates of Klebsiella pneumoniae from Pakistan. Pakistan Journal of Pharmaceutical Sciences. 2018;**31**:1591-1596

[8] Knapp KM, English BK. Carbapenems. In: Seminars in Pediatric Infectious Diseases. Elsevier; 2001. pp. 175-185

[9] Sumita Y, Fakasawa M. Potent activity of meropenem against *Escherichia coli* arising from its simultaneous binding to penicillin-binding proteins 2 and 3. Journal of Antimicrobial Chemotherapy. 1995;**36**:53-64

[10] Bonfiglio G, Russo G, Nicoletti G. Recent developments in carbapenems. Expert Opinion on Investigational Drugs. 2002;**11**:529-544

[11] Queenan AM, Bush K. Carbapenemases: The versatile β-lactamases. Clinical Microbiology Reviews. 2007;**20**:440-458

[12] Bush K. Recent developments in β-lactamase research and their implications for the future. Clinical Infectious Diseases. 1988;**10**:681-690

[13] Rasmussen BA, Bush K. Carbapenem-hydrolyzing beta-lactamases. Antimicrobial Agents and Chemotherapy. 1997;**41**:223

[14] Ambler R, Coulson A, Frere J-M, et al. A standard numbering scheme for the class A beta-lactamases. Biochemical Journal. 1991;**276**:269

[15] Walsh T. The emergence and implications of metallo-β-lactamases in Gram-negative bacteria. Clinical Microbiology and Infection. 2005;**11**:2-9

[16] Codjoe F, Donkor E. Carbapenem resistance: A review. Medical Science. 2018;**6**:1

[17] Logan LK, Weinstein RA. The epidemiology of carbapenem-resistant Enterobacteriaceae: The impact and evolution of a global menace. The Journal of Infectious Diseases. 2017;**215**:S28-S36

[18] Meletis G. Carbapenem resistance: Overview of the problem and future perspectives. Therapeutic Advances in Infectious Disease. 2016;**3**:15-21

[19] Bush K, Fisher JF. Epidemiological expansion, structural studies, and clinical challenges of new β-lactamases from gram-negative bacteria. Annual Review of Microbiology. 2011;**65**:455-478

[20] Bush K, Jacoby GA. Updated functional classification of β-lactamases. Antimicrobial Agents and Chemotherapy. 2010;**54**:969-976

[21] Bedenić B, Plečko V, Sardelić S, Uzunović S, Godič Torkar K. Carbapenemases in gram-negative bacteria: Laboratory detection and clinical significance. BioMed Research International. 2014;**2014**

[22] Walsh C. Molecular mechanisms that confer antibacterial drug resistance. Nature. 2000;**406**:775

[23] Tang Y, Li G, Liang W, Shen P, Zhang Y, Jiang X. Translocation of carbapenemase gene blaKPC-2 both internal and external to transposons occurs via novel structures of Tn1721 and exhibits distinct movement patterns. Antimicrobial Agents and Chemotherapy. 2017;**61**:1-10

[24] Jeon M-H, Choi S-H, Kwak YG, et al. Risk factors for the acquisition of carbapenem-resistant *Escherichia coli* among hospitalized patients. Diagnostic Microbiology and Infectious Disease. 2008;**62**:402-406

[25] Alekshun MN, Levy SB. Molecular mechanisms of antibacterial multidrug resistance. Cell. 2007;**128**:1037-1050

[26] Muniesa M, Garcia A, Miro E, et al. Bacteriophages and diffusion of β-lactamase genes. Emerging Infectious Diseases. 2004;**10**:1134

[27] Hidron AI, Edwards JR, Patel J, et al. Antimicrobial-resistant pathogens associated with healthcare-associated infections: Annual summary of data reported to the National Healthcare Safety Network at the Centers for Disease Control and Prevention, 2006-2007. Infection Control & Hospital Epidemiology. 2008;**29**:996-1011

[28] Guh AY, Limbago BM,Kallen AJ. Epidemiology and prevention of carbapenem-resistant Enterobacteriaceae in the United States. Expert Review of Anti-infective Therapy. 2014;**12**:565-580

[29] Lutgring JD, Limbago BM. The problem of carbapenemase-producing-carbapenem-resistant-Enterobacteriaceae detection. Journal of Clinical Microbiology. 2016;**54**:529-534

[30] Logan LK. Carbapenem-resistant enterobacteriaceae: An emerging problem in children. Clinical Infectious Diseases. 2012;**55**:852-859

[31] Wanger A, Chavez V, Huang R, Wahed A, Actor J, Dasgupta A. Antibiotics, antimicrobial resistance, antibiotic susceptibility testing, and therapeutic drug monitoring for selected drugs. Microbiology and Molecular Diagnosis in Pathology. 2017:119-153. DOI: 10.1016/B978-0-12-805351-5.00007-7

[32] Guh AY, Bulens SN, Mu Y, et al. Epidemiology of carbapenem-resistant Enterobacteriaceae in 7 US communities, 2012-2013. Jama. 2015;**314**:1479-1487

[33] Cantón R, Akóva M, Carmeli Y, et al. Rapid evolution and spread of carbapenemases among Enterobacteriaceae in Europe. Clinical Microbiology and Infection. 2012;**18**:413-431

[34] Molton JS, Tambyah PA, Ang BS, Ling ML, Fisher DA. The global spread of healthcare-associated multidrug-resistant bacteria: A perspective from

Asia. Clinical Infectious Diseases. 2013;**56**:1310-1318

[35] Bergen PJ, Landersdorfer CB, Zhang J, et al. Pharmacokinetics and pharmacodynamics of 'old' polymyxins: What is new? Diagnostic Microbiology and Infectious Disease. 2012;**74**:213-223

[36] Ellington MJ, Livermore DM, Pitt TL, Hall LM, Woodford N. Mutators among CTX-M β-lactamase-producing *Escherichia coli* and risk for the emergence of fosfomycin resistance. Journal of Antimicrobial Chemotherapy. 2006;**58**:848-852

Chapter 5

Mobile Genetic Elements in Vancomycin-Resistant *Enterococcus faecium* Population

Gastón Delpech, Leonardo García Allende and Mónica Sparo

Abstract

Horizontal gene transfer constitutes a key driving force in bacterial evolution. The ability to acquire mobile genetic elements encoding antimicrobial resistance has contributed to the emergence of *Enterococcus faecium* as one of the main human nosocomial opportunistic pathogens. The deep analysis of the vancomycin-resistant *E. faecium* (VREfm) population's mobilome, as the architecture and evolution of the core genome enables to observe VREfm plasticity and power of adaptation in animals, plants, environment and food. The persistence of VREfm is facilitated by the exchange of plasmids, phages and conjugative transposons that have allowed them to achieve a rapid adaptation to changes in environmental conditions. They can acquire resistance determinants from several species and transfer resistance genes to other potentially pathogenic bacteria such as methicillin-resistant *Staphylococcus aureus* strains.

Keywords: *E. faecium*, vancomycin, resistance, mobile genetic elements

1. Introduction

Enterococcus faecium is a main bacterial agent of healthcare-associated infections in immunocompromised as well as severely ill patients, with a worldwide distribution [1, 2].

In 1988, vancomycin-resistant *E. faecium* (VREfm) was reported for the first time. Along the 1990s, a fast increase of infectious diseases due to this bacterium was detected in the United Kingdom hospitals, linking its emergence with the employment of the glycopeptide avoparcin in animal husbandry for food industry. In addition, at U.S hospitals it was observed the emergence of VREfm, but with not proved association to the use of avoparcin in animals [3–5].

The World Health Organization's global priority pathogens list of antibiotic-resistant bacteria has categorized VREfm as of high priority. For infectious diseases produced by VREfm, it has been reported that the therapeutic options are more limited, altogether with higher mortality rates and financial costs for the Health system when compared with vancomycin-susceptible enterococci [6, 7].

Food chain can be considered as one possible way of VREfm spread or for the transfer of its antimicrobial resistance genes to humans, as it has been reported for cattle, pork and poultry meat [5, 8].

In the European Union, despite the avoparcin ban 18 years ago, VREfm circulation in the environment has continued. A likely cause of vancomycin-resistance plasmid genes persistence is the co-selection of other antimicrobials used in animals, such as macrolides or narasin, as it has been suggested by the presence of *ermB* type transporter genes (macrolide-lincosamide-streptogramin B resistance), as well as ABC type transporter genes and the presence of a toxin/anti-toxin system. Other possibilities which can relate with VREfm spread is their persistence in food farms, slaughterhouses or their environments due to poor hygienic conditions or through avian transmission [9–11].

It is important to highlight that, enterococci, as part of human and animal intestinal microbiota, are able to acquire resistance genes from other commensal bacteria, which can be spread as well to other pathogenic bacteria [12, 13].

Evolution of *E. faecium* from intestinal commensal bacteria to opportunistic pathogen is a complex and sequential process, in which seem to have been involved different factors, such as resistance and virulence determinants acquisition and persistence. Their expression is assumed to give an adaptive advantage since these factors facilitate the colonization of different epithelial cells (urinary, oral or intestinal), and at the same time, the bacterial adhesion to a wide variety of extracellular matrix proteins.

2. VREfm: natural and acquired antimicrobial resistance

E. faecium is intrinsically resistant to penicillin, ampicillin, cephalosporins and other β-lactams by mutations in the penicillin-binding protein PBP5 that is encoded by a horizontally transferred gene. Globally, enterococci are *in vivo* resistant to clindamycin (efflux pumps), trimethoprim-sulfamethoxazole (missing target) and the majority of aminoglycosides (enzymatic degradation). Furthermore, *E. faecium* has been acquiring resistance to quinolones, rifampicin and chloramphenicol, through mutations or by horizontal gene transfer [14–18].

In regard with vancomycin resistance, only the *vanA* and *vanB* genotypes are epidemiologically relevant in clinical isolates. In this sense, the *vanA* cluster is the most prevalent glycopeptide resistance determinant in clinical settings. Recently, the presence of *vanB* cluster has increased in Europe, while is the main vancomycin resistance mechanism in Australia. These genotypes are associated with mobile genetic elements. The *vanA* gene cluster is generally part of the Tn*3*-family transposon Tn*1546*. Among *vanB* cluster, $vanB_2$ is the most frequent subtype and constitutes an integral part of the integrative conjugative element Tn*1549*/*5382* [19–23].

2.1 The VREfm-mobilome

The mobilome is defined as all the mobile genetic elements (MGEs) able to move around within or between genomes. MGEs contribute to genome plasticity and dissemination of antimicrobial resistance and pathogenicity bacterial genes. In *E. faecium*, the acquisition of exogenous DNA is involved in the change of a commensal bacterium for becoming a pathogenic strain [24].

Horizontal gene transfer (HGT) allows the exchange of genetic material between bacteria. The most important HGT mechanism is conjugation, where the type IV secretion systems create channels between bacterial cells for transferring DNA.

The others mechanisms involved in HGT are transformation, in which bacteria are able to internalize naked DNA located in their immediate environment, and transduction, in which DNA is trapped within bacteriophages that have infected a bacterial cell and, then, is released and inserted into the genome of a new cell after bacteriophage transmission. Other gene transfer mechanisms as nanotubes, micro-vesicles and gene-transfer agents have not been described in enterococci yet [25, 26].

There have been described three mechanisms of attack and defense interacting with HGT, toxin-antitoxin (T/A) systems, restriction/modification (R/M) systems and Clustered Regularly Interspaced Short Palindromic Repeats (CRISPR)-Cas enzymes.

- T/A systems are small elements conformed by a toxin gen and its related antitoxin. Plasmid-encoded TA elements are important for plasmid maintenance. There are five types of T/A systems but only type 2 is prevalent in enterococci. In *E. faecium*, type 2 T/A systems comprise Axe/Txe and omega/epsilon/zeta. These plasmid-located T/A systems are enriched in clinical multi-drug resistant isolates [26–28].

- R/M systems, in which a restriction enzyme cleaves in a specific unmethylated DNA site and other enzyme links a methyl group to the same site; thus, DNA cleavage is blocked. This system contributes with the regulation of gene exchange in *E. faecium* [29].

- CRISPR-Cas systems constitute endogenous barriers to HGT in bacteria. A set of genes (*cas*) encoding nucleases are located near the CRISPR. Nosocomial clade of *E. faecium* is in great measure deficient of the CRISPR-Cas systems [30]. This fact is associated with the increased presence of MGEs. Conversely, commensal *E. faecalis* contain type II CRISPR-Cas systems, but multi-drug resistant (MDR) strains not carry complete CRISPR systems. Thus, MDR *E. faecalis* are prone for acquiring antibiotic resistance genes [31].

Among enterococci, different types of DNA arrangements and/or MGEs can be found, such as insertion sequences (IS), pathogenicity islands (PAIs), transposons (Tn) and plasmids.

IS are DNA segments (0.5–2 kb) able to autonomously move and to be found integrated in any replicon, in chromosomes as well as in plasmids. When IS appear in the middle of genes, they can interrupt the encoding sequence and inactivate the gene expression.

PAIs are fractions of a microorganism's genomic DNA linked with encoding sequences for virulence traits, such as adhesins, host immune evasion factors, toxins, cell components lytic enzymes, among others. Usually, PAIs are included in plasmids and their origin is associated with horizontal transfer of genetic material.

Tn are genetic elements that are directly movable as DNA and can harbor adaptive functions such as an antimicrobial resistance mechanism.

Plasmids are small extrachromosomal DNAs that can replicate independently (replicons). In enterococci, these genetic elements are wide-spread. Plasmid size is variable and is reflected in the number of genes they contain and the range of encoded functions. Plasmids are able to include antimicrobial resistance genes, stability modules and conjugation modules. In addition, are termed conjugative plasmids when they encode the type IV secretion system (T4SS) and are mobilizable if they contain the origin of transfer (*oriT*) and the relaxase protein, type IV coupling protein T4CP. In enterococci, plasmid replication proteins may be classified by mode of replication, sequence similarity and subdomains present within the translated gene. Replication proteins replicate the plasmids by unidirectional leading strand Rolling Circle Replication (RCR) and by bi-directional Theta (q) replication. RCR plasmids are frequently small, cryptic and unstable over a 10–15 kb size.

A plasmid typing method based on the replication regions from various plasmid incompatibility groups was described in enterococci and other Gram-positive bacteria, and 19 replicon families (*rep*-family) and some unique replicons were found [32].

The q plasmids are subdivided into replicon families: Rep_3, Inc18 and RepA_N:

- Rep_3 plasmids: narrow host range of similar size to RCR plasmids and often cryptic.

- Inc18 plasmids: often conjugative (25–50 kb) broad host-range plasmids; most of them harbor resistance determinants.

- RepA_N plasmids (10–300 kb): prevalent in low G + C content Gram positive bacteria with a narrow host range.

This scheme can be modified by recombination, leading to mosaic structures [26, 33–36].

The pheromone-responsive plasmids have been described mainly in *E. faecalis*; pAD1 and pCF10 were the first described.

Different plasmid diversity between VREfm and *E. faecalis* strains producers of nosocomial infections can be observed. VREfm, mainly CC17, show many *rep* types as rep_{11} (pB82), rep_{14} (pRI1), rep_{18} (pEF418), rep_{unique} (pC1Z2) rep_1 and rep_2 (Inc18), rep_{17} (pRUM), rep_{unique} (pHTβ) were found. Vancomycin-resistant *E. faecalis* carry a lower diversity of plasmid, generally associated with rep_9 type (pheromone responsive pAD1), rep_1 and rep_2 (Inc-18 type) as well [34].

The presence of big transferable plasmids, also known as megaplasmids (>150 kb) is common among clinical isolates of *E. faecium*, and can have a role related with their virulence. Often, these plasmids contain genes linked with different carbohydrates metabolism, such as hyl_{Efm} gene. Initially, it was suggested that this gene encoded for a hyaluronidase. Nevertheless, more recent sequencing studies showed that, actually, this gene encodes for a glycosyltransferase which allows the utilization of complex carbohydrates. Furthermore, it has been proven that the transfer of these MGEs to non-carrying plasmid commensal strains of *E. faecium*, will increase their virulence and their gastrointestinal colonization capability [19, 33, 37].

Worldwide, most of the VREfm strains recovered in clinical settings were included into the clonal complex 17 (CC17). Afterwards, they were divided into three lineages (17, 18 and 78), using multilocus sequence typing studies (MLST). More recently, the Bayesian Analysis of Population Structure (BAPS), applied to MLST data established two nosocomial groups: 2–1 (lineage 78) and 3–3 (lineages 17/18). All CC17 *E. faecium* strains contain many exogenously acquired genes such as IS, phages, and Tn encoding antimicrobial resistance. Furthermore, hospital-adapted VREfm are ciprofloxacin and ampicillin-resistant, with virulence traits also found in theirs genomes. VREfm strains have cell surface protein genes, regulatory genes, putative PAIs, plasmids, IS and integrated phages, which promote their adaptation to the healthcare-associated environment. The IS*16* and the *esp* gene are carried by an integrative conjugative element (ICEEfm1) with the *intA* integrase gene, and are considered as markers of nosocomial *E. faecium* strains [5, 17, 29].

In *E. faecium* CC17, the location of hyl_{Efm} gene was described in a large conjugative plasmid, pLG1 (281.02 kb), in association with the *vanA* operon, the *ermB* gene (macrolide-lincosamide-streptogramin B resistance) and the *tcrYAZB* operon (heavy metal resistance). The hyl_{Efm} gene, an important factor involved in enterococcal colonization and adhesion, it has also been described as part of a genomic island. The dissemination of the multi-resistant megaplasmid pLG1, carrying hyl_{Efm} could explain the spread of the so frequently isolated hospital-associated *E. faecium* CC17 genotype [33].

Transposable elements contribute with the genome plasticity by different mechanisms. They are substrates for homologous recombination within and between different DNA elements and rearrangements are carried out in chromosome and plasmid DNA [38].

In glycopeptide-resistant enterococci, vancomycin resistance is classified into eight acquired gene clusters: *vanA*, *vanB*, *vanD*, *vanE*, *vanG*, *vanL*, *vanM* and *vanN*. VanA- and VanB-type vancomycin-resistant enterococci (VRE) constitute the majority of VRE in clinical settings. VanA-type VRE shows high-level resistance to vancomycin (Minimum Inhibitory Concentration, MIC = 64–100 mg/L) and teicoplanin (MIC = 16–512 mg/L), while VanB-type VRE is susceptible to teicoplanin (MIC = 0.5–1 mg/L) and expresses different levels of resistance to vancomycin (MIC = 4–1000 mg/L). Also, it can be mentioned the intrinsic *vanC* genotype, found in *E. gallinarum* and *E. casseliflavus* [39, 40]. The main phenotypic and genotypic features of glycopeptide resistant enterococci are shown in **Table 1**.

The *vanB* gene cluster consists of a two-component regulatory system (*vanRB*, *vanSB*) and five resistance genes (*vanYB*, *vanW*, *vanHB*, *vanB*, *vanXB*). Conversely to the highly conserved resistance genes, the amino acid sequences of VanSB and VanRB show less similarity to those of VanSA and VanRA. These differences could be responsible for the characteristics of VanB-type resistance [41].

Furthermore, low-level vancomycin resistant *E. faecium* can turn into high-level vancomycin resistant during antibiotic therapy. This variant was named vancomycin-variable *E. faecium* [20, 39, 42]. A schematic diagram of *van* operon is shown in **Figure 1**.

Tn*1546* carries the *vanA* gene, and is often located on a plasmid belonging to the broad host range Inc18 family, involved in the *vanA* transfer from enterococci to *Staphylococcus aureus*. Typically, the *vanA* operon is associated with Tn, such as Tn*1546*, implicating two genes for the transposition of the element (*orf1* and *orf2*), and one gene involved with teicoplanin resistance (*vanZ*). The *vanA* gene cluster includes seven open reading frames transcribed from two separate promoters. The regulatory apparatus is encoded by the *vanR* (response regulator) and *vanS* (sensor kinase) two-component system. Both are transcribed from one promoter, while the remaining genes are transcribed from other promoter. *vanH* (dehydrogenase that converts pyruvate to lactate) and *vanA* (ligase that forms D-Ala-D-Llac dipeptide) modify the production of peptidoglycan precursors. The production of the normal ending D-Ala-D-Ala of the pentapeptide does not continue. The products of *vanX* (encodes a dipeptidase that cleaves D-Ala-D-Ala) and *vanY* (encodes a D, D-carboxypeptidase) genes hydrolyze, interrupt the production of the pentapeptides and cleave the pentapeptides that can still be produced. The variations in the composition of this vancomycin resistance operon is due to the insertion of IS elements. The *vanB* operon is carried by Tn*1547*, Tn*1549* and Tn*5382*. Tn*1549* conjugative Tn, is mainly located in large conjugative chromosomal elements and less frequently integrated in conjugative plasmids. This conjugative *vanB* Tn is widely prevalent among *VanB* type enterococci and other Gram-positive bacteria. The *vanB* operon has a similar genetic organization to the *vanA* because *vanB* operon contains two distinct promoters transcribing seven open reading frames. But *vanB* encodes a two-component signaling system (named *vanRB* and *vanSB*) that is considerably different from that encoded in *vanA*. Furthermore, *vanB* encodes homologs of *vanH* and the D-Ala-D-Ala ligase, and the peptidases (*vanX* and *vanY*). In addition, *vanB* lacks a homolog of *vanZ*, and instead encodes a protein *vanW*, with a role not totally explained. *vanB* gene (ligase) has been divided in three subtypes,

Phenotype	VanA	VanB	VanD	VanE	VanG	VanL	VanM	VanN	VanC
Resistance	Acquired	Acquired	Acquired	Acquired	Acquired	Acquired	Acquired	Acquired	Intrinsic
MICvan	16–>1000	4–>1000	16–128	8–32	16	8	>256	16	2–32
MIC_{tei}	16–512	0.25–2	2–64	0.5	0.5	0.5	96	0.5	0.12–2
Expression	I	I	C	I	I	I	I	C	I, C
Mobiltiy	Yes	Yes	No	No	Yes	No	Yes	Yes	No
Precursor	Ala-Lac	Ala-Lac	Ala-Lac	Ala-Ser	Ala-Ser	Ala-Ser	Ala-Lac	Ala-Ser	Ala-Ser
Operon	*vanA*	*vanB*	*vanD*	*vanE*	*vanG*	*vanL*	*vanM*	*vanN*	*vanC*
Subtypes	N/A	B1-B3	D1-D5	N/A	G1-G2	N/A	N/A	N/A	C1-C4
Required genes for expression	*vanH*, *vanA*, *vanX*	$vanH_B$, *vanB*, $vanX_B$	$vanH_D$, *vanD*, $vanX_D$	*vanE*, $vanXY_E$, $vanT_E$	*vanG*, *vanXYG*, *vanTG*	*vanL*, $vanXY_1$, $vanT_m$, $vanTr_1$	$vanH_M$, *vanM*, $vanX_M$	*vanN*, $vanXY_N$, $vanT_N$	*vanC*, $vanXY_C$, *vanT*

MIC, minimum inhibitory concentration (mg/L); van, vancomycin; tei, teicoplanin; I, inducible; C, constitutive; N/A, not applies. Adapted from [39, 40].

Table 1.
Main phenotypic and genotypic features of glycopeptide-resistant enterococci.

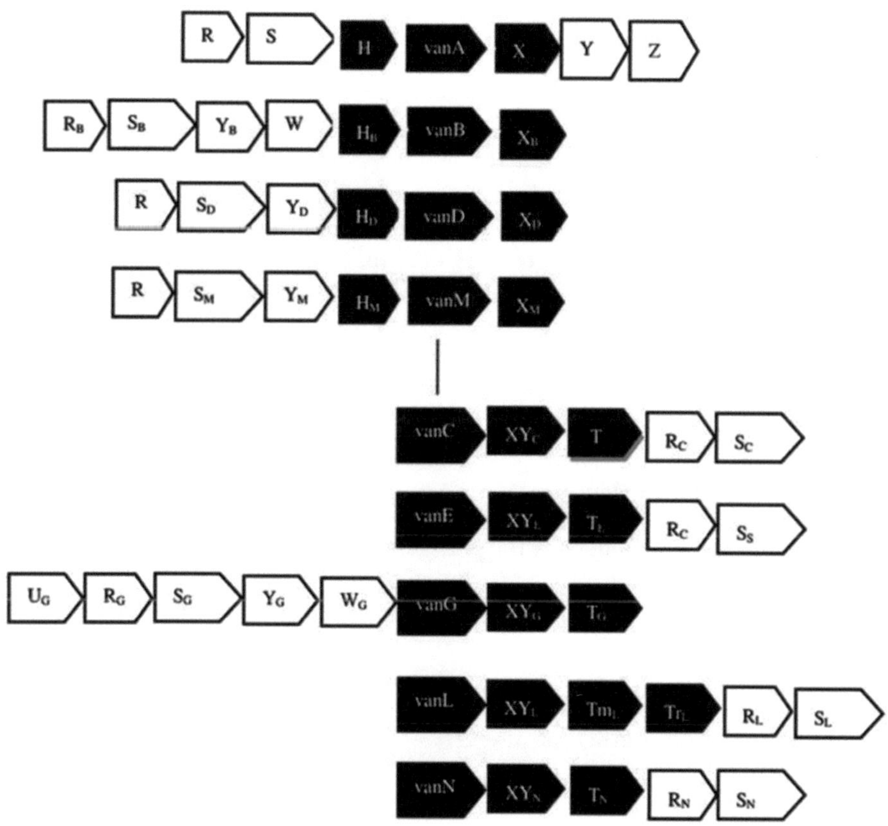

Figure 1.
Schematic diagram of van operon. Adapted from [40].

vanB1-3, based on nucleotidic sequence differences. *vanB2* subtype is the most commonly spreaded in clinical enterococci. Also, is part of conjugative Tn, Tn*1549*/Tn*5382*-like. The first description of a *vanB2*-Tn*1549*-like element in pheromone responsive plasmids (pCF10-like) carried by *E. faecalis* was reported at Japan. *vanB1* has only been described for certain isolates as part of composite Tn or an integrative conjugative element [21, 43–48].

It has been described that some *vanA* genotype isolates had a new type F Tn*1546* Tn associated with two insertion sequences: IS*1216V* and IS*1251* [49].

The *vanM* cluster is similar to *vanA*, *vanB*, and *vanD*, while *vanL* and *vanN* are similar to *vanC*. The *vanM* operon has been described in VREfm isolates and showed a close genetic arrangement to *vanD* and *in vitro* transferable resistance by conjugation. The *vanN* operon is the most recently identified gene cluster described in *E. faecium* and is a similar operon to *vanG*, but can be transferred by MGEs only in this enterococcal species. The IS elements can produce structural alterations in the genes, and as a consequence leads to changes of resistance phenotype. The *vanA* gene cluster is prone to IS-mediated alterations, modifying sometimes the vancomycin resistance phenotype, as being susceptible to glycopeptides but with possibility to revert to a resistant phenotype. These bacteria were named vancomycin-variable enterococci (VVE), which could cause serious clinical issues because of their possibility to escape of detection and surveillance as well as facilitating the horizontal transfer of vancomycin resistance [50–52].

An additional operon (*vanF*) was described but only in *Paenibacillus popilliae*. This *vanF* cluster has a high similarity to the amino acid sequences of the *vanA* operon, and *P. popilliae* has been proposed as a possible origin for vancomycin resistance in enterococci [53].

3. Conclusions

The massive use of glycopeptides (vancomycin and teicoplanin) and non-glycopeptide agents such as extended-spectrum cephalosporins in clinical settings have been implicated in the emergence of VREfm. Delayed effective antimicrobial therapy more than 48 h after the beginning of VRE bacteremia is associated with higher mortality rates.

The core genes bring a phylogenomic reconstruction of the *E. faecium* population structure; the main contribution of accessory genes includes the adaptation of this species to nosocomial environments. It was observed that the plasmid component drives host specificity, while their whole genome and chromosome share a common evolutionary history.

The clinical isolates' mobilome are quite different from the other hosts. In VREfm the plasmid component of the pan-genome plays an important role in adaptation and its emergence as a nosocomial pathogen.

Conflict of interest

Authors declare no conflicts of interest.

Author details

Gastón Delpech[1], Leonardo García Allende[2] and Mónica Sparo[1*]

1 School of Medicine, Universidad Nacional del Centro de la Provincia de Buenos Aires, Olavarría, Argentina

2 Hospital Privado de Comunidad, Mar del Plata, Argentina

*Address all correspondence to: monicasparo@gmail.com

References

[1] Cattoir V, Leclercq R. Twenty-five years of shared life with vancomycin-resistant enterococci: Is it time to divorce? Journal of Antimicrobial Chemotherapy. 2013;**68**:731-742. DOI: 10.1093/jac/dks469

[2] Sievert DM, Ricks P, Edwards JR, Schneider A, Patel J, Srinivasan A, et al. Antimicrobial-resistant pathogens associated with healthcare-associated infections: Summary of data reported to the National Healthcare Safety Network at the Centers for Disease Control and Prevention, 2009-2010. Infection Control and Hospital Epidemiology. 2013;**34**:1-14. DOI: 10.1086/668770

[3] Leclercq R, Derlot E, Duval J, Courvalin P. Plasmid-mediated resistance to vancomycin and teicoplanin in *Enterococcus faecium*. New England Journal of Medicine. 1988;**319**:157-161

[4] Reacher MH, Shah A, Livermore DM, Wale MCJ, Graham C, Johnson AP, et al. Bacteraemia and antibiotic resistance of its pathogens reported in England and Wales between 1990 and 1998: Trend analysis. BMJ. 2000;**320**:213-216

[5] Sparo M, Delpech G, García Allende N. Impact on public health of the spread of high-level resistance to gentamicin and vancomycin in enterococci. Frontiers in Microbiology. 2018;**9**:3073. DOI: 10.3389/fmicb.2018.03073

[6] Cheah AL, Spelman T, Liew D, Peel T, Howden BP, Spelman D, et al. Enterococcal bacteraemia: Factors influencing mortality, length of stay and costs of hospitalization. Clinical Microbiology and Infection. 2013;**19**:E181-E189. DOI: 10.1111/1469-0691.12132

[7] World Health Organization. Global Priority List of Antibiotic-Resistant Bacteria to Guide Research, Discovery, and Development of New Antibiotics. Geneva, Switzerland: WHO; 2017

[8] Delpech G, Pourcel G, Schell C, De Luca M, Basualdo J, Bernstein J, et al. Antimicrobial resistance profiles of *Enterococcus faecalis* and *Enterococcus faecium* isolated from artisanal food of animal origin in Argentina. Foodborne Pathogens and Disease. 2013;**9**:939-944. DOI: 10.1089/fpd.2012.1192

[9] Aarestrup FM. Characterization of glycopeptide-resistant *Enterococcus faecium* (GRE) from broilers and pigs in Denmark: Genetic evidence that persistence of GRE in pig herds is associated with coselection by resistance to macrolides. Journal of Clinical Microbiology. 2000;**38**:2774-2777

[10] Garcia-Migura L, Liebana E, Jensen LB, Barnes S, Pleydell E. A longitudinal study to assess the persistence of vancomycin-resistant *Enterococcus faecium* (VREF) on an intensive broiler farm in the United Kingdom. FEMS Microbiology Letters. 2007;**275**:319-325

[11] Nilsson O, Greko C, Bengtsson B, Englund S. Genetic diversity among VRE isolates from Swedish broilers with the coincidental finding of transferrable decreased susceptibility to narasin. Journal of Applied Microbiology. 2012;**112**:716-722. DOI: 10.1111/j.1365-2672.2012.05254.x

[12] Sparo M, Urbizu L, Solana MV, Pourcel G, Delpech G, Confalonieri A, et al. High-level resistance to gentamicin: Genetic transfer between *Enterococcus faecalis* isolated from food of animal origin and human microbiota. Letters in Applied Microbiology. 2012;**54**:119-125. DOI: 10.1111/j.1472-765X.2011.03182.x

[13] Weigel LM, Clewell DB, Gill SR, Clark NC, McDougal LK, Flannagan SE,

et al. Genetic analysis of a high-level vancomycin-resistant isolate of *Staphylococcus aureus*. Science. 2003;**302**:1569-1571

[14] Deshpande LM, Fritsche TR, Moet GJ, Biedenbach DJ, Jones RN. Antimicrobial resistance and molecular epidemiology of vancomycin resistant enterococci from North America and Europe: A report from the SENTRY antimicrobial surveillance program. Diagnostic Microbiology and Infectious Disease. 2007;**58**:163-170

[15] Miller WR, Munita JM, Arias CA. Mechanisms of antibiotic resistance in enterococci. Expert Review of Anti-Infective Therapy. 2014;**12**:1221-1236. DOI: 10.1586/14787210.2014.956092

[16] Murray BE. The life and times of the *Enterococcus*. Clinical Microbiology Reviews. 1990;**3**:46-65

[17] Pourcel G, Sparo M, Corso A, Delpech A, Gagetti P, de Luca MM, et al. Molecular genetic profiling of clinical and foodborne strains of enterococci with high level resistance to gentamicin and vancomycin. Clinical Microbiology Open Access. 2017;**6**:1. DOI: 10.4172/2327-5073.1000272

[18] Werner G, Fleige C, Ewert B, Laverde-Gomez JA, Klare I, Witte W. High-level ciprofloxacin resistance among hospital-adapted *Enterococcus faecium* (CC17). International Journal of Antimicrobial Agents. 2010;**35**:119-125. DOI: 10.1016/j.ijantimicag.2009.10.012

[19] Arias CA, Murray BE. The rise of the *Enterococcus*: Beyond vancomycin resistance. Nature Reviews Microbiology. 2012;**10**:266-278. DOI: 10.1038/nrmicro2761

[20] Courvalin P. Vancomycin resistance in gram-positive cocci. Clinical Infectious Diseases. 2006;**42**(suppl 1): S25-S34

[21] Dahl KH, Lundblad EW, Røkenes TP, Olsvik O, Sundsfjord A. Genetic linkage of the *vanB2* gene cluster to Tn*5382* in vancomycin-resistant enterococci and characterization of two novel insertion sequences. Microbiology. 2000;**146**:1469-1479

[22] Guzman Prieto AM, van Schaik W, Rogers MR, Coque TM, Baquero F, Corander J, et al. Global emergence and dissemination of enterococci as nosocomial pathogens: Attack of the clones? Frontiers in Microbiology. 2016;**7**:788. DOI: 10.3389/fmicb.2016.00788

[23] Howden BP, Holt KE, Lam MM, Seemann T, Ballard S, Coombs GW, et al. Genomic insights to control the emergence of vancomycin-resistant enterococci. MBio. 2013;**4**:e00412-e00413. DOI: 10.1128/mBio.00412-13.

[24] Willems RJ, van Schaik W. Transition of *Enterococcus faecium* from commensal organism to nosocomial pathogen. Future Microbiology. 2009;**4**:1125-1135. DOI: 10.2217/fmb.09.82

[25] Clewell DB, Francia MV. Conjugation in gram-positive bacteria. In: Funnal BE, Philips GJ, editors. Plasmid Biology. Washington, DC: ASM Press; 2004. pp. 227-256

[26] Clewell DB, Weaver KE, Dunny GM, Coque TM, Francia MV, Hayes F. Extrachromosomal and mobile elements in enterococci: Transmission, maintenance, and epidemiology. In: Gilmore MS, Clewell DB, Ike Y, Shankar N, editors. Enterococci: From Commensals to Leading Causes of Drug Resistant Infection. Boston, MA: Massachusetts Eye and Ear Infirmary; 2014. pp. 1-84

[27] Grady R, Hayes F. Axe-Txe, a broad-spectrum proteic toxin-antitoxin system specified by a multidrug-resistant,

clinical isolate of *Enterococcus faecium*. Molecular Microbiology. 2003;**47**:1419-1432

[28] Meinhart A, Alonso JC, Sträter N, Saenger W. Crystal structure of the plasmid maintenance system epsilon/zeta: Functional mechanism of toxin zeta and inactivation by epsilon 2 zeta 2 complex formation. Proceedings of the National Academy of Sciences of the United States of America. 2003;**100**:1661-1666

[29] Lebreton F, van Schaik W, McGuire AM, Godfrey P, Griggs A, Mazumdar V, et al. Emergence of epidemic multidrug-resistant *Enterococcus faecium* from animal and commensal strains. MBio. 2013;**4**:e00534-e00513. DOI: 10.1128/mBio.00534-13

[30] Palmer KL, Gilmore MS. Multidrug-resistant enterococci lack CRISPR-cas. MBio. 2010;**1**:e00227-e00210. DOI: 10.1128/mBio.00227-10

[31] Hullahalli K, Rodrigues M, Thy Nguyen U, Palmer K. An attenuated CRISPR-Cas system in *Enterococcus faecalis* permits DNA acquisition. MBio. 2018;**9**:e00414-e00418. DOI: 10.1128/mBio.00414-1

[32] Jensen LB, Garcia-Migura L, Valenzuela AJ, Lehr M, Hasman H, Aarestrup FM. A classification system for plasmids from enterococci and other gram-positive bacteria. Journal of Microbiological Methods. 2010;**80**:25-43. DOI: 10.1016/j.mimet.2009.10.012

[33] Freitas AR, Tedim AP, Novais C, RuizGarbajosa P, Werner G, Laverde-Gomez JA, et al. Global spread of the *hyl*$_{\mathrm{Efm}}$ colonization-virulence gene in megaplasmids of the *Enterococcus faecium* CC17 polyclonal subcluster. Antimicrobial Agents and Chemotherapy. 2010;**54**:2660-2665. DOI: 10.1128/AAC.00134-10

[34] Freitas AR, Tedim AP, Francia MV, Jensen LB, Novais C, Peixe L, et al. Multilevel population genetic analysis of *vanA* and *vanB Enterococcus faecium* causing nosocomial outbreaks in 27 countries (1986-2012). Journal of Antimicrobial Chemotherapy. 2016;**71**:3351-3366

[35] Lilly J, Camps M. Mechanisms of theta plasmid replication. Microbiology Spectrum. 2015;**3**:PLAS-0029-2014. DOI: 10.1128 /microbiolspec.PLAS-0029-2014

[36] Smillie C, Garcillán-Barcia MP, Francia MV, Rocha EP, de la CruzF.Mobilityofplasmids.Microbiology and Molecular Biology Reviews. 2010;**74**:434-452. DOI: 10.1128/MMBR.00020-10

[37] Rice LB, Laktičová V, Carias LL, Rudin S, Hutton R, Marshall SH. Transferable capacity for gastrointestinal colonization in *Enterococcus faecium* in a mouse model. Journal of Infectious Diseases. 2009;**199**:342-349. DOI: 10.1086/595986

[38] Rice LB, Carias LL, Marshall S, Rudin SD, Hutton-Thomas R. Tn*5386*, a novel Tn*916*-like mobile element in *Enterococcus faecium* D344R that interacts with Tn*916* to yield a large genomic deletion. Journal of Bacteriology. 2005;**187**:6668-6677

[39] Cercenado E. *Enterococcus*: Phenotype and genotype resistance and epidemiology in Spain. Enfermedades Infecciosas y Microbiología Clínica. 2011;**29**(Suppl. 5):59-65. DOI: 10.1016/S0213-005X(11)70045-3

[40] Ahmed MO, Baptiste KE. Vancomycin-resistant enterococci: A review of antimicrobial resistance mechanisms and perspectives of human and animal health. Microbiology Drug Resistance. 2018;**24**:590-606. DOI: 10.1089/mdr.2017.0147

[41] Evers S, Courvalin P. Regulation of VanB-type vancomycin resistance gene expression by the VanS(B)-VanR(B) two-component regulatory system in *Enterococcus faecalis* V583. Journal of Bacteriology. 1996;**178**:1302-1309

[42] Coburn B, Low DE, Patel SN, Poutanen SM, Shahinas D, Eshaghi A, et al. Vancomycin-variable *Enterococcus faecium*: *In vivo* emergence of vancomycin resistance in a vancomycin-susceptible isolate. Journal of Clinical Microbiology. 2014;**52**:1766-1767. DOI: 10.1128/JCM.03579-13

[43] Arthur M, Molinas C, Depardieu F, Courvalin P. Characterization of Tn*1546*, a Tn*3*-related transposon conferring glycopeptide resistance by synthesis of depsipeptide peptidoglycan precursors in *Enterococcus faecium* BM4147. Journal of Bacteriology. 1993;**175**:117-127

[44] Clark NC, Cooksey RC, Hill BC, Swenson JM, Tenover FC. Characterization of glycopeptide-resistant enterococci from U.S. hospitals. Antimicrobial Agents and Chemotherapy. 1993;**37**:2311-2317

[45] Garnier F, Taourit S, Glaser P, Courvalin P, Galimand M. Characterization of transposon Tn*1549*, conferring VanB-type resistance in *Enterococcus* spp. Microbiology. 2000;**146**:1481-1489

[46] Perichon B, Courvalin P. VanA-type vancomycin-resistant *Staphylococcus aureus*. Antimicrobial Agents and Chemotherapy. 2009;**53**:4580-4587. DOI: 10.1128/AAC.00346-09

[47] Sarti M, Campanile F, Sabia C, Santagati M, Gargiulo R, Stefani S. Polyclonaldiffusion of Beta-lactamase-producing *Enterococcus faecium*. Journal of Clinical Microbiology. 2012;**50**:169-172. DOI: 10.1128/JCM.05640-11

[48] Zheng B, Tomita H, Inoue T, Ike Y. Isolation of VanB-type *Enterococcus faecalis* strains from nosocomial infections: First report of the isolation and identification of the pheromone-responsive plasmids pMG2200, encoding VanB-type vancomycin resistance and a Bac41-type bacteriocin, and pMG2201, encoding erythromycin resistance and cytolysin (Hly/Bac). Antimicrobial Agents and Chemotherapy. 2009;**53**:735-747. DOI: 10.1128/AAC.00754-08

[49] Khan MA, Shorman M, Al-Tawfiq JA, Hays JP. New type F lineage-related Tn*1546* and a *vanA*/*vanB* type vancomycin-resistant *Enterococcus faecium* isolated from patients in Dammam, Saudi Arabia during 2006-2007. Epidemiololgy and Infection. 2013;**141**:1109-1114. DOI: 10.1017/S0950268812001574

[50] Kohler P, Eshaghi A, Kim HC, Plevneshi A, Green K, Willey BM, et al. Toronto invasive bacterial diseases network (TIBDN). Prevalence of vancomycin-variable *Enterococcus faecium* (VVE) among *vanA*-positive sterile site isolates and patient factors associated with VVE bacteremia. PLoS ONE. 2018;**13**:e0193926. DOI: 10.1371/journal.pone.0193926

[51] Lebreton F, Depardieu F, Bourdon N, Fines-Guyon M, Berger P, Camiade S, et al. D-Ala-d-Ser VanN-type transferable vancomycin resistance in *Enterococcus faecium*. Antimicrobial Agents and Chemotherapy. 2011;**55**:4606-4612. DOI: 10.1128/AAC.00714-11

[52] Partridge SR, Kwong SM, Firth N, Jensen SO. Mobile genetic elements associated with antimicrobial resistance. Clinical Microbiology Reviews. 2018;**31**:e00088-e00017. DOI: 10.1128/CMR.00088-17

[53] Patel R, Piper K, Cockerill FR 3rd, Steckelberg JM, Yousten AA. The biopesticide *Paenibacillus popilliae* has a vancomycin resistance gene cluster homologous to the enterococcal VanA vancomycin resistance gene cluster. Antimicrobial Agents and Chemotherapy. 2000;**44**:705-709

Chapter 6

Enterococci: An Important Nosocomial Pathogen

Sonia Bhonchal Bhardwaj

Abstract

Enterococci, particularly *Enterococcus faecalis* and *Enterococcus faecium*, are an important cause of nosocomial infections and have become a major issue worldwide. Nosocomial infections due to vancomycin resistant Enterococci (VRE) occur frequently. A significant increase in prevalence of VRE has been reported recently in many countries. Enterococci are second most frequent cause of nosocomial urinary tract infection, bacteremia and infective endocarditis. They are also related to etiology of intra-abdominal an pelvic infections, gastrointestinal infections and oral infections. The ability of Enterococci to survive in adverse conditions, presence of virulence factors and possession of intrinsic and acquired antibiotic resistance traits poses a therapeutic challenge. Due to high level of multidrug resistance in VRE, Enterococcus has become an important organism in health based settings.

Keywords: Enterococci, vancomycin, resistance, nosocomial, infections

1. Introduction

Enterococci are Gram-positive, non-spore forming and facultative anaerobic cocci. They are indigenous flora of the intestinal tract, oral cavity and vagina in healthy persons. The genus comprises 54 species which are ubiquitously present in nature [1]. Enterococci have emerged as an important nosocomial pathogens second to Staphylococci which is the leading cause of nosocomial infections worldwide [2]. Enterococci are important nosocomial pathogens causing up to 10% of all infections in the hospitalized patients [3]. In these Enterococci infections approximately 60% of infections are caused by *Enterococcus faecalis* and *Enterococcus faecium* causes the remaining [4]. In the last decade both *E. faecalis* and *E. faecium* have emerged as important nosocomial pathogens. Other Enterococcal species causing nosocomial human infections are *E. avium*, *E. gallinarum*, *E. casseliflavus*, *E. durans*, *E. raffinosus* and *E. mundtii*. Majority of clinical isolates (63–81%) are identified as *E. faecalis*, around 13–23% as *E. faecium* and other enterococcal species comprise around 3–4% of the clinical isolates [5].

2. Prevalence of vancomycin resistant Enterococci (VRE)

Nosocomial infections particularly by vancomycin resistant Enterococci (VRE) have become a major problem since last few years though VRE are organisms of low virulence and pathogenicity. Nosocomial infections caused by VRE are highly

IntechOpen

prevalent in intensive care units of hospitals. These infections are particularly high in presence of underlying health factors like diabetes, liver transplantation, neutropenia, diabetes mellitus and renal dysfunction. Recently it has also been seen that VRE bloodstream infections have higher mortality rates as compared to vancomycin susceptible Enterococci (VSE) [6, 7]. Data from countries like Germany shows an increase of VRE from less than 5% in 2001 to 14.5% in 2013 mainly vancomycin resistant *E. faecium* [8]. In Europe of all the nosocomial infections reported 9.6% were of Enterococci [9]. In USA 3% of the nosocomial infections are due to VRE [10]. VRE nosocomial infections cause greater number of invasive treatment resulting in extended stay in hospital and cost [11]. Hospitals in some countries have now established VRE screening in high risk areas and isolation of patients to prevent spread of the resistant pathogen [12]. A study has shown the prevalence of VRE colonization in patients who had history of previous administration of antibiotics for more than 2 weeks were 10 times more likely of getting VRE colonization [13]. Other studies have also reported similar findings which show antibiotic exposure can cause colonization of VRE in hospital settings because of their resistance to commonly used antibiotics, virulence factors and ability to acquire genes [14].

3. Genetic factors and antibiotic resistance in VRE

The genes Van A, Van B, Van C, Van D and Van E are responsible for vancomycin resistance in Enterococci. Van M has been identified which is also an important vancomycin resistant determinant among different *E. faecium* lineages in hospitals in Shanghai, China [15]. Each Van operon has different ecological origin, Van A has originated from soil organisms, van B, Van G and Van D from gut microbiota [16]. Vancomycin resistance in Enterococci is of two types (a) Intrinsic resistance—Enterococci spp. like *E. gallinarum* and *E. casseliflavus* show an inherent low level resistance to Vancomycin. They have Van C genes that produce Vancomycin minimum inhibitory concentration (2–32 μg/ml) [17] A hospital wide outbreak of vancomycin resistant *E. gallinarum* has been reported in Colombia showing that uncommon species of Enterococci are capable of spreading in the hospital environment and producing nosocomial infections [18]. The second type is (b) acquired resistance—Enterococci species acquire resistance genes and become resistant to vancomycin. This is seen in *E. faecium* and *E. faecalis* and to some extent in *E. raffinosus*, *E. avium*, *E. durance* and other enterococcal species. The most common isolated Enterococci species which is VRE in hospital settings is *E. faecium*. It has been seen that *E. faecium* produces high vancomycin minimum inhibitory concentration (64–1000 μg/ml) [19]. There has been a significant increase in VRE prevalence. The emergence and rapid spread of VRE has led to the use of new antibiotics like linezolid, daptomycin and tigecycline. Linezolid is a oxazolidinone antibiotic. An oxazolidinone resistance gene optr A has been identified in *E. faecalis* and *E. faecium* isolates of human and animal origin [20] .Linezolid resistance is still less prevalent reported as 1.1 and 1.8% in *E. faecium* and *E. faecalis* isolates from 19 US hospitals [21]. Daptomycin resistance is more prevalent in *E. faecium* than *E. faecalis* isolates. Around 3.9 and 0.2% of *E. faecium* and *E. faecalis* isolates have been reported in various hospital settings [22]. Tigecycline is a semisynthetic derivative of tetracycline. Tigecycline resistance in *E. faecium* and *E. faecalis* is rare and reported as 0.3%. It is being used to treat bacteremia caused by MDR enterococci. The increased use of antibiotics in hospitals is causing gut dysbiosis and enterococci possess surviving ability take over the niche in the gastrointestinal tract and this could be the primary source of enterococcal infections [23].

4. Lineages of nosocomial Enterococci

The ability of *E. faecium* to exchange mobile genetic elements carrying antimicrobial resistance genes and virulence determinants has resulted in hospital adapted clones [24]. Esp was the first adaptive element found in hospital strains of *E. faecium*. The *E. faecium* esp. gene has been linked to biofilm formation, UTI and endocarditis [24]. New determinants have been now linked to hospital isolates of *E. faecium*. A genomic analysis study of *E. faecium* hospital strains identified gain and loss of gene clusters in clinical and non-clinical isolates of *E. faecium* [25]. Genomic studies of nosocomial *E. faecium* infection have confirmed the transmission of *E. faecium* Clad A115. Recently it has been seen a significant presence of hospital associated VRE fm lineages in the wastewater and need of controlling healthcare associated dissemination of VRE fm [26]. However studies on *E. faecalis* ecotypes have shown no appearance of distinct *E. faecalis* strains over a significant period of time. Virulence factors like antibiotic resistance and virulence genes, esp., capsule polysaccharide genes and genes determining gelatinase, aggregation factor, cytolysin and ace are identified in *E. faecalis* isolates [27]. The non-emergence of distinct ecotypes of *E. faecalis* and multiplicity of closely related ecotypes is not seen in *E. faecalis* as compared to *E. faecium*. A genomic analysis of 168 *E. faecalis* hospital isolates showed no genes and non-synonymous single nucleotide polymorphisms in the three lineages of hospital strains [28]. A recent study has also demonstrated that the acquisition of mobile genetic elements in *E. faecalis* V583, makes it unable to coexist with commensal enterococci in humans [29].

5. Nosocomial infections by VRE

Nosocomial infections by Enterococci are Urinary tract infection, endocarditis, bacteremia, catheter related infections, wound infections, intra- abdominal and pelvic infections and recently even oral infections have been reported.

6. Urinary tract infection (UTI)

Enterococci cause both uncomplicated and complicated health care associated UTI. *E. faecalis*. Vancomycin resistant *E. faecalis* and vancomycin resistant *E. faecium* have been mainly implicated in Enterococcal UTI. VRE is fast becoming a major cause of health care associated UTI. The treatment of UTI involves the use of broad spectrum antibiotics which is a major cause of resistant strains to vancomycin (VRE). The complications range from uncomplicated cystitis, pyelonephritis, perinephric abscess, and prostatitis. These organisms are responsible for nosocomial infection of urinary tract particularly in intensive care units (ICU). Enterococci have been particularly reported in catheter associated urinary tract infections, CAUTI (28.4%). Enterococci species are capable of producing biofilms, which are a population of cells attached irreversibly on various biotic and abiotic surfaces. CAUTI are associated with multispecies biofilms. Biofilms are difficult to remove and result in many chronic infections. Bacteria in biofilms colonize medical devices such as catheters, pacemakers, prosthetic heart valves and orthopedic appliances [30]. These multispecies biofilms have synergistic or antagonistic effects of interspecies interaction. Many studies have shown the association of biofilm producing enterococci and urinary catheter [31, 32]. Enterococci biofilms which are formed on catheter in CAUTI are resistant to immune clearance, urination force and even antibiotics. These enterococci utilize fibrinogen formed on catheter surface and form resistant biofilms. *E. faecalis* attachment in biofilm formation seen in vitro is partially inhibited by uropathogenic *E. coli* (UPEC) but biofilm formation by *K. pneumoniae* or UPEC are not affected by *E. faecalis* but *E. faecalis* increased *E. coli* biofilm mass accumulation and it has been seen that co-culture of an *E. faecium* probiotic strain with enteropathogenic *E. coli* increased the antibiotic sensitivity of *E. coli* to aminoglycosides, B-lactams and quinolones [33]. Biofilm formation confers the organism resistance to phagocytosis and antimicrobial agents. UTI by *E. faecalis* is mediated by virulence factors of the genes esp., srtC, ebp A, ebpC, ace, epaB, msrA, msr B, sigV, efbA, and grvR/etaR. *E. faecium* also displays similar genes related to virulence. Both *E. faecalis* and *E. faecium* isolated from nosocomial UTIs show kidney tropism. It is important to study factors in enterococcal causing pyelonephritis [33].

7. Bacteremia

There is a high prevalence of blood stream infections caused by Gram-positive bacteria and 45% are caused by Enterococci. Bacteremia is a common manifestation of vancomycin resistant Enterococci. Due to use of intravascular and urinary catheters these nosocomial infections are acquired. *E. faecium* in the blood stream is associated with increased mortality due to high levels of resistance. Risk factors identified with VRE bacteremia include intestinal colonization, long term antibiotic use, severity of illness, bone marrow transplant, hematologic malignancy, indwelling urinary catheters, corticosteroid treatment, chemotherapy and parenteral nutrition [34]. Studies have shown that bacteremia caused by vancomycin resistant Enterococci strains carry higher mortality rates (2.5-fold increase) as compared to bacteremia caused by vancomycin sensitive strains. In one such study the prognosis of VRE bacteremia was not much changed even with the availability of antimicrobial agents with greater potency. *E. faecalis* sigma factor Sig V that regulates gene expression in response to stress conditions has been implicated in enterococci survival and colonization in systemic infection. Absence of sig V in systemic infection in mice resulted in attenuation of bacterial translocation reducing colonization of kidney and liver. Virulence factors like Bgs A and Bgs B have also been implicated in colonization of endocarditic lesions and bacteremia. BgsA and Bgs B are now being used to treat enterococcal infections by using them as drug targets [35]. Similarly gene Asr has been implicated in *E. faecium* pathogenesis in systemic infections. Nosocomial enterococcal bacteremia have been associated with urinary catheters, intra-abdominal, burn wound, pelvic, biliary and bone sources. VRE bacteremia results in 2.5-fold increase in mortality as compared to vancomycin sensitive (VSE) bacteremia [18].

8. Infective endocarditis

Enterococci are the second most cause of infective endocarditis. Endocarditis caused by VRE faecalis causes GI or GU manipulation, damaged mitral or aortic valve infections, liver transplantation whereas VRE faecium endocarditis is associated with infection of tricuspid valve [36]. *E. faecalis* is also associated with community acquired endocarditis. Characteristic signs of infection include fever or a new murmur. Typical stigmata of endocarditis like petechiae, osler spots are rare and occur with sub-acute infections. Genitourinary infection or instrumentation often precedes the onset of enterococcal endocarditis. In published series of enterococcal endocarditis men often outnumber women and mostly it occurred in elderly individuals. In the current therapeutic regimes, the mortality rate of enterococcal endocarditis remains around 20%.

9. Intra-abdominal and pelvic infections

VRE has been isolated from intra-abdominal and pelvic infections. The usual infections include abscesses wounds or peritonitis. Often it is a part of polymicrobial infection with Gram negative or anaerobic organisms. Usually infecting strains originate from patients intestinal flora and cause intra-abdominal infection. Enterococci are able to cause monomicrobial peritonitis infections particularly in patients undergoing chronic peritoneal dialysis or liver cirrhosis.

10. Gastrointestinal infections (GI)

GI related enterococcal infections are opportunistic infections particularly occurring during colorectal surgery and colorectal cancer. Pre-colonization with VRE in patients can result in bacteremia following antibiotic induced disruption of gut microbiota [37]. Reg IIIy, a c type lectin is secreted by intestinal epithelial and paneth cells that removes Gram positive bacteria from the gut. Antibiotic treatment causes Reg IIIy down-regulation [38]. Therapeutic strategies have been devised to prevent intestinal colonization of resistant enterococci, introducing probiotic *E. faecalis* pheromone induced killing of drug resistant *E. faecalis* reactivating Reg IIIy introducing obligate anaerobic commensal bacteria containing *Barnesiella* species which prevents *E. faecium* gut colonization and bacteremia [39]. High collagenase producing *E. faecalis* strains have been found to be associated with colorectal anastomotic leak by activating tissue matrix metalloproteinase 9 that cleaves host extracellular matrix [40]. Enterococci produce menaquinone and extracellular superoxide in intestine. This results in high oxidative stress which is linked with colorectal cancer as high genomic instability of intestinal tumor cells as around 80% of colon cancers are caused due to genetic mutations.

11. Central nervous system infections (CNS)

Although CNS infections have been reported rarely with VRE but occur in elderly patients having underlying health issues like malignancies, pulmonary and cardiac complications [41]. In them VRE *faecium* is reported at 82% and less so of VRE *faecalis*. These infections present as fever, mental disorientation, focal CNS deficits and petechial rash. CSF investigations show pleocytosis, low glucose and increased protein levels.

12. Skin and skin structure infections (SSSE)

Enterococci are part of polymicrobial infections which are found to be associated with SSSE [42]. Enterococci are frequently isolated from diabetic foot ulcers and 2–5% of patients undergoing inpatient surgery. In studies using animal models it has been seen that *E. faecalis* capsular polysaccharide in SSSI predominantly is related to the persistence of the organism. A gene cpsI encodes the carbohydrate for capsular polysaccharide.

13. Oral infections

Enterococci are inhabitants of the oral cavity and as opportunistic pathogen cause oral diseases like caries, endodontic infections, periodontitis and peri-implantitis. In endodontic infections the failure of root canal treatment by endodontic infections is now well evidenced. Enterococci have high resistance to endodontic medicaments and forms resistant biofilms. This is implicated in root canal treatment failure [43, 44]. Enterococci prevalence is also seen in gingivitis and periodontitis (3.7–35%) [45]. Oral Enterococci constitute the highest percentage of virulent genes and ability to form resistant biofilms. The oral cavity may hence be an important reservoir of virulent antibiotic resistant enterococci strains. VRE colonization occurs mainly in GI tract, skin, genitourinary tract and oral cavity. Enterococci can persist from months to years. The hands of health care workers are the most common source of transmission in nosocomial infections [46]. The need of oral care is particularly important in nosocomial settings. The spread of the nosocomial VRE occurs and when the immunity is lowered VRE multiply to cause disease. Few studies have shown that antibiotic resistant enterococci is transmitted by food [47–49] but recently Vidana et al. [50] have said there is no food related transmission of enterococci. Enterococci are now showing a high degree of resistance to tetracycline, chloramphenicol, erythromycin besides vancomycin pose a threat for spread of nosocomial infection particularly in patients of ICUs and on mechanical ventilators [51]. Vancomycin resistance is an independent predictor for the overall increase of hospital costs for the patient but also for the individual hospital [52].

14. Conclusion

VRE and have now become an important nosocomial pathogen globally. VRE causes range of infections from UTI, bacteremia, infective endocarditis, intra-abdominal and pelvic infections, central nervous system infections and even oral infections. The ability of enterococci to form recalcitrant biofilms, colonize and express virulence factors, genome plasticity, resistant to antibiotics, survival ability makes it an important nosocomial pathogen to which new therapeutic strategies have to be devised for the treatment of VRE. A periodic surveillance of VRE in hospitals is essential for limiting the spread of antibiotic resistance. Future therapy should be targeted to prevent VRE colonization of patients with immunosuppression.

Author details

Sonia Bhonchal Bhardwaj
Department of Microbiology, Dr. Harvansh Singh Judge Institute of Dental Sciences and Hospital, Panjab University, Chandigarh, India

*Address all correspondence to: sbbhardwaj2002@yahoo.com

References

[1] De Perio MA, Yarnold PR, Warrent J, et al. Risk factors and outcomes associated with non *Enterococcus faecalis*, non *E. faecium* enterococcal bacteremia. Infection Control and Hospital Epidemiology. 2006;**27**(1):28-33

[2] Sakka V, Tsiodras S, Galani L, et al. Risk factors and predictors of mortality in patients colonized with vancomycin resistant enterococci. Clinical Microbiology and Infection. 2008;**14**(1):14-21

[3] Schmidt-Hieber M, Blau IW, Schwartz S, et al. Intensified strategies to control vancomycin-resistant Enterococci in immunocompromised patients. International Journal of Hematology. 2007;**86**(2):158-162

[4] Diaz Granados CA, Zimmer SM, Klein M, et al. Comparison of mortality assosciated with vancomycin resistant and Vancomycin susceptible enterococcal bloodstream infections: A meta analysis. Clinical Infectious Diseases. 2005;**41**(3):327-333

[5] Ruoff KL, de La ML, Murtagh MJ, Spargo JD, Ferraro MJ. Species identities of enterococci isolated from clinical specimens. Journal of Clinical Microbiology. 1990;**28**(3):435

[6] Billington EO, Phang SH, Gregson DB, Pitout JD, Ross T, Church DL, et al. Incidence, risk factors and outcomes for enterococcus species, blood stream infections. A population based study. International Journal of Infectious Diseases. 2014;**26**:76-82

[7] Gaca AO, Gilmore MS. Killing of VRE *E. faecalis* by commensal strains: Evidence for evolution and accumulation of mobile elements in the absence of competition. Gut Microbes. 2016;**7**:90-96. DOI: 10.1080/19490976.2015.1127482

[8] Gastmeier P, Schroder C, Behnke M, Meyer M, Meyer E, Geffers C. Dramatic increase in vancomycin resistant enterococci in Germany. The Journal of Antimicrobial Chemotherapy. 2014;**69**(6):1660-1664

[9] Data from the ECDC Surveillance. Atlas-Antimicrobial Resistance. 2017. Available from: http//ecdc.europa.eu/en/healthtopics/antimicrobial_resistance/database/pages/table-reports.aspx

[10] Simner PJ, Adam H, Baxter M, McCracken M, Golding G, Karlowsky JA, et al. Canadian antimicrobial resistance. Epidemiology of vancomyccin resistant enterococci in Canadian hospitals (CANWARD study, 2007 to 2013). Antimicrobial Agents and Chemotherapy. 2015;**59**(7):4315-4317

[11] Lodise TP, Mckinnon PS, Tam VH, Rybak MJ. Clinical outcomes for patients with bacteremia caused by vancomycin-resistant enterococcus in a level 1 trauma center. Clinical Infectious Diseases. 2002;**34**(7):922-929

[12] Vonberg RP, Chberny IF, Kola A, Mattner F, Borgmann S, Dettenkofer M, et al. Prevention and control of the spread of vancomycin resistant enterococci: Results of a workshop held by the German Society of Hygiene and Microbiology. Anaesthesist. 2007;**56**(2):151-157

[13] Fortun J, Coque TM, Davila PM. Risk factors associated with ampicillin resistance in patients with bacteremia caused by *E. faecium*. Journal of Antimicrobial Chemotherapy. 2002;**50**(6):1003-1009

[14] Abebe W, Endris M, Tiruneh M, Moges F. Prevalence of vancomycin resistant enterococci and associated risk factors among clients with and without HIV in Northwest Ethiopia: A cross

sectional study. BMC Public Health. 2014;**14**:185

[15] Chen C, San J, Guo Y, Lin D, Guo Q, Hu F, et al. High prevalence of Van M in Vancomycin-resistant *E. faecium* isolates from Shanghai, China. Antimicrobial Agents and Chemotherapy;**59**:239-244. DOI: 10.1128/AAC.04174-14

[16] Howden BP, Hot KE, Lam MMC, Seeemanm T, Ballard S, Coombs GW, et al. Genomic insights to control the emergence of vancomycin resistant enterococci. MBio;**4**:e412-e413. DOI: 10.1128/mbio.00412-13

[17] Erlandson KM, Sun J, Iwen PC, et al. Impact of the more potent antibiotics quinupristin delfopristin and linezolid on outcome measures of patients with VRE bacteremia. Clinical Infectious Diseases. 2008;**46**(1):30-36

[18] O'Driscoll T, Crank CW. Vancomycin resistant enterococcal infections. Epidemiology, clinical manifestations and optimal management. Infection and Drug Resistance. 2015;**8**:217-230

[19] Contreras GA, Granadoz CA, Cortes L, Reyes J, Vanegas S, Panesso D, et al. Nosocomial outbreak of *Enterococcus gallinarum*: Untaming of rare species of enterococci. Journal of Hospital Infection. 2008;**70**:346-352

[20] Wang Y, Lv Y, Cai J, Schwarz S, Cui L, Hu Z, et al. A novel gene optr A, that confers transferable resistance to oxazolidinones and phenicols and its presence in *E. faecalis* and *E. faecium* of human and animal origin. The Journal of Antimicrobial Chemotherapy;**70**:2182-2190. DOI: 10.1093/jac/dkv116

[21] Edelsberg J, Weycher D, Baron R, Li X, Wu H, Ostr G, et al. Prevalence of antibiotic resistance in US hospitals. Diagnostic Microbiology and Infectious Disease. 2013;**78**:255-262. DOI: 10.1016/j.diagmicrobio.2013.11.011

[22] Lubbert C, Rodloff AC, Hamed K. Real world treatment of enterococcal infections with daptomycin: Insights from a large European registry (EU-CORE). Infectious Disease and Therapy;**4**:259-279. DOI: 10.1007/s40121-015-0072-z

[23] Agegne M, Abera B, Derbie A, Yismaw G, Shiferaw MB. Magnitude of Vancomycin resistant Entercocci (VRE) colonization among HIV infected patients attending ART Clinic in West Amahara government hospitals. International Journal of Microbiology. 2018;**2018**:7. DOI: 10.1155/2018/7510157

[24] Van Schaik W, Top J, Riley DR, Boekhorst J, Vrijenhoek JE, Schapendonk CM, et al. Pyrosequencing-based comparative genome analysis of the nosocomial pathogen *E. faecium* and identification of a large transferable pathogenicity island. BMC Genomics. 2010;**11**:239. DOI: 1186/1471-2164-11-239

[25] Kim EB, Marco ML. Non clinical and clinical *E. faecium* but not *E. faecalis* have distinct structural and functional genomic features. Applied and Environmental Microbiology. 2013;**80**:154-165

[26] La Rosa SL, Snipen LG, Murray BE, Williams RJ, Gilmore MS, Diep DB, et al. A genomic virulence reference map of *E. faecalis* reveals an important contribution of phage 03 like elements in nosocomial genetic lineages to pathogenicity in *Caenorhabditis elegans* infection model. Infection and Immunity. 2015;**83**:2156-2167. DOI: 10.1128/IAL02801-14

[27] Van Hal SJ, Ip CL, Ansari MA, Wilson DJ, Espedido BA, Jesen SO, et al. Evolutionary dynamics of *E. faecium*

reveals complex genomic relationships between isolates with independent emergence of vancomycin resistance. Microbial Genomics. 2016;**2**:48. DOI: 10.1099/mgen000048

[28] Gouiliouris T, Raven KE, Moradigara VD, et al. Detection of Vancomycin-resistant *E. faecium* hospital adapted lineages in municipal wastewater treatment plants indicates widespread distribution and release into the environment. Genome Research. 2019;**29**:1-9. DOI: 10.1101/gr.232629.117

[29] Raven KE, Reutr S, Goulioris T, Reynolds R, Russell JE, Brown NM, et al. Genome based characterization of hospital-adapted *Enterococcus faecalis* lineages. Nature Microbiology. 2016;**1**:15033. DOI: 10.1038/nmicrobio.2015.33

[30] Costerton JW, Stewart PS, Greenberg EP. Bacterial biofilms: A common cause of persistent infections. Science. 1999;**284**:1318-1322

[31] Akhter J, Ahmed S, Saleh AA, Anwar S. Antimicrobial resistance and in vitro biofilm forming ability of Enterococci species isolated from urinary tract infection in a tertiary care hospital in Dhaka. Bangladesh Medical Research Council Bulletin. 2014;**40**:6-9

[32] Galvan EM, Mateya C, Ielpi L. Role of interspecies interactions in dual species biofilms developed in vitro by uropathogens isolated from urinary catheter associated bacteriuria. Biofouling. 2016;**32**:1067-1077. DOI: 10.1080/08927014.2016.1231300. PMID: 27642801

[33] Al K, Martin SM, Lyon W, Hayes W, Caparo MG, Hultgren SJ. *E. faecalis* tropism for the kidneys in the urinary tract of C57BL/6J mice. Infection and Immunity. 2005;**73**:2461-2468. DOI: 10.1128/IAI.732471-24682005. PMID: 15784592

[34] Le Jeune A, Torelli R, Sanguinetti M, Giard JC, Hartke A, Auffray Y, et al. The extracytoplasmic function sigma factor sig V plays a key role in the original mode of lysozyme resistance and virulence of *E. faecalis*. PLoS One. 2010;**5**:e9658. DOI: 10.1371/journal.pone.009658. PMID: 20300180

[35] Theilacker C, Sanchez-Carballo P, Oma I, Fabretti F, Sava I, Kropec A, et al. Glycolipids are involved in biofilm accumulation and prolonged bacteremia *E. faecalis*. Molecular Microbiology. 2009;**71**:1055-1069. DOI: 10.1111/j.1365-2958.2008.06587x. PMID: 19170884

[36] Forrest GN, Arnold RS, Gammie JS, Gilliam BL. Single center experience of a vancomycin resistant enterococcal endocarditis cohort. The Journal of Infection. 2011;**63**(6):420-428

[37] Kommineni S, Brett DJ, Lam V, Chakraborty R, Hayword M, Simpson P, et al. Bacteriocin production augments niche competition by enterococci in the mammalian gastrointestinal tract. Nature. 2015;**526**:719-722. DOI: 10.1038/nature15524. PMID: 26479034

[38] Cash HL, Witham CV, Behrendt CL, Hooper LV. Symbiotic bacteria direct expression of an intestinal bacteria and lectin. Science. 2006;**313**:1126-1130. DOI: 10.1126/Science1127119. PMID: 16931762

[39] Ubeda C, Bucci V, Cabal Hero S, Dju Kovic A, Toussaint NC, Equinda M, et al. Intestinal microbiota containing *Baresiella* species cures vancomycin resistant *E. faecium* colonization. Infection and Immunity. 2013;**81**:965-973. DOI: 10.1128/A.01197-12. PMID: 23319552

[40] Shogan BD, Belogortseva N, Luong PM, Zaborin A, Lax S, Bethel C, et al. Collagen degradation and MMP9 activation by *E. faecalis* contribute to intestinal anastomatic leak. Science

Translational Medicine. 2015;7:286ra68. DOI: 10.1126/scitraslmed.3010658. PMID: 25947163

[41] Wang JS, Muzewich K, Edmond MB, Bearman G, Stevens MP. Central nervous system infections due to VRE: Case series and review of literature. International Journal of Infectious Diseases. 2014;**25**:26-31

[42] Arias CA, Murray BE. *Enterococcus* species, *Streptococccus gallolyticus* group and *Leuconostoc* species. In: Bennett JE, Dolin R, Blaser MJ, editors. Marebel Douglas and Bennetts: Principles and Practice of Infectious Diseases. Philadelphia: Saunders; 2015. pp. 2328-2339

[43] Duggan JM, Sedgley CM. Biofilm formation of oral and endodontic *E. faecalis*. Journal of Endodontia. 2007;**33**:815-818. PMID: 17804318

[44] Appelbe OK, Sedgley CM. Effects of prolonged exposure to alkaline pH on *E. faecalis* survival and specific gene transcripts. Oral Microbiology and Immunology. 2007;**22**:169-174. PMID: 17488442

[45] Sun J, Sundsford A, Song X. *E. faecalis* from patients with chronic periodontitis: virulence and antimicrobial resistance traits and determinants. European Journal of Clinical Microbiology & Infectious Diseases. 2012;**31**:267-272. DOI: 10.1007/s10096-011-1305-zPMID21660501

[46] Snyder GM, Thom KA, Faruno JP, et al. Detection of methicillin resistant *Staphylococcus aureus* and vancomycin resistant enterococci on the gowns and gloves of healthcare workers. Infection Control and Hospital Epidemiology. 2008;**29**(7):583-589

[47] Anderson AC, Jonas D, Huber I, Karygianni L, Nolber J, Hellwig E, et al. *E. faecalis* from food, clinical specimens and oral sites. Prevalence of virulence factors in association with biofilm formation. Frontiers in Microbiology. 2016;**6**:1534. DOI: 10.3389/fmcib.2015.01534. PMID: 01534; PMID: 26793174

[48] Linden PK. Optimising therapy for vancomycin resistant enterococci (VRE). Seminars in Respiratory and Critical Care Medicine. 2007;**28**(6):632-645

[49] Miranda JM, Franco CM, Vasquez BI, Fente CA, Barros-Velazquez J, Cepeda A. Evaluation of chromooccult enterococci agar for the isolation and selective enumeration of Enterococcus species in broilers. Letters in Applied Microbiology. 2005;**41**:153-156

[50] Vidana R, Rashid MU, Ozenci V, Weintraub A, Lund B. The origin of *E. faecalis* explored by comparison of virulence factor patterns and antibiotic resistance to that of isolates from stool samples, blood cultures and food. International Endodontic Journal. 2016;**49**(4):343-351. DOI: 10.1111/iej.1264 PMID 25950381

[51] Komiyama EY, Lepesqueur LSS, Yassuda CG, Samaranayake LP, Parahitiyawa NB, Balducci I, et al. Enterococci species in the oral cavity: Prevalence, virulence factors and antimicrobial susceptibility. PLoS One;**11**(9):e0163001. DOI: 10:1371/journal.Pone.0163001

[52] Puchter L, Chaberny IF, Schwab F, Vonberg RP, Bange FC, Ebadi E. Economic burden of nosocomial infections caused by vancomycin resistant enterococci. Antimicrobial Resistance and Infection Control. 2018;**7**:1. DOI: 10.1186/s13756-017-0291-z

Section 3

Infections

Chapter 7

Pathology of Gangrene

Yutaka Tsutsumi

Abstract

Pathological features of gangrene are described. Gangrene is commonly caused by infection of anaerobic bacteria. Dry gangrene belongs to noninfectious gangrene. The hypoxic/ischemic condition accelerates the growth of anaerobic bacteria and extensive necrosis of the involved tissue. Clostridial and non-clostridial gangrene provokes gas formation in the necrotic tissue. Acute gangrenous inflammation happens in a variety of tissues and organs, including the vermiform appendix, gallbladder, bile duct, lung, and eyeball. Emphysematous (gas-forming) infection such as emphysematous pyelonephritis may be provoked by *Escherichia coli* and *Klebsiella pneumoniae*. Rapidly progressive gangrene of the extremities (so-called "flesh-eating bacteria" infection) is seen in fulminant streptococcal, *Vibrio vulnificus*, and *Aeromonas hydrophila* infections. Fournier gangrene is an aggressive and life-threatening gangrenous disease seen in the scrotum and rectum. Necrotizing fasciitis is a subacute form of gangrene of the extremities. Of note is the fact that clostridial and streptococcal infections in the internal organs may result in a lethal hypercytokinemic state without association of gangrene of the arms and legs. Uncontrolled diabetes mellitus may play an important role for vulnerability of the infectious diseases. *Pseudomonas*-induced malignant otitis externa and craniofacial mucormycosis are special forms of the lethal gangrenous disorder.

Keywords: anaerobic bacteria, clostridial gas gangrene, flesh-eating bacteria, necrotizing fasciitis, non-clostridial gas gangrene

1. Introduction

Gangrene is a lesion of ischemic tissue death. Typically, the acral skin of the hand and foot accompanies numbness, pain, coolness, swelling, and the skin color changes to reddish black. When severe infection is associated, fever and sepsis may follow. Risk factors of gangrene include diabetes mellitus, atherosclerosis, smoking, major trauma, alcoholism, liver cirrhosis, renal insufficiency, immunosuppression, acquired immunodeficiency syndrome (AIDS), drug abuse, malnutrition, and pernio. Clinically, the disease is divided into dry gangrene, wet gangrene, gas gangrene, internal gangrene, and necrotizing fasciitis. In all cases except for dry gangrene, the necrotic tissue is infected. Treatments include surgery, antibiotics administration, and efforts to control the underlying cause. Hyperbaric oxygen therapy can be tried. Amputation and debridement are performed as surgical treatments. Maggot therapy (artificial implantation of maggots in cavitated lesions) may be performed for digesting tissue debris of diabetic wet gangrene on the extremities.

In the present review article, pathologic features of varied gangrenous lesions are illustrated. In addition to gross findings, microscopic features are presented mainly with hematoxylin and eosin (H&E) and Gram stains. When needed, immunohistochemical approach is combined [1, 2]. Immunostaining using rabbit antisera raised against *Bacillus* Calmette-Guérin (BCG; *Mycobacterium bovis*), *Bacillus cereus*, *Treponema pallidum*, and *Escherichia coli* is employed. These low-specificity (widely cross-reactive) antimicrobial antisera effectively yield clear high-sensitivity signals with a low background [3, 4]. Please visit the author's Website at https://pathos223.com/en/ [5].

2. Dry gangrene

Dry gangrene represents coagulative necrosis of ischemic tissue, caused by inadequate blood supply due to peripheral artery disorders. The term dry gangrene is used only for necrosis of the acral limb [6, 7]. Patients with atherosclerosis, hypercholesterolemia, and diabetes mellitus are susceptible to dry gangrene, particularly when they smoke. The low local oxygen level provokes putrefaction without bacterial growth. The affected portions become dry, solidified, and reddish black (**Figure 1**). Once gangrene has developed, the affected tissue is no longer salvageable. The boundary of the dried lesion is sharply demarcated from the nonischemic skin so that autoamputation may follow [8]. Because of the lack of infection, dry gangrene is not so emergent as wet gangrene and gas gangrene. However, dry gangrene may develop to wet gangrene when the secondary infection happens. Diabetes mellitus is a serious and the most important risk factor for developing both dry and wet gangrenes.

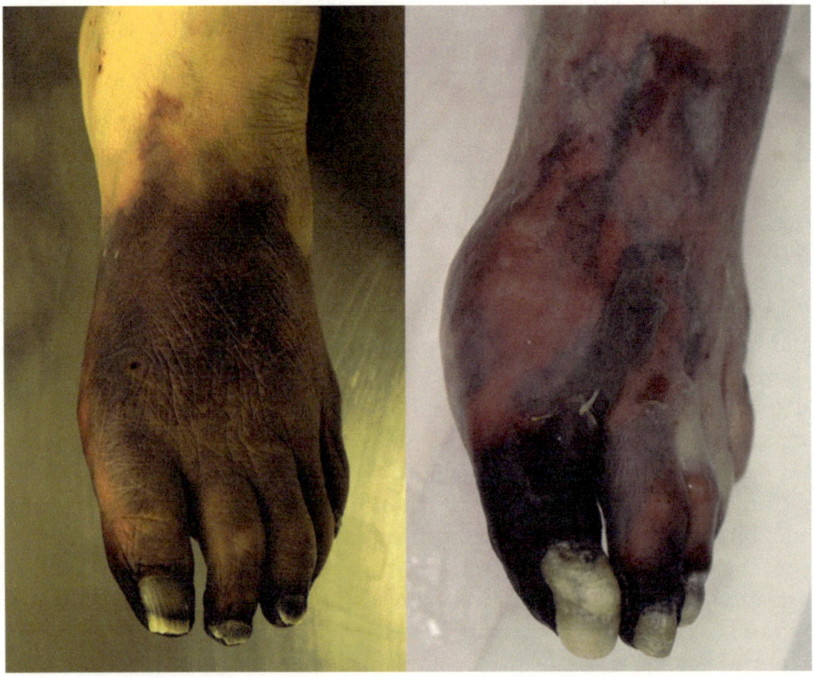

Figure 1.
Dry gangrene (gross appearance of two cases). Atherosclerosis-induced dry gangrene is seen in the foot (left). The border of necrotic lesion is relatively sharp. In the right panel, the toes of a diabetic patient are dry and black-colored, and wet gangrene with red swelling and epidermal blister formation followed (the courtesy of Drs. Mitsuhiro Tachibana and Yasuhito Kaneko at Department of Diagnostic Pathology and Dermatology, Shimada Municipal Hospital, Shimada, Japan).

3. Wet gangrene

Wet or infected gangrene is featured by bacterial infection of the necrotic tissue, and secondary sepsis accompanies a poor prognosis when compared with dry gangrene [9–11]. The affected part becomes markedly edematous, soft, rotten, and dark. Blisters filled with turbid fluid are formed on the discolored and cold-on-touch skin (**Figure 2**). Secondary infection of Gram-positive cocci is common. Infection of saprogenic (anaerobic) bacteria causes a foul smell. Gas formation is often associated, eliciting crepitation on touch. Causative bacteria are polymicrobial or monobacterial. In case of monobacterial infection by *Clostridium perfringens*, we call the status as clostridial gas gangrene. Wet gangrene rapidly progresses via the blockage of blood flow, and the hypoxic stagnant blood promotes rapid growth of anaerobic bacteria that often release exotoxins. The mortality rate of wet gangrene is high so that emergency salvage amputation is often necessary. Disseminated infection (sepsis) eventually leads the patient to death. The predisposing disorders for developing wet gangrene include diabetes mellitus, arteriosclerosis obliterans (atherosclerotic arterial obstruction), and calciphylaxis/calcific uremic arteriolopathy or "gray scale" (painful and intractable ulcers caused by arteriolar wall calcification in patients with chronic renal failure under dialysis).

Several lethal conditions described below are encompassed in the category of wet gangrene. These include polymicrobial necrotizing fasciitis, gas gangrene, Fournier's gangrene, fulminant streptococcal infection, *Vibrio vulnificus* infection, and *Aeromonas hydrophila* infection.

4. Pernio (frostbite or chilblains)

Pernio (frostbite or chilblains) is a vascular disease affecting small vessels of the peripheral skin. Persistent low temperature (cooling) or freezing of the skin causes pernio. Persistent hypoxia of the tissue eventually results in necrosis and ulceration. In a chronic stage, scleroderma-like change may follow. Histopathological features of pernio include mild inflammation around small vessels, peri-eccrine inflammation, and necrosis of the subcutaneous fat tissue with formation of multinucleated giant cells. The epidermis may reveal spongiosis, basal vacuolation, and

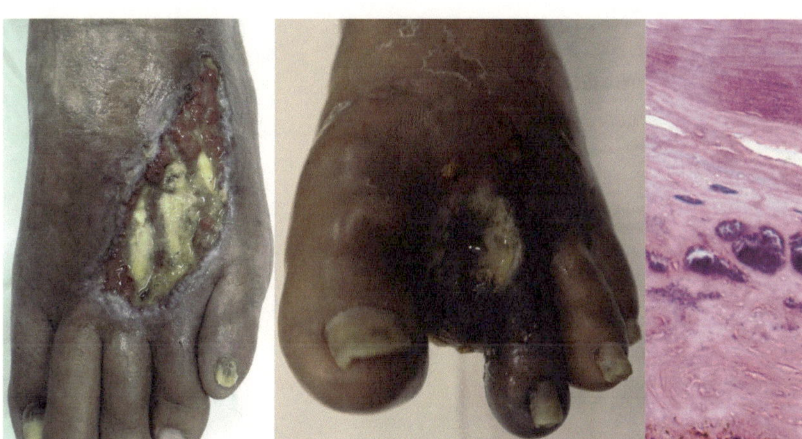

Figure 2.
Wet gangrene (gross appearance of two cases and H&E). Infected deep irregular ulcers are formed in the back of the foot (left) and the base of the second toe after autoamputation (right). Histologically, Gram-positive cocci in the necrotic upper dermis are observed in the debridement specimen (the courtesy of Dr. Yasuhito Kaneko at Department of Dermatology, Shimada Municipal Hospital, Shimada, Japan).

keratinocyte necrosis [12–14]. Representative features are displayed in **Figure 3**. These histopathologic pictures are seen in other vascular disorders, provoking a chronic irritative process of the skin.

5. Decubitus (pressure ulcer or bedsore)

Decubitus (pressure ulcer or bedsore) is formed as a result of long-term pressure, completely or partially blocking the skin blood flow [15]. The sites on a bony prominence are commonly affected, including the skin overlying the sacrum, the greater trochanter, the heel, and the scalp. Decubitus commonly develops in individuals who are on chronic bedrest or consistently use a wheelchair. Factors influencing the skin tolerance against pressure include malnutrition, skin wetness, diseases reducing the blood flow to the skin such as atherosclerosis, and diseases reducing the skin sensation such as paralysis or neuropathy. The advanced age, smoking, complicated diseases (atherosclerosis, diabetes mellitus, and secondary infection), and the use of anti-inflammatory drugs may hamper healing of decubitus.

There is a preceding erythematous stage before ulceration. The late stage presents as a black eschar form. The ulcer often deeply reaches the periosteal tissue. When a pocket is formed, secondary infection may become serious (**Figure 4**). Infection provokes slow or stalling healing and pale granulation tissue [16, 17]. Infected wounds may have a gangrenous odor. Bacterial biofilm formation leads to delayed healing of the decubital ulcer. Infected decubitus may progress to wet gangrene or clostridial/non-clostridial gas gangrene (**Figure 5**) [18], as described in Section 7. The colonization of *Staphylococcus aureus*, particularly methicillin-resistant *Staphylococcus aureus* (MRSA), in the decubitus must be the important target of infection control [19]. It should be noted that the eradication of MRSA can be achieved only after healing of ulceration.

The National Pressure Ulcer Advisory Panel (NPUAP) in the United States proposed the staging of decubital ulcer [20].

Stage I: Intact skin with non-blanchable redness of a localized area usually over a bony prominence.

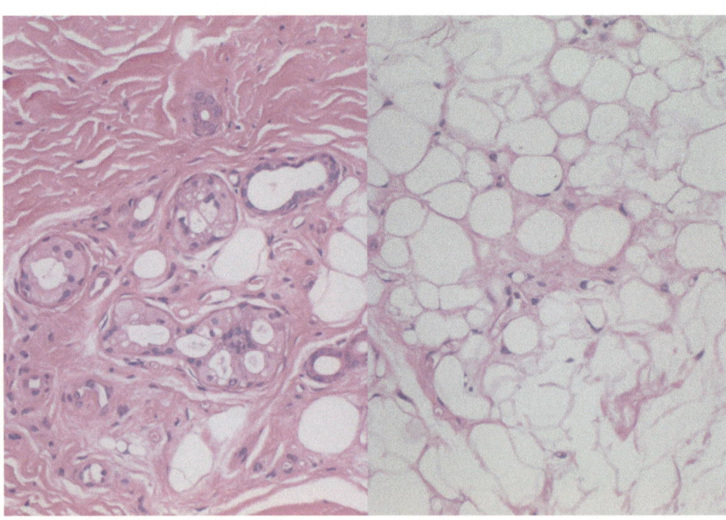

Figure 3.
Pernio (H&E). Biopsy from the skin of sole demonstrates angiectasia of capillary vessels around the eccrine sweat gland (left) and fat necrobiosis (right). Loss of fat cell nuclei, membranous deposition in the cytoplasm, and focal stromal hyalinizing fibrosis are observed.

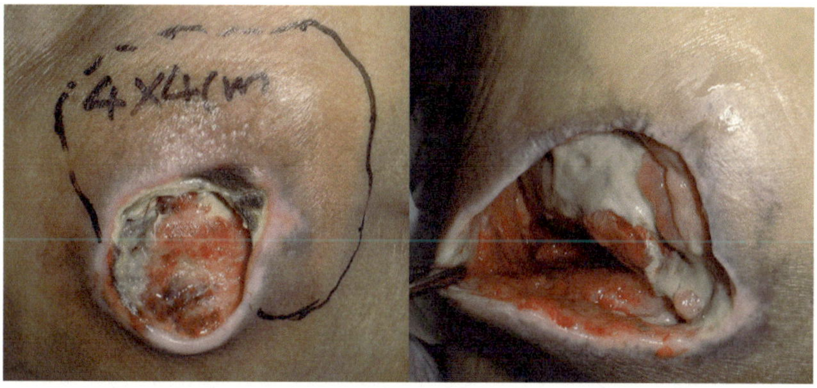

Figure 4.
Large and deep decubiti with purulent exudation and pocket formation at the sacral region (two cases). Deep mining ulcers, so-called pockets, are noted. In the left case, the 40 × 40 mm pocket is indicated by dotted lines on the skin. Secondary infection is inevitable (the courtesy of Dr. Sandai Ohnishi at Hakuhokai Home Healthcare Clinic, Nagoya, Japan).

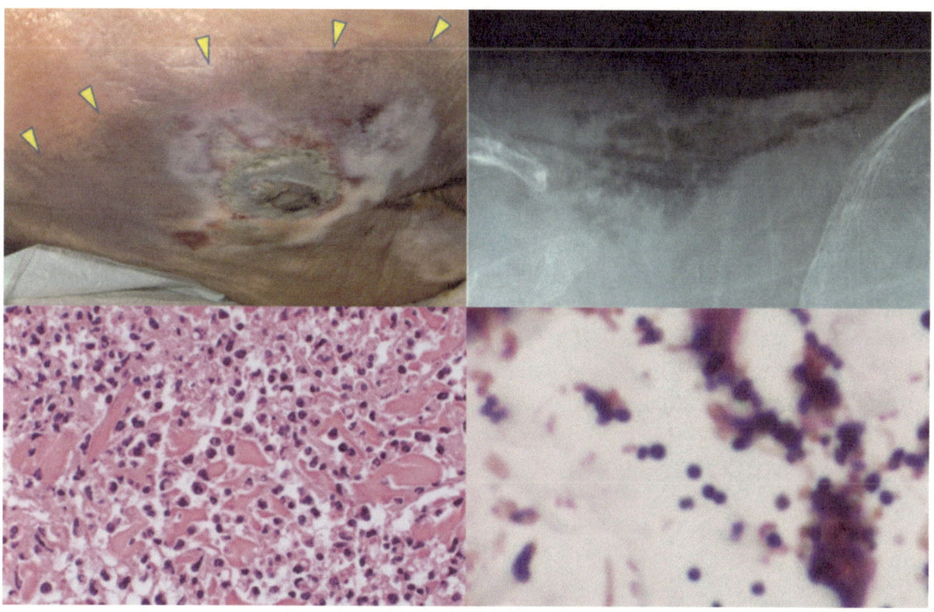

Figure 5.
Non-clostridial gas gangrene caused by group A β-hemolytic streptococcal infection (gross and radiologic findings, H&E and Gram). Infected sacral decubitus seen in a 72-year-old diabetic female patient with a history of brain infarction progressed to gas gangrene. The arrowheads indicate the red-colored skin area with crepitation on touch. X-ray examination discloses gas formation in the soft tissue. Debridement tissue reveals massive gangrenous inflammation with infection of Gram-positive cocci.

Stage II: A shallow open ulcer with a red, pink wound bed, without slough.

Stage III: Full thickness tissue loss. Subcutaneous fat may be visible but bone, tendon, and muscle are not exposed. Slough may be present but does not obscure the depth of tissue loss.

Stage IV: Full thickness tissue loss with exposed bone, tendon, or muscle. Slough or eschar may be present on some parts of the wound bed. Undermining/tunneling (pocket formation) is often seen.

For the treatment purpose, the eschar stage decubitus can surgically be removed and skin-grafted (**Figure 6**). Histologically, the advanced lesion shows the full-thickness dermal necrosis with deep ulceration and abscess/gangrene formation.

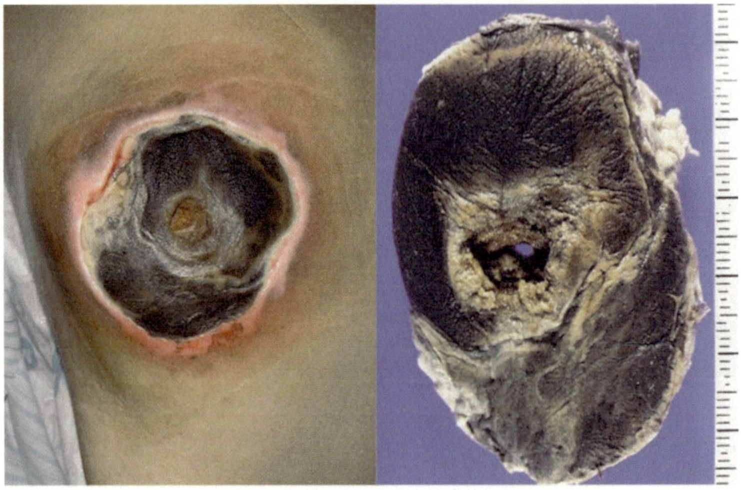

Figure 6.
Surgical removal of a large decubital ulcer covered with black eschar at the trochanter region. Surgical treatment was effective in this intractable ulceration (the courtesy of Dr. Sandai Ohnishi, Nagoya, Japan).

Patterns of bacterial infection are often unique. *Staphylococcus aureus*, including MRSA, mainly colonizes the superficial layer (**Figure 7**), while Gram-negative rods, including *Pseudomonas aeruginosa* and *Escherichia coli*, are observed in the deep layer (**Figures 8** and **9**) [2].

The phenotype of MRSA can be demonstrated immunohistochemically in routinely formalin-fixed, paraffin-embedded lesions [21]. *S. aureus* is immunoreactive not only for staphylococcal antigens but also for protein A, an immunoglobulin-binding protein specifically expressed on the cell wall of *S. aureus*. The multidrug resistance of MRSA, determined by the expression of penicillin-binding protein 2′ (PBP2′) encoded by the *mecA* gene, can be immunophenotyped with monoclonal antibodies. Representative findings are demonstrated in **Figure 10**.

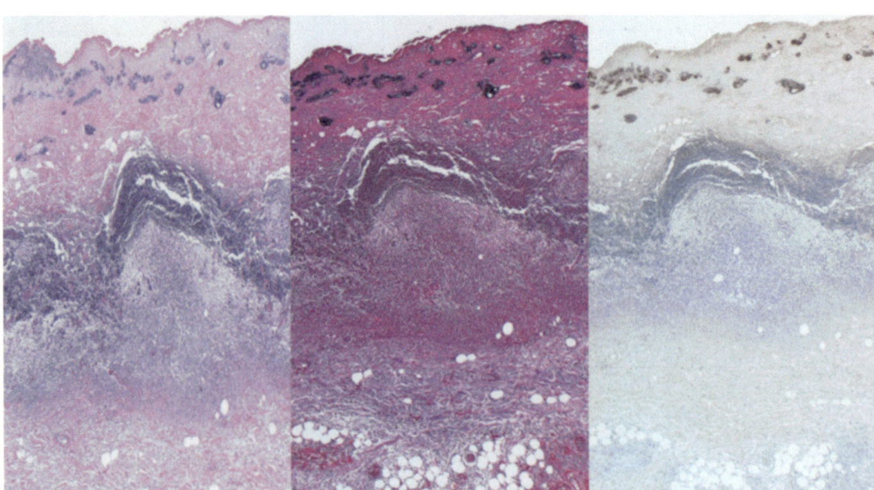

Figure 7.
Microscopic double-layered appearance of the resected decubital ulcer (H&E, Gram and immunostain). Colonization of Staphylococcus aureus, *probably MRSA, is observed along the eroded surface and clearly illustrated by Gram stain and immunostaining for staphylococcal antigens. Abscess formation with ischemic gangrene is noted in the deep zone.*

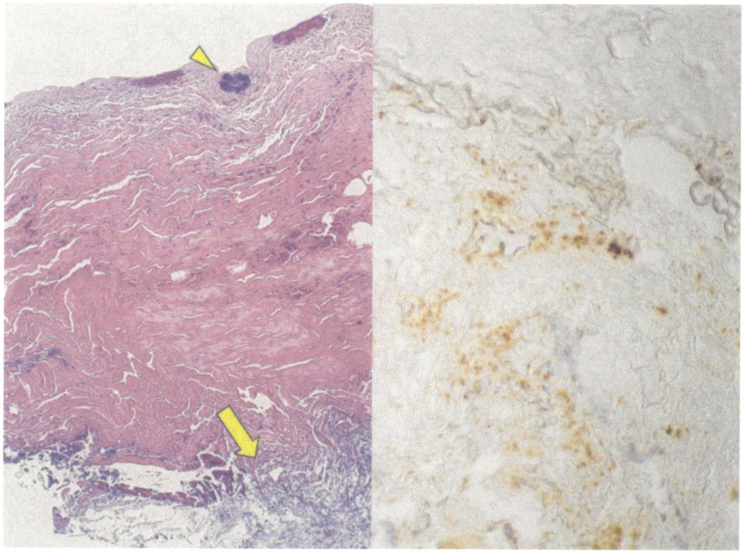

Figure 8.
Pseudomonas *infection in the deep part of the decubitus (H&E and immunostain). In the subcutaneous abscess (arrow), immunostaining using a monoclonal antibody against* Pseudomonas aeruginosa *demonstrates phagocytized microbes in neutrophils.* E. coli *antigens are negative. Arrowhead indicates superficially colonized* Staphylococcus aureus, *as shown in* ***Figure 7****.*

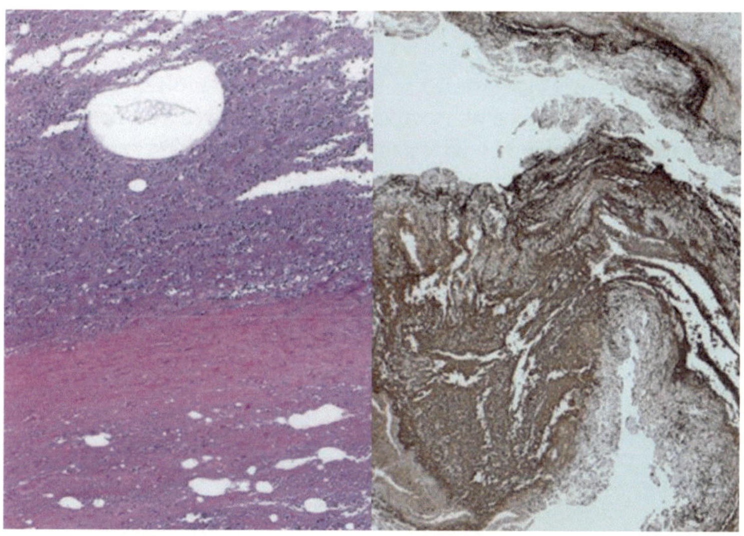

Figure 9.
Another decubital ulcer with massive colonization of Gram-negative rods in the deep gangrenous tissue (H&E and immunostain). Gas formation is associated. A monoclonal antibody (J5) against lipopolysaccharide common in Enterobacteriaceae *illustrates an advanced infective process in the decubitus.*

6. Gas gangrene (clostridial myonecrosis)

6.1 Traumatic gas gangrene

Gas gangrene caused by infection of *Clostridium perfringens* (formerly called *C. welchii*) is a life-threatening emergency, as a representative and grave form of wet gangrene [22–25]. *C. perfringens* is an obligate anaerobic Gram-positive bacillus forming spores on culture plates. Traumatic skin invasion of the microbe results in

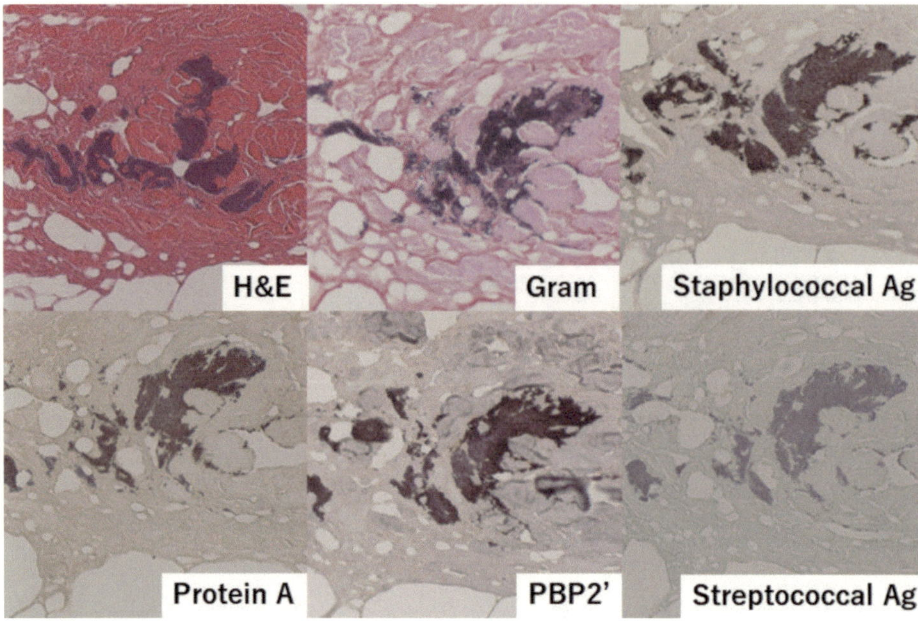

Figure 10.
Immunohistochemical identification of MRSA in formalin-fixed, paraffin-embedded sections (H&E, Gram and immunostain). The Gram-positive coccal colonies in the gangrenous decubital lesion express staphylococcal antigens, protein A (staphylococcal IgG Fc-binding protein) and penicillin-binding protein 2′ (PBP2′), confirming the nature of MRSA. Streptococcal antigens are negative.

massive ischemic necrosis (gangrene) of the soft tissue involving the striated muscle. Gas production is quite characteristic, and the involved tissue thus reveals crepitation on touch (**Figure 11**). The gas is composed of 5.9% hydrogen, 3.4% carbon dioxide, 74.5% nitrogen, and 16.1% oxygen. As the bacteria grow under an anaerobic condition, the degree of ischemia in the involved tissues and organs becomes advanced. Tissue necrosis is accelerated by α-toxin production of the microbe. Putrid odor is associated. Intravascular hemolysis is a common event due to bacterial production of hemolysin (α-toxin). The prognosis is very poor. The disease is also called as clostridial histotoxic syndrome. Gram-positive rods are microscopically localized adjacent to gas bubbles (see below).

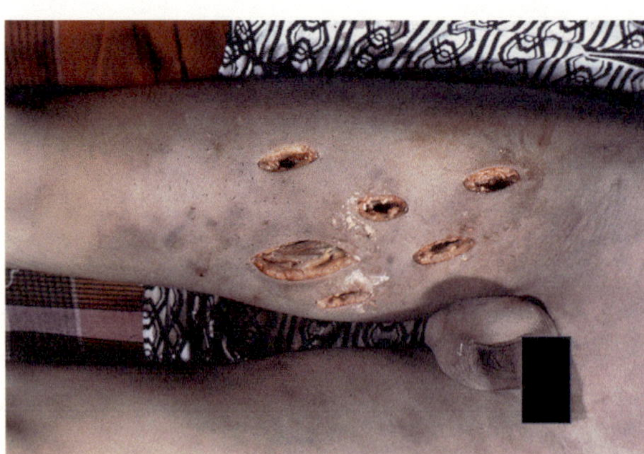

Figure 11.
Traumatic gas gangrene of the right thigh (gross appearance). Gas-forming gangrenous process of the soft tissue results in marked swelling of the thigh. Crepitation was palpable on touch. Surgical debridement has been performed for the treatment purpose.

DOI: http://dx.doi.org/10.5772/intechopen.93505

6.2 Nontraumatic gas gangrene

C. perfringens commonly resides in the gut lumen of healthy individuals, so that the nontraumatic gas gangrene is encountered in the internal organs such as the gut, bile duct, and pancreas [26, 27]. Representative autopsy cases are presented below.

The pancreas is occasionally assaulted by *C. perfringens* [28–30]. An autopsy case of fulminant pancreatitis (emphysematous pancreatitis) in a 66-year-old diabetic man, presenting just a two-day clinical course, is demonstrated. Diabetes mellitus was poorly controlled. The patient suffered sudden abdominal and back pain, and acute pancreatitis was diagnosed by a markedly elevated serum amylase level. Abdominal computed tomography scan demonstrated gas retention in the pancreatic head, intrahepatic branches of the bile duct, and in the abdominal cavity. At autopsy, features of acute hemorrhagic and necrotizing pancreatitis with infiltration of neutrophils were observed (**Figure 12**). Clusters of rods were identified in necrotic, gas-forming areas, and the bacteria grew also along the pancreatic duct. Neutrophilic reaction was sparse in the hypoxic area showing bacterial growth. Not all of the bacteria were stained blue with Gram stain (some remain unstained), and the formation of spores was abortive within the living body (**Figure 13**). These microscopic features were consistent with infection of *C. perfringens*.

Another case of pancreatic gas gangrene in a diabetic male patient aged 70's showed numerous Gram-positive rods around the gas-filled space formed in the necrotic pancreas, confirming the diagnosis of *C. perfringens* infection. Gross and microscopic findings of the foamy liver are illustrated in **Figure 14**. The cut surface of the formalin-fixed liver shows numerous gas-filled spaces, giving characteristic spongy/foamy appearance.

Nontraumatic gas gangrene may be associated with colon cancer [31, 32]. An 81-year-old female patient with rectal cancer became acutely ill with abdominal pain and paralytic ileus. The patient soon died of septic shock. Autopsy clarified nontraumatic gas gangrene of the colorectum caused by clostridial infection in rectal adenocarcinoma. The growth of Gram-positive, gas-forming rods was observed in the cancer tissue, crypts of the noncancerous colorectal mucosa, and also in the liver. Gangrenous inflammation was observed in the entire layer of the colorectal wall. Acute tubular necrosis represented the shock kidney. The microscopic appearance is displayed in **Figure 15**.

Gastric gas gangrene is infrequently experienced [33]. A 65-year-old diabetic male patient underwent endoscopic mucosal resection of intramucosal gastric

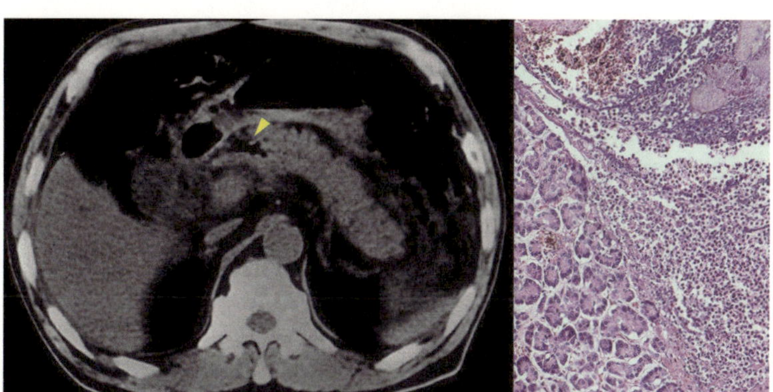

Figure 12.
Clostridial acute hemorrhagic and necrotizing pancreatitis (CT scan and H&E). Computed tomography scan demonstrates gas formation in the pancreatic head (arrowhead). At autopsy, neutrophils infiltrate the pancreatic parenchyma, giving features of severe acute pancreatitis.

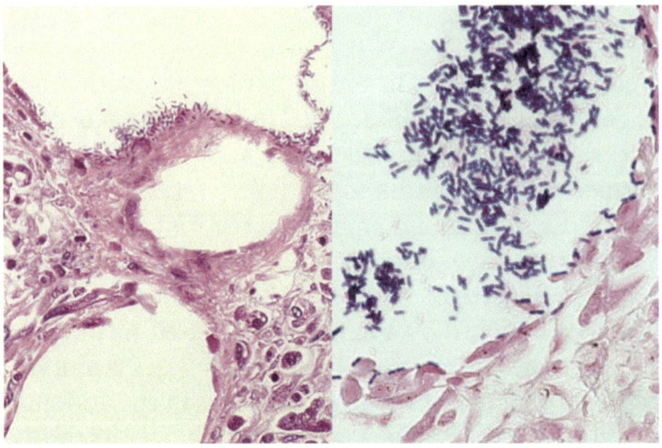

Figure 13.
C. perfringens *grown in acute necrotizing pancreatitis (H&E and Gram). Gas bubbles are observed in the necrotic pancreatic tissue with sparse inflammatory infiltration. The rods growing in the bubble are unevenly Gram-positive (some bacilli are not stained blue). Spores (representing unstained dots in the bacterial body) are only focally recognizable in the living body.*

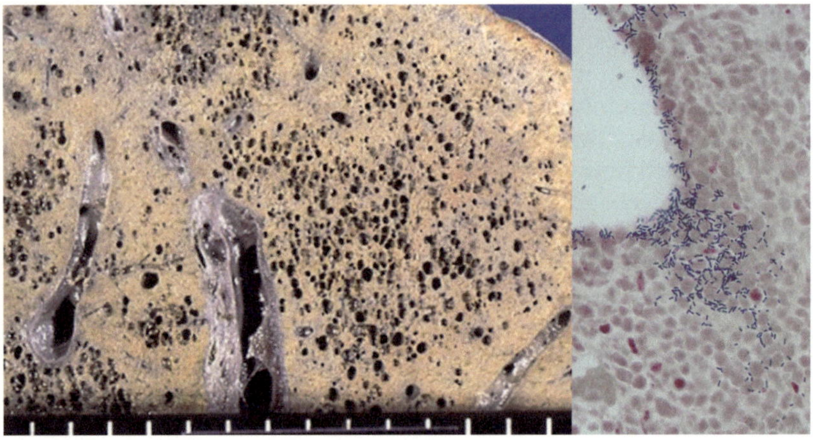

Figure 14.
Gas gangrene involving the liver (gross and Gram). Numerous gas bubbles replace the liver parenchyma, giving foamy or spongy appearance. The hepatocytes reveal ischemic changes, and Gram-positive rods are clustered around the gas bubble. Note that the condition allowing the growth of obligate anaerobic Clostridium perfringens *must be highly hypoxic.*

adenocarcinoma located at the gastric angle. Next day, he became acutely ill with abdominal pain and distention, and circulatory collapse soon followed. At autopsy, colonization of Gram-positive rods was noted at the base of ulcer caused by the endoscopic operation (**Figure 16**). The liver revealed multifocal foamy appearance due to gas formation by Gram-positive rods growing among the liver cell cord. The final diagnosis was gas gangrene caused by clostridial infection on the iatrogenic gastric mucosal trauma.

C. septicum may cause spontaneous, nontraumatic gas gangrene [34], and *C. sordellii* may induce gas gangrene of the uterus, as a consequence of spontaneous abortion, normal vaginal delivery, and traumatic injury [35]. As illustrated in **Figure 17**, *C. butyricum* happened to infect the stomach, resulting in fulminant death of a male patient aged 60's. *C. butyricum*, a resident of healthy human gut, uniquely produces butyric acid as a metabolite, hence named. Foamy appearance of the gastric wall was quite characteristic. The liver also appeared foamy/spongy. The formation of spores inside the rugby ball-shaped Gram-positive rod bodies is microscopically characteristic of *C. butyricum*. This is in sharp contrast to poor spore

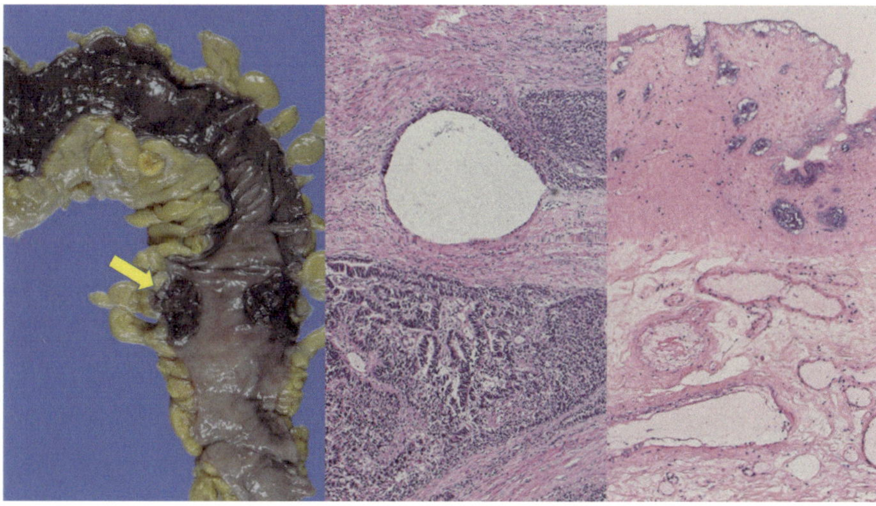

Figure 15.
Rectal cancer-associated nontraumatic gas gangrene in an 81-year-old female patient (gross and H&E). The rectal cancer (arrow) and the edematous proximal colon reveal hemorrhagic necrosis. Gas formation is observed in the tissue of rectal adenocarcinoma with marked ischemic change. Large-sized rods colonize the crypts of the non-cancerous necrotic colorectal mucosa (the courtesy of Dr. Hirokazu Kurohama, Regional Pathological Diagnosis Support Center, Atomic Bomb Disease Institute, Nagasaki University, Nagasaki, Japan).

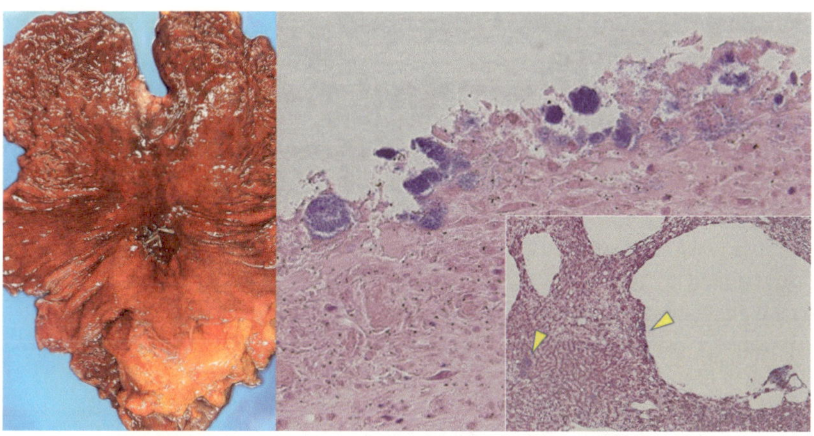

Figure 16.
Gastric gas gangrene in a 65-year-old diabetic male patient (gross and H&E). Red-swollen stomach after endoscopic mucosal resection for early gastric cancer has an artificial ulcer at the gastric angle. Rods colonize the ulcer base. Gas formation is evident in the liver tissue (inset). Arrowheads indicate bacterial colonies (the courtesy of Dr. Chunlin Ye, Emergency Department, Saishukan Hospital, Kitanagoya, Japan).

formation by *C. perfringens*. Surgically curable *C. butyricum*-induced intestinal gas gangrene is described in the Section 14.3.

7. Non-clostridial gas gangrene

Gas gangrene is commonly caused by clostridial infection, but non-clostridial bacteria may also provoke gas gangrene mostly in the extremities [36–38]. Early diagnosis and therapy are required, because the disease rapidly progresses to fatal toxemia. This unique dermatologic emergency is featured by the detection of nontraumatic subcutaneous emphysema of the leg with or without association of

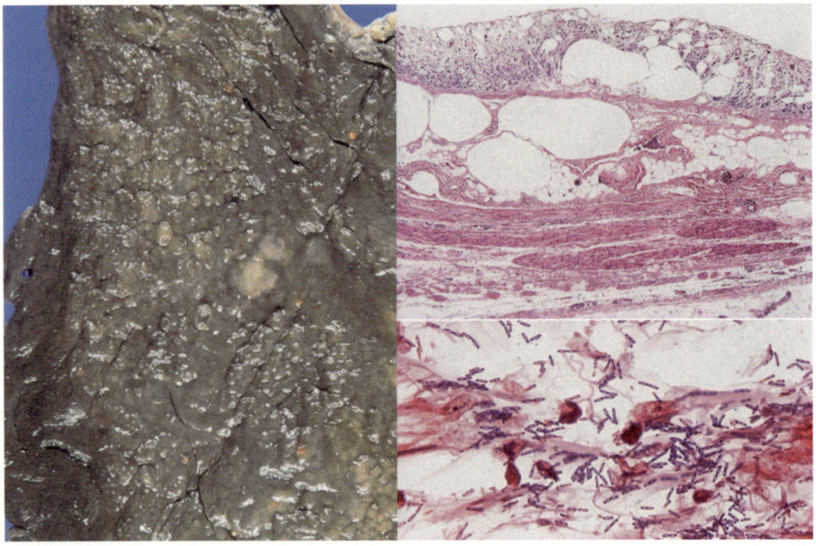

Figure 17.
Clostridium butyricum-*induced lethal gastric gas gangrene (gross, H&E and Gram). The gastric wall demonstrates formation of gas bubbles both grossly and microscopically. The growing Gram-positive rods exhibit distinct spore formation in rugby ball-shaped bacterial bodies, morphologically consistent with* C. butyricum *(the courtesy of Dr. Mayu Fukushima, a pathologist at Hamamatsu Medical University Hospital, Hamamatsu, Japan).*

erythema, tenderness, or bullous lesions. Non-clostridial gas gangrene most often results from polymicrobial infection of mixed kinds of microbes, and it is mainly seen in diabetic patients [39–41]. The causative gas-producing bacteria include *Escherichia coli*, *Klebsiella pneumoniae*, *Enterobacter cloacae*, *Pseudomonas aeruginosa*, *Aeromonas hydrophila*, *Bacteroides* spp., and *Streptococcus anginosus* group (former *S. milleri* group) [42]. Groups A, B, and G streptococci also cause gas gangrene, as a form of fulminant streptococcal infection [43], as described in the Section 10.1. **Figure 5** illustrates the gas-forming fulminant group A β-hemolytic streptococcal infection, caused by a deeply ulcerated (pocket-forming) decubitus at the sacral region of a 72-year-old diabetic woman. Another diabetic male patient aged 70's with advanced rectal adenocarcinoma suddenly manifested nontraumatic and non-clostridial gas gangrene in the abdominal cavity. Massive transportal infection of gas-forming *E. coli* resulted in the formation of foamy liver (**Figure 18**). Intrahepatic vascular-invasive growth of Gram-negative rods was observed under a microscope, and infection of *E. coli* was immunohistochemically confirmed.

Emphysematous (gas-producing) inflammation may be encountered in a variety of organs and tissues, as described in the next section.

8. Gangrenous inflammation of internal organs

Gangrenous inflammation may occur in a wide variety of internal organs, such as the vermiform appendix, gallbladder, bile duct, pancreas, lung, kidney, eyeball, etc. The lesion may be localized within the organ, but it often extends to the surrounding tissues, so as to be fatal. When the anaerobic pathogens produce gas, we call the serious condition as "emphysematous" inflammation (as a form of localized gas gangrene).

8.1 Gangrenous appendicitis

Acute appendicitis is featured by sudden onset epigastric pain radiating to pelvis and high tachycardiac fever. When perforated, generalized abdominal tenderness

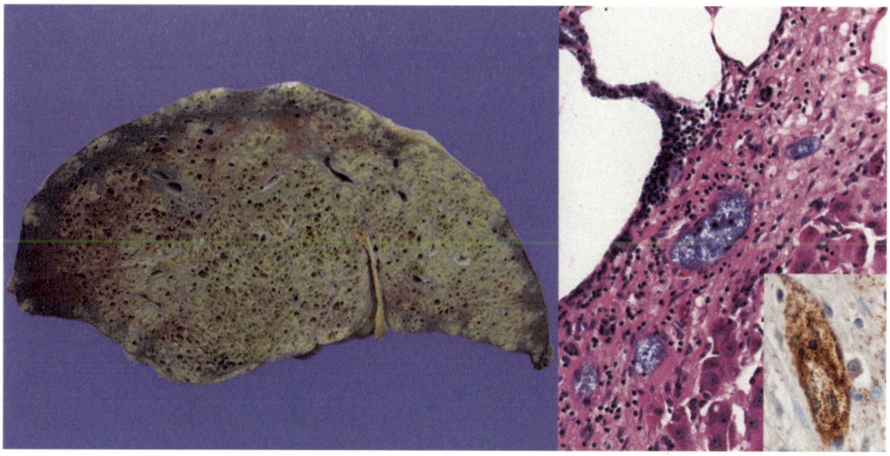

Figure 18.
E. coli-*induced non-clostridial gas gangrene accompanying foamy liver seen in a diabetic male patient aged 70's with advanced rectal adenocarcinoma (gross and H&E, inset: immunostain). Transportal infection of* E. coli *provoked foamy liver due to gas formation by the infected bacteria. The rods embolic in capillary vessels of the liver are immunoreactive with a monoclonal antibody against* E. coli *antigen (inset).*

and peritonism occur. Acute appendicitis is caused by the blockage of the appendiceal lumen (most commonly by fecalith impaction). The blockage results in increased luminal pressure, impaired blood flow, and invasive infection of bacterial flora. When the gangrenous process proceeds, rupture of the appendix can result [44–46]. Mixed bacterial infection is proven. Causative pathogens include *Escherichia coli*, *Bacteroides fragilis*, *B. splanchnicus*, *B. intermedius*, *Peptostreptococcus*, *Pseudomonas*, *Lactobacillus*, *Bilophila wadsworthia*, *Fusobacterium nucleatum*, *Eggerthella lent*a, and *Streptococcus anginosus* (or *milleri*) group. An average of 10.2 different microorganisms have been isolated from the infected lesion. Microscopically, the appendiceal wall reveals marked transmural collection of neutrophils and massive necrosis with the disappearance of the proper muscle layer. Colonization of cocci and rods is easily observed within the gangrenous lesion. Fibrinopurulent peritonitis is associated. Medium-sized blood vessels are thrombosed, accelerating the gangrenous change. Representative findings are displayed in **Figure 19**.

8.2 Gangrenous and emphysematous cholecystitis

Gangrenous cholecystitis is defined as infection-associated transmural necrosis and perforation of the gallbladder wall, as a result of secondary ischemia due to vascular thrombosis. Mural necrosis (infarction) provokes perforation in 25% of cases. Gangrenous cholecystitis represents a form of acute acalculous cholecystitis (**Figure 20**), and the pathology and epidemiology differ from chronic cholecystitis induced by gallstones [47–49]. *Enterobacteriaceae* and anaerobic bacteria are frequently cultured from the bile. The mortality rate is high between 15 and 50%. Risk factors for the development of gangrenous cholecystitis include male sex, advanced age, delayed surgery, cardiovascular diseases, and diabetes mellitus.

Emphysematous cholecystitis is a fulminant and sinister form of acute gangrenous cholecystitis, and it is characterized by the presence of gas both in the lumen (pneumobilia) and wall of the gallbladder. Gas may be extended to the biliary tree or adjacent structures. Either clostridial or non-clostridial etiology is encountered [50]. In case of non-clostridial infection, mixed infection of rods and cocci is often proven microscopically (**Figure 21**). Emphysematous cholecystitis, a form of gas

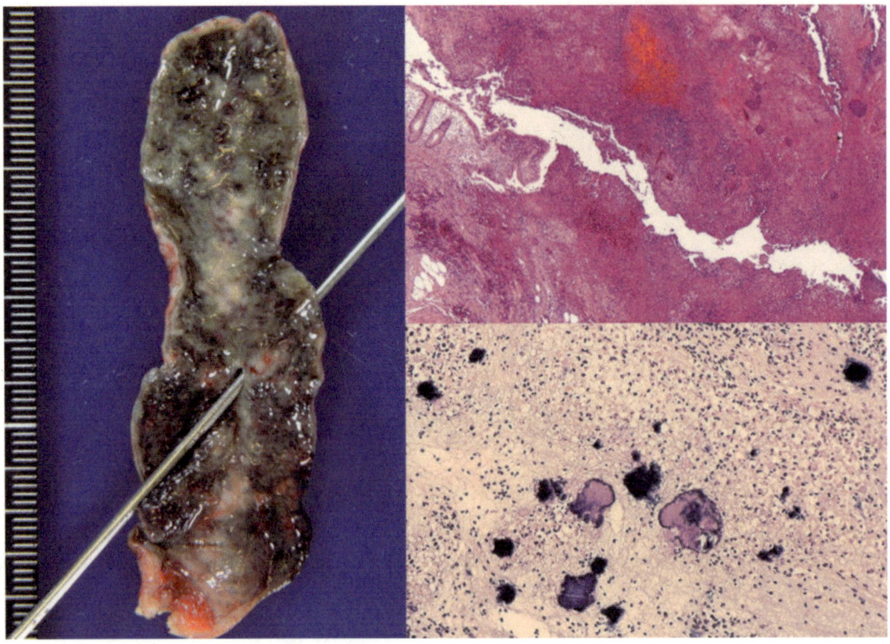

Figure 19.
Gangrenous appendicitis (gross, H&E and Gram). Massive necrotizing inflammation of the appendiceal wall results in perforated purulent peritonitis. A probe is inserted at the site of perforation. Gram-positive bacterial colonies are scattered in the necrotic exudation.

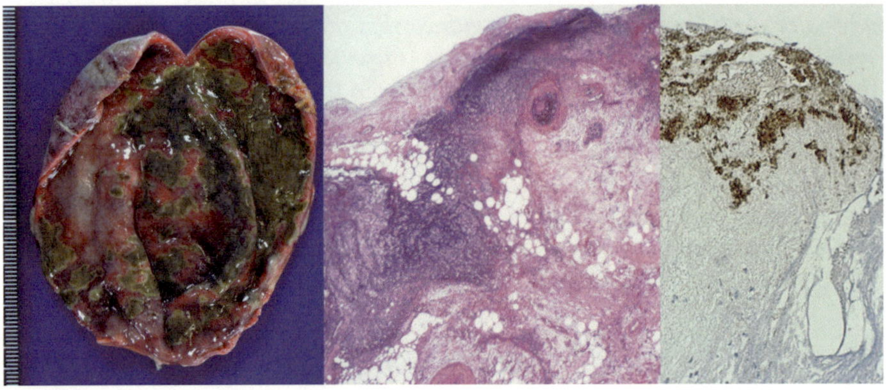

Figure 20.
Gangrenous cholecystitis (gross, H&E and immunostain). Surgically removed gallbladder reveals marked necrotizing inflammation with bile-stained (green-colored) multiple mucosal ulceration. Bacterial colonies growing in the necrotic exudation are strongly immunoreactive for E. coli *antigens.*

gangrene of gallbladder origin, carries a very high mortality rate. Those who suffer from diabetes mellitus or immunosuppression are especially susceptible to this serious condition.

8.3 Gangrenous cholangitis

Gangrenous cholangitis is a severe form of acute cholangitis without biliary stones [47, 51, 52]. Varied pathogens such as *Enterococcus*, *Escherichia coli*, and *Pseudomonas aeruginosa* cause ascending biliary tract infection [53].

A 70-year-old man complained of epigastralgia, vomiting, and difficulty in walking. Abdominal computed tomography scan suggested panperitonitis.

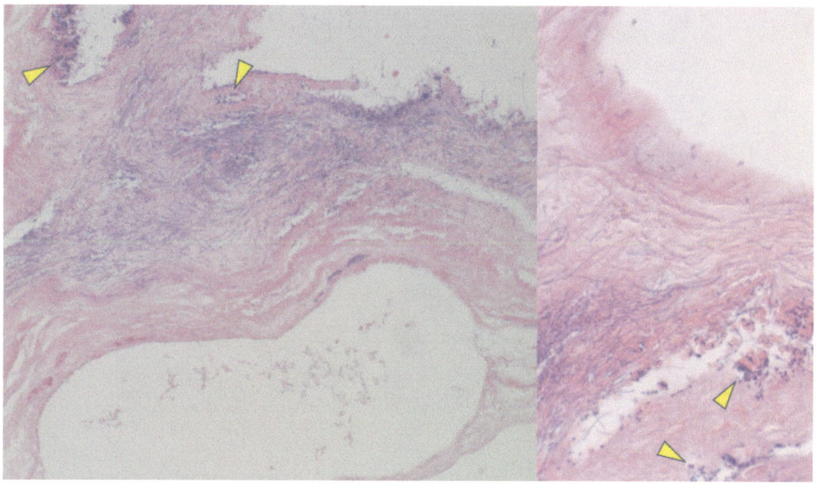

Figure 21.
Emphysematous cholecystitis (gas gangrene of the gallbladder) (H&E). The gallbladder wall accompanies gas bubbles released from thin long rods growing in the necrotic tissue. Co-infection of cocci (arrowheads) is noted. There is little cellular reaction in this highly hypoxic tissue.

Emergency laparotomy indicated an 8 mm-sized perforation in the common bile duct in association with biliary peritonitis. Gallbladder was dilated, but without gallstones. Cholecystectomy and partial resection of the common bile duct was performed. T-tube drainage and pazufloxacin administration were effective to control the infection. Surgical specimens of the common bile duct and gallbladder microscopically showed transmural necrosis with perforation/ulceration and massive infection of Gram-positive cocci. The cocci were immunoreactive for enterococcal antigens, and culture of the bile demonstrated *Enterococcus faecalis* (**Figure 22**). Neutrophilic reaction was mild in the gangrenous lesion. Scanning electron microscopy demonstrated clustered cocci at the site of perforation (**Figure 23**).

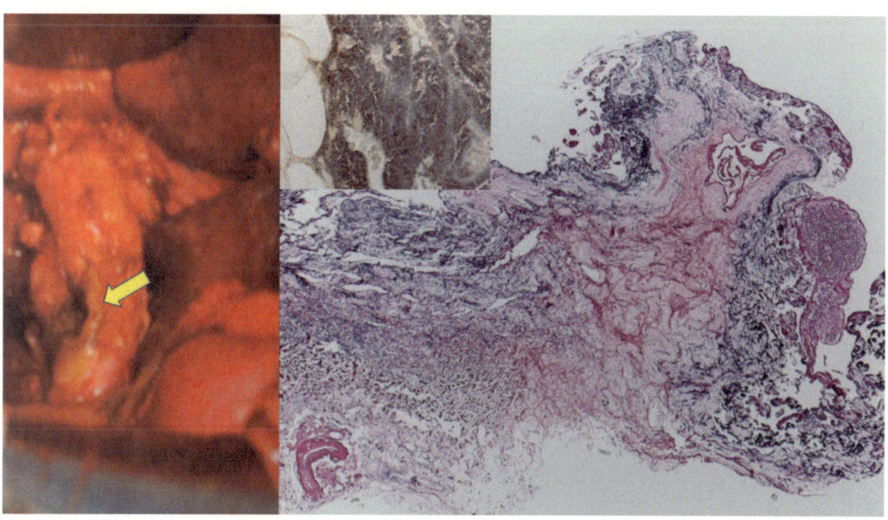

Figure 22.
Perforated enterococcal cholangitis (gross, H&E and immunostain). Massive infection of Enterococcus faecalis *provokes transmural necrosis and perforation of the common bile duct (arrow). Enterococcal antigens (inset) are immunohistochemically demonstrated in the cocci overwhelmingly growing throughout the destroyed bile duct.*

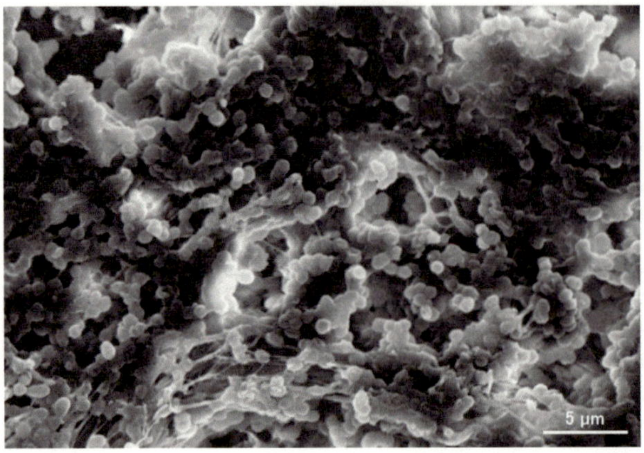

Figure 23.
Scanning electron microscopy of perforated enterococcal cholangitis. Numerous cocci, 0.7 µm in size, are clustered at the site of perforation. Bar indicates 5 µm.

A diabetic lady aged 40's complaining of severe abdominal and back pain visited an emergency suite. Diabetes mellitus had been poorly controlled. Mild obstructive dilatation of the bile duct and gallbladder were associated. Endoscopic retrograde biliary drainage was performed, but the patient soon died of septic shock. Autopsy demonstrated severe gangrenous and acalculous cholangitis and cholecystitis. Necrotic change with active growth of Gram-negative rods was proven in the biliary tree. Immunostaining using a monoclonal antibody disclosed the *Pseudomonas aeruginosa* antigen in the invasive bacilli (**Figure 24**). Neutrophilic reaction was relatively mild. The lower (intrapancreatic) part of the common bile duct remained intact. The association of diabetes mellitus was evident: the pancreatic islets revealed pronounced deposition of amyloid substances, and the kidney showed diabetic glomerulosclerosis with nodular lesions.

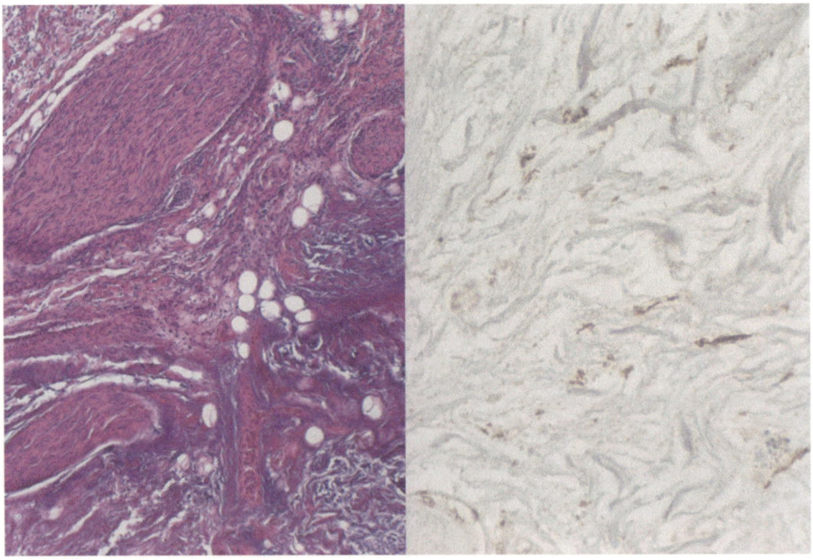

Figure 24.
Acute Pseudomonas *cholangitis (H&E and immunostain). Diabetes mellitus accelerated severe necrotizing (gangrenous) inflammation of the extrahepatic biliary tree. Neutrophilic reactions are limited. The rods are immunoreactive for a* Pseudomonas aeruginosa *antigen visualized with a monoclonal antibody. Acalculous necrotizing cholecystitis was associated.*

Luminal obstruction of the bile duct by pancreatobiliary malignancy is often associated with bactibilia and provokes secondary (ascending) bacterial infection. Enterococci often colonize the cancer tissue, and obstructive cholangitis and liver abscess may follow [54]. They are responsible for postoperative septic complications. The surgical specimen of cholangiocellular carcinoma in a female patient aged 80's showed necrotizing inflammation of the intrahepatic bile duct, as illustrated in **Figure 25**. Gram-positive cocci infected the necrotic cancer tissue. Culture of the bile was positive for *Enterococcus faecalis*.

8.4 Pulmonary gangrene

Pulmonary gangrene is a rare form of acute and severe necrotizing pneumonia [55–57]. A necrotic process with cavity formation is observed in a pulmonary segment or lobe. The term pulmonary gangrene is applied when a large amount of lung tissue is sloughed off. The extent of necrosis is far extensive in pulmonary gangrene when compared with usual pulmonary abscess (**Figure 26**). The lesion is often located in the upper lobe of the lung. Thrombosis of large and small vessels plays a significant role in the ischemic pathogenesis. *Klebsiella pneumoniae* is often isolated from the gangrenous lesion. Infection of anaerobes should be the cause of foul smell. The anaerobes may secondarily infect the lung slough under the progressively anaerobic environment.

8.5 Emphysematous pyelonephritis and renal papillary necrosis

Emphysematous pyelonephritis is a severe, multifocal, necrotizing, and gas-forming form of acute ascending bacterial infection of the renal parenchyma. Extracapsular extension is common. The disease is most often seen in patients with poorly controlled diabetes mellitus. The common causative pathogens are *Enterobacteriaceae*, particularly *Escherichia coli* and *Klebsiella pneumoniae* [58–60].

E. coli-induced emphysematous pyelonephritis in a male patient aged 60's is demonstrated. The patient suffering from alcoholic cirrhosis manifested lumbar pain and high fever. Septic shock killed the patient. The total clinical course was

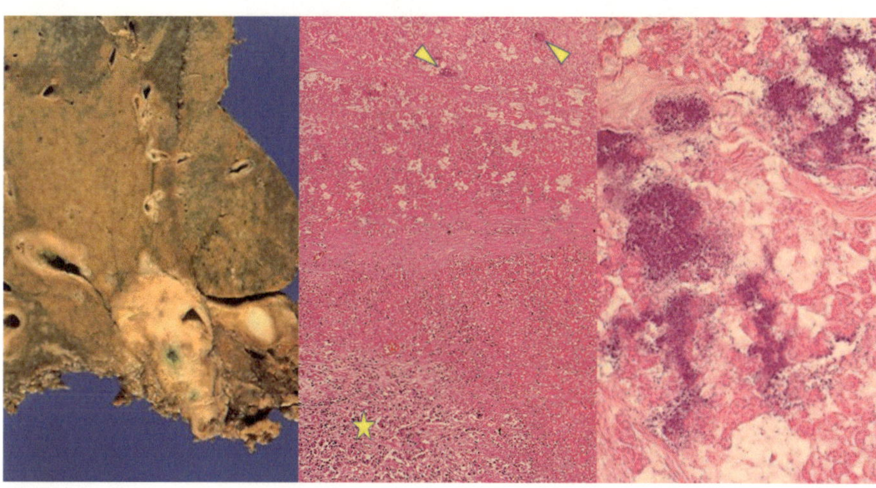

Figure 25.
Enterococcal intrahepatic cholangitis superimposed on cholangiocellular carcinoma in the surgically resected liver (gross and H&E). Colonization of culture-proven Enterococcus faecalis *is demonstrated in the necrotic cancer tissue (arrowheads), provoking acute intrahepatic cholangitis. Asterisk indicates poorly differentiated adenocarcinoma. High-powered H&E picture of the cocci is shown in the right panel.*

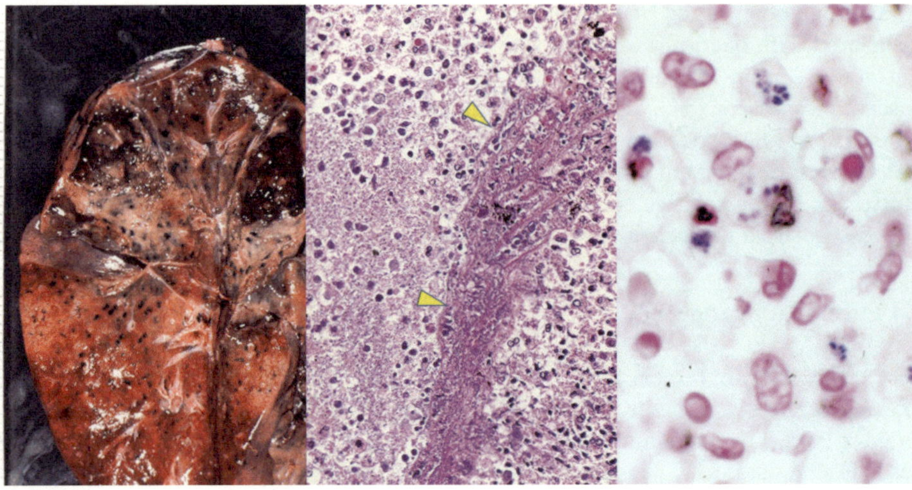

Figure 26.
Pulmonary gangrene (gross, H&E, Gram). Necrotizing (cavity-forming) pneumonia is noted in bilateral upper lobes of the lung. Foul smell was characteristic. Gangrenous inflammation is evident histologically. Microbial culture from the lung lesion identified Bacteroides, Pseudomonas aeruginosa *and* Peptostreptococcus. *Pseudomonal infection is indicated by arrowheads, and Gram-positive cocci (probably representing* Peptostreptococcus*) are phagocytized by neutrophils.*

9 days. At autopsy, both kidneys were enlarged and accompanied multifocal gangrenous changes in association with small foamy bubbles. Foul smell was not associated. Microscopically, gas formation was evident in the necrotic renal parenchyma, in association with diffuse neutrophilic infiltration (**Figure 27**). Numerous Gram-negative rods immunohistochemically expressing *E. coli* antigens are clustered within the necrotic renal tubules and around gas-filled bubbles. Microbial culture confirmed infection of *E. coli*. The condition can be categorized in non-clostridial gas gangrene.

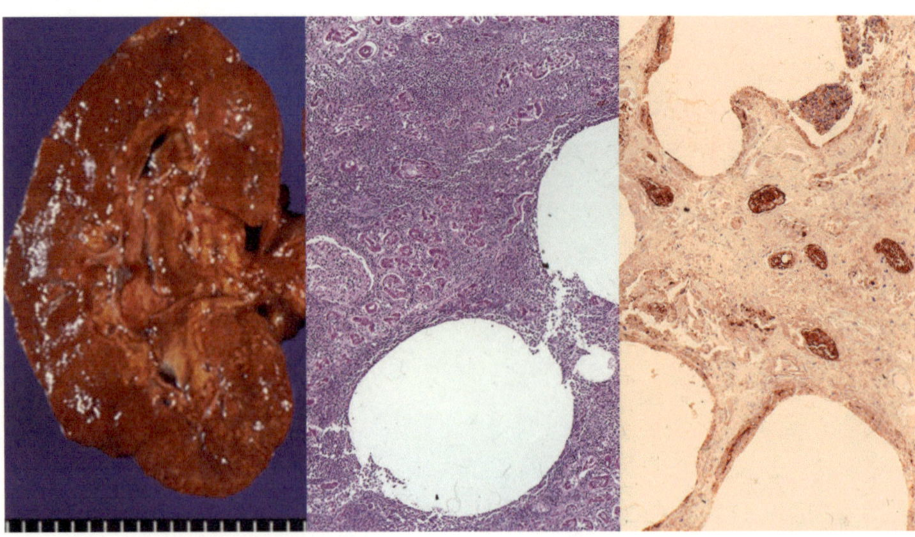

Figure 27.
E. coli-*infected emphysematous pyelonephritis in a diabetic male patient aged 70's (gross, H&E and immunostain for* E. coli *antigens). The enlarged kidney shows multifocal gangrenous changes with formation of small bubbles. Gas-forming infection of* E. coli *is evident both histologically and immunohistochemically in severe acute purulent pyelonephritis.*

Renal papillary necrosis is another form of lethal renal infection of *E. coli* seen in poorly controlled diabetic patients (**Figure 28**). The disease is characterized by coagulation necrosis of the renal medullary pyramid: the renal papillae are anatomically vulnerable to ischemic changes [61]. *E. coli* septicemia often follows, and the prognosis is poor.

8.6 Endophthalmitis

Endophthalmitis represents bacterial or fungal infection of the eyeball, as an acute illness (medical emergency) having up to a few days duration [62–64]. Patients complain of blurred vision, red eye, pain, and lid swelling. Due to progressive vitritis, hypopyon can be seen at the time of presentation. Exogenous organisms invade the eyeball via trauma, surgery, or corneal infection. When infection spreads to the adjacent orbital soft tissue, it is called as panophthalmitis. Endophthalmitis is localized to the eye, and it does not result in bacteremia or fungemia. Patients with Hansen's disease (leprosy) are highly susceptible to traumatic eyeball infection. Streptococcal infection may be proven in the surgical specimen. Prolonged inflammation results in ophthalmophthisis (**Figure 29**). Gram-positive cocci, including *Staphylococcus epidermidis* and *Streptococcus viridans*, are commonly isolated after surgery for cataract or intravitreal injection. Gram-negative bacteria such as *Pseudomonas aeruginosa*, *Hemophilus influenzae*, and *Moraxella catarrhalis* infrequently cause endophthalmitis. *Bacillus cereus* and fungi, particularly *Fusarium* spp., are the major cause of post-traumatic endophthalmitis [65]. **Figure 30** illustrates a surgical specimen of a *Fusarium*-infected eyeball. Traumatic corneal infection extended to the surrounding tissues such as the lens, palpebra, and orbit to provoke panophthalmitis. The fungal colonies on the surface microscopically reveal several-celled (chained or beaded), fusiform to sickle-shaped macroconidia (hyphae).

Endocarditis-associated endogenous endophthalmitis is usually caused by *Staphylococcus aureus* and streptococci. *Klebsiella pneumoniae* is another important

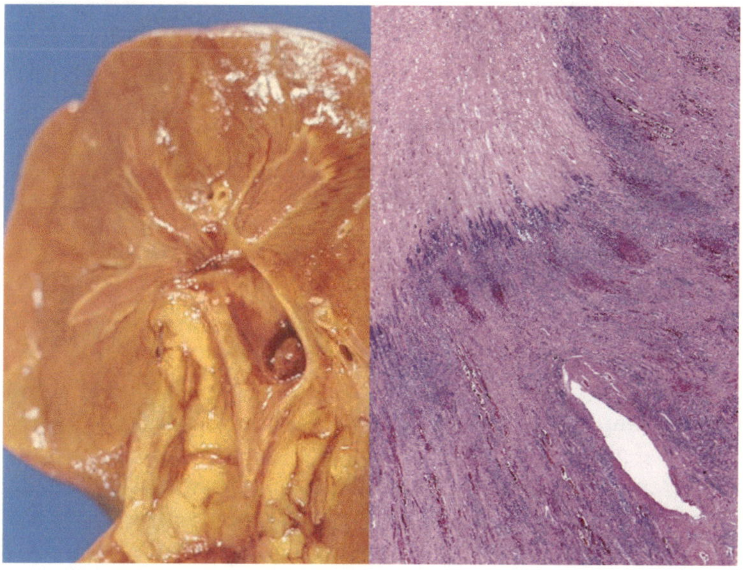

Figure 28.
Renal papillary necrosis in a male patient aged 60's with uncontrolled diabetes mellitus (gross and H&E). The patient manifested symptoms of acute pyelonephritis and died of acute renal failure. At autopsy, the renal papillae are necrotic and demarcated with yellowish zones. Ascending infection of E. coli *was associated.*

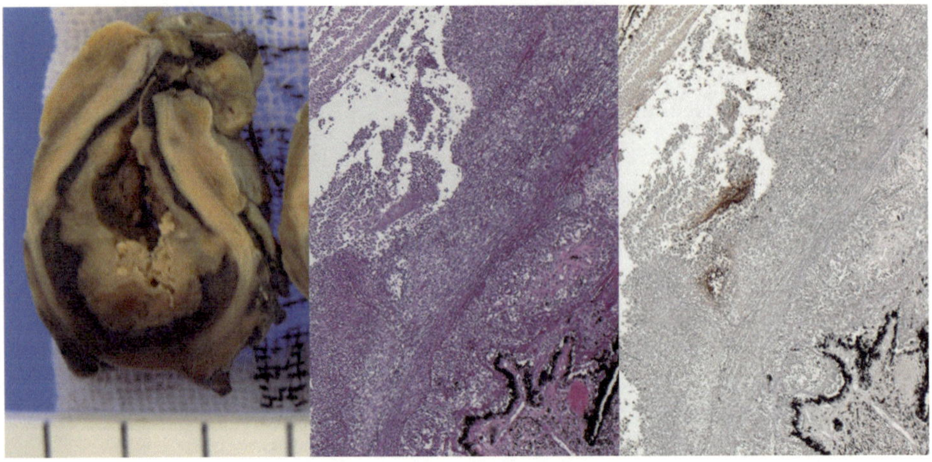

Figure 29.
Endophthalmitis in a leprosy patient (gross, H&E and immunostain). The eyeball is totally collapsed and deteriorated. Traumatic infection resulted in ophthalmophthisis. Gram-positive cocci inside the eyeball are immunoreactive for streptococcal antigens. Black melanin pigment in the iris is shown in the right bottom corner.

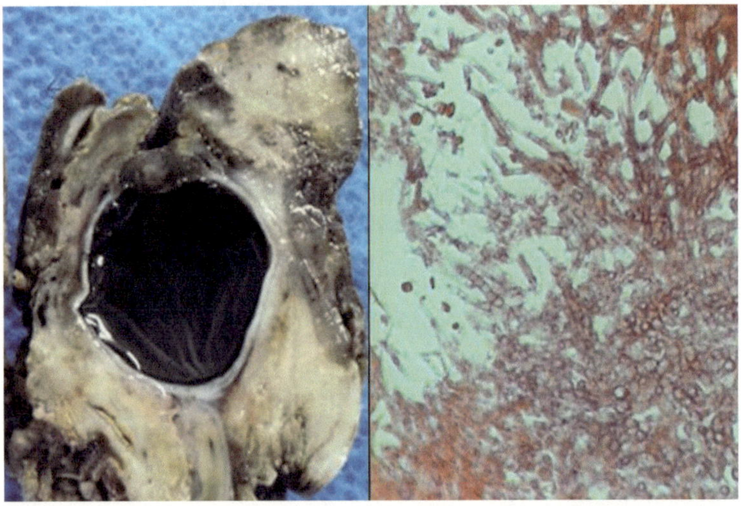

Figure 30.
Traumatic ophthalmitis caused by Fusarium *infection in a Cambodian teenager (gross and H&E). The corneal fungal infection extended to the lens, palpebra and orbital connective tissue. Chained or beaded (several-celled) appearance of hyphae is characteristic of* Fusarium *spp. (the courtesy of Dr. Chhut Vanthana, a pathologist at Sihanouk Hospital Center of HOPE, Phnom Penh, Cambodia).*

pathogen for endogenous endophthalmitis. Hyperalimentation may lead to endophthalmitis caused by *Candida albicans*.

8.7 Gangrenous/emphysematous inflammation in other organs

Gangrenous/emphysematous inflammation may occur in the stomach [33, 66] (see **Figures 16** and **17**), esophagus [67], colorectum [68] (see **Figures 15** and **18**), urinary bladder [69, 70], ureter [71], urethra [72], penis [73], epididymis/testis [74, 75], endometrium [76], vagina [77], breast [78], bone [79], striated muscle [80], aorta [81], mediastinum [82], and endocardium [83]. Most cases are categorized in the non-clostridial etiology. Clostridial infection is seen in the gastrointestinal tract and pancreas, including emphysematous pancreatitis [28], as described in the Section 6.2.

9. Vincent angina and noma (cancrum oris, gangrenous stomatitis)

Vincent angina, named after the French physician Jean H. Vincent (1862–1950), represents acute necrotizing ulcerative gingivitis caused by fusiform bacteria and spirochetes [84, 85]. It is also called as trench mouth or fusospirochetosis. The patients complain of progressive painful swelling and hemorrhagic ulceration of the gum. The punched-out ulcer, 2–4 mm in size, is seen in the interdental papilla, and is covered with white pseudomembranes. Bad breath is associated. The infection can effectively be treated with penicillin. Infrequently, Vincent angina may spread to involve the mouth and throat to be diagnosed as acute necrotizing periodontitis.

Noma is a rapidly progressive and necrotizing infection of the soft and hard tissues around the oral cavity, as an advanced clinical form of Vincent angina [86, 87]. It is also called as fusospirochetal gangrene. It represents gangrenous stomatitis or necrotizing fasciitis of the oral cavity. The preferred age of the patients is below 10 years, and the disease mostly occurs in malnourished children of African poverty. The prognosis is poor. In developed countries, severely immunosuppressed patients (including acquired immunodeficiency syndrome) with poor oral hygiene may suffer from this critical condition. It begins in the form of Vincent angina, and is rapidly followed by painless and extensive necrosis of the oral cavity. Eventually, the extensive involvement of the cheek, nose, palate, and maxillary bones results in serious facial destruction. Hence, the name of "cancrum oris" (meaning oral cancer). Gas formation may be associated. In noma neonatorum, the disease manifests massive orofacial (mucocutaneous) gangrene in the neonate [88]. A similar disorder may be encountered in the genitalia and is called as noma pudendi.

The polymicrobial etiology is known in both conditions. Gram stain smeared from the ulcer easily identifies both fusiform bacteria and long spiral-shaped spirochetes (**Figure 31**). The key players are anaerobic, Gram-negative fusiform pathogens, *Fusobacterium nucleatum* (older term: *Bacillus fusiformis*) and *Prevotella intermedia*. The spiral microbes are identified as *Borrelia vincentii*. Many other bacteria have been co-isolated, including *Porphyromonas gingivalis* (an anaerobic, Gram-negative, porphyrin-producing bacillary pathogen of periodontitis), *Tannerella forsynthesis*, *Treponema denticola*, *Staphylococcus aureus*, and nonhemolytic streptococci.

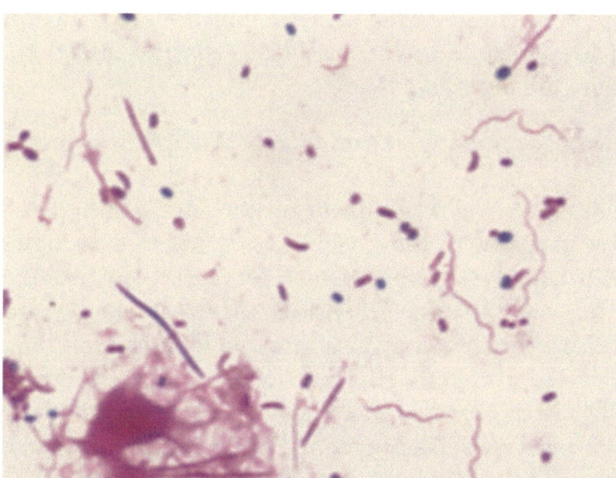

Figure 31.
Vincent angina (Gram). Gram-stained smear prepared from a painful gingival ulcer demonstrates mixed bacterial infection, including Gram-negative fusiform bacilli and filamentous spiral microbes. Gram-positive cocci and long rods are also intermingled.

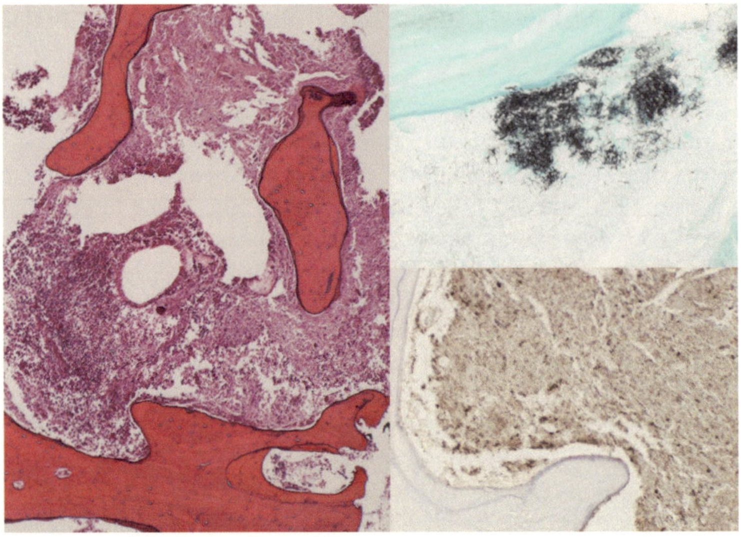

Figure 32.
Noma-like condition in a diabetic male patient aged 80's (progressive ulcerative gingivitis) (H&E, Grocott and immunostain). A gas-forming, necrotizing lesion is observed in the biopsied maxillary bone. Grocott methenamine silver stain identifies colonies of filamentous bacteria in the lesion, probably representing Actinomyces *colonization. The Gram-negative bacteria around the gas bubble are immunoreactive with a commercial antiserum against* Escherichia coli, *which shows wide cross-reactivity to Gram-negative bacteria (the courtesy by Dr. Tatsuru Ikeda, Pathology Center, Hakodate Goryoukaku Hospital, Hakodate, Japan).*

Figure 32 demonstrates a diabetic male patient aged 80's, suffering from noma-like condition (progressive ulcerative gingivitis with massive maxillary necrosis). Numerous bacilli accompanying gas formation and immunoreactive with *E. coli* antiserum grew in the maxillary bone. Colonies of filamentous bacteria, representing anaerobic *Actinomyces* spp., were coinfected.

10. Flesh-eating bacteria infection

A variety of microbes cause progressive and often lethal gangrenous lesions in the soft tissue, particularly on the extremities. The mass media often call this frightening condition as "flesh-eating bacteria infection." Three representative forms, fulminant streptococcal infection, *Vibrio vulnificus* infection, and *Aeromonas hydrophila* infection, are described below.

10.1 Fulminant streptococcal infection (streptococcal myonecrosis)

Streptococcal myonecrosis, a fulminant form of necrotizing fasciitis, presents a rapidly progressive gangrene of the extremities caused by infection of *Streptococcus pyogenes* (group A β-hemolytic *Streptococcus*), representing a prototype of "flesh-eating bacteria infection" [89, 90]. The disease affects persons of any age. Groups B and G β-hemolytic *Streptococcus* may also cause an identical fulminant condition [91, 92]. In some cases, protein S deficiency may be responsible for the necrotizing inflammation. It has been reported that vimentin, an intracellular intermediate filament of nonepithelial cells, is upregulated in the injured skeletal muscle cells and functions as the major skeletal-muscle protein binding to streptococci [93]. The life-threatening gangrene follows the subacute form of necrotizing fasciitis or occurs suddenly without preexisting ulceration. As shown in **Figure 5**, an advanced, deep pocket-forming decubitus in the sacral region may cause the lethal gangrenous lesion categorized in non-clostridial gas gangrene [18].

Clinically, high fever, pain at the site of infection, and skin necrosis (gangrene) with hemorrhagic bulla formation are associated. Scarlatiniform rash may be noted. Finally, massive gangrenous necrosis involves the extremity.

Microscopically, pronounced myonecrosis with foci of infection of Gram-positive cocci is observed. Gram-positive cocci grow within the lesion of advancing gangrenous necrosis of soft tissue. Cellular reactions are minimal, because of the ischemic (anaerobic) state with poor blood flow. In the cultured blood, short chains of Gram-positive cocci, morphologically typical of *Streptococcus*, are seen (**Figure 33**). Streptococcal septicemia provokes streptococcal toxic shock-like syndrome [94]. The bacterial exotoxins (superantigens) such as streptococcal pyrogenic exotoxins-A, B, C, F, and streptococcal superantigen provoke a severe cytokine storm. Hypercytokinemia activates hemophagocytosis by macrophages. Activation of NLRP3 inflammasome may be an essential event for the cytokine storm in streptococcal toxic shock-like syndrome [95].

The bacteria are commonly sensitive to penicillin and its derivatives, but the intravenous antibiotics administration is clinically ineffective, principally because of the absence of blood flow. The drug can hardly reach the site of infection.

10.2 *Vibrio vulnificus* infection

Progressive gangrene of the extremities caused by infection of *Vibrio vulnificus* is characteristically seen in patients with liver cirrhosis or hemochromatosis [96–99]. High iron concentration in the serum is essential for the bacteria to grow in the body. The genus *Vibrio* is categorized in the "halophilic" bacteria preferring to a high salt concentration for growth on plates. In contrast to *V. cholerae* and *V. parahaemolyticus* growing at the salt concentration of sea water (3–3.5%), *V. vulnificus* prefers to a lower salt concentration of the brackish (estuarine) water at the mouse of the river. *V. vulnificus* resides in the sea fish and oyster, particularly during the summertime. The bacteria proliferate in the gut of the sea creature when the temperature is high. Two transmission pathways of the pathogen are known: transenteric infection and traumatic skin infection. The former septicemic condition is often fatal, initiating a painful skin lesion on the arm or leg resembling honeybee bite. Gangrenous changes of the extremity progress rapidly.

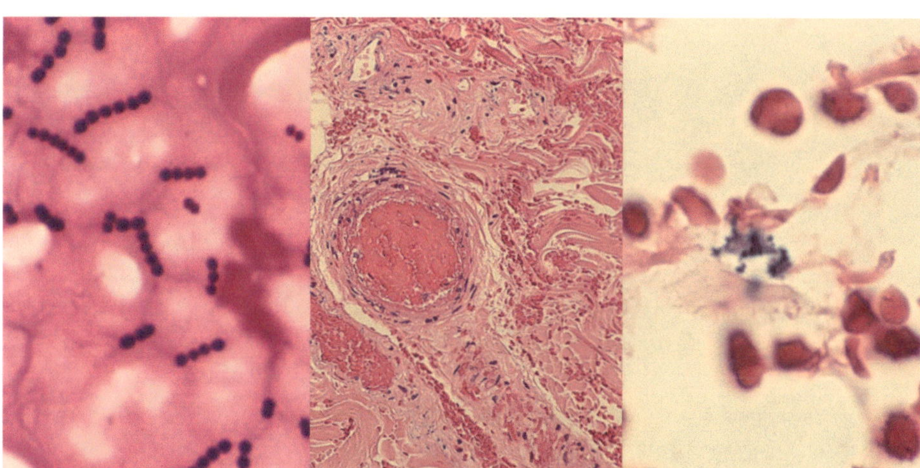

Figure 33.
Fulminant streptococcal infection (streptococcal myonecrosis) (Giemsa, H&E and Gram). Numerous chained cocci are demonstrated in the cultured blood. Vessels are thrombosed, and the striated muscle fibers show coagulation necrosis. Colonies of Gram-positive cocci are scattered in the ischemic tissue.

Gas formation is not associated. The traumatic infection of *V. vulnificus* is caused by an accidental trauma of the hand or fingers during cooking raw fish (preparing sashimi) or injuring the foot on the rocky seacoast. The prognosis is better than the former. The incidence of infection of the halophilic pathogen nicknamed "flesh-eating bacteria" is high in Japan.

Microscopically, perivascular cuffing of Gram-negative bacteria, showing a coccoid change, is noted in the involved ischemic/necrotic skin and soft tissue, while the cellular reaction is minimal (**Figure 34**).

10.3 *Aeromonas hydrophila* infection

Lethal gangrene of the extremities or face is also caused by *Aeromonas hydrophila* in patients under an immunocompromised condition, with diabetes mellitus or on hemodialysis, as a form of opportunistic infection [100–104]. The bacteria invade the skin via a minor trauma. **Figure 35** illustrates gross features of lethal gangrene of the right upper arm caused by *A. hydrophila*. Vesicles are formed on the necrotic skin. *A. hydrophila* belongs to the family *Vibrio* and widely distributes in fresh water and soil. *A. hydrophila* can grow at low temperature to cause food poisoning (watery or bloody diarrhea) due to production of heat-labile enterotoxins. An outbreak of *A. hydrophila* wound infection has also been reported among the participants for mud football games in Australia [105]. There were many infected scratches and pustules distributed over the bodies.

Microscopically, the lesion shows clusters of Gram-negative rods around necrotic subcutaneous tissue. Cellular reaction is poor. Gas formation may be associated. In the case as shown in **Figure 36**, necrotizing foci of infection were disseminated in the rectum, epididymis, prostate, liver, and kidneys.

11. Fournier's gangrene

Fournier's gangrene is a special form of fulminant cellulitis (fatal gangrene) involving the male scrotum and perineum [106–109]. The necrotizing change rapidly progresses to the surrounding soft tissue, eventually resulting in septicemia.

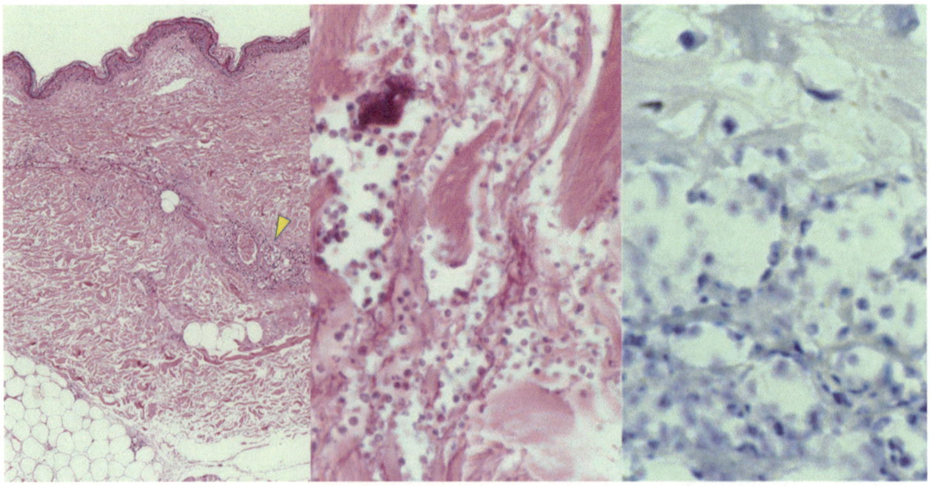

Figure 34.
Vibrio vulnificus *infection in a cirrhotic male patient (H&E and Giemsa). In a biopsy specimen sampled in an emergency suite, perivascular cuffing by infected microbes is observed around small vessels and sweat glands (arrowhead) in the deep dermis through subcutis. Coccoid transformation is recognized in H&E and Giemsa stained preparations. Inflammatory reaction is sparse. Gram stain showed negativity.*

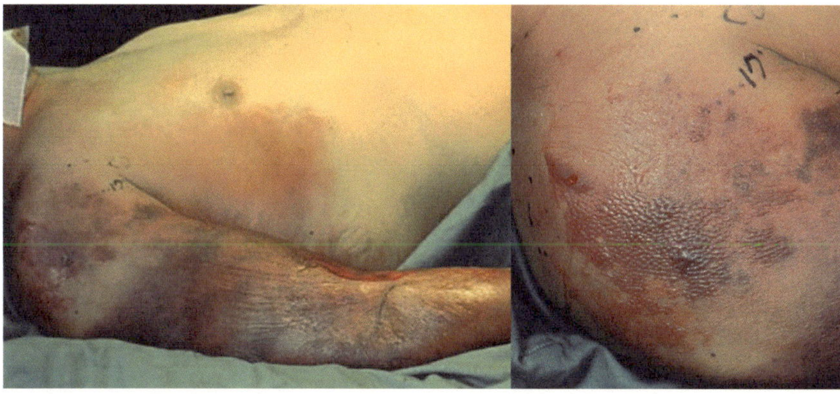

Figure 35.
Aeromonas hydrophila *infection in a diabetic male patient aged 50's (gross appearance). Lethal gangrene is observed on the right upper arm. Vesicular skin change is evident. Autopsy confirmed that septicemia caused multiorgan abscess formation (see* ***Figure 36****).*

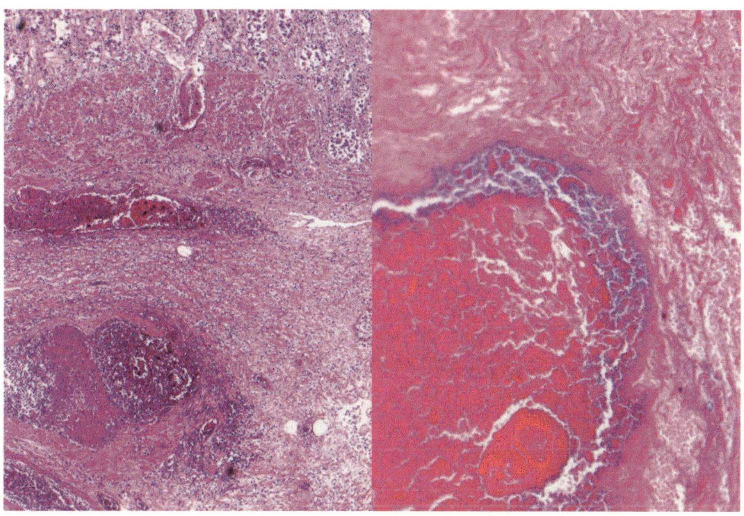

Figure 36.
Aeromonas hydrophila *infection (H&E). Septic and necrotic/hemorrhagic lesions are seen in the rectal submucosa (left) and epididymis (right). Septic embolism is noted in the rectum, while Gram-negative rods are clustered around the dilated and thrombosed vascular structure in the epididymis, where inflammatory reaction is sparse.*

The prognosis is poor. The scrotum is markedly swollen and becomes reddish-black in color (**Figure 37**). The penis is either involved or spared. The physiological lack of subcutaneous fat tissue in the scrotum and penis accelerates the bacterial spread. Gas production and malodor may be associated. It belongs to non-clostridial gas gangrene when gas production is noted. The preferred age ranges from 50 to 80 years. Male patients of Fournier's gangrene often have a history of diabetes mellitus. Immunocompromised condition also accelerates the disease. Perianal abscess should be a risk factor of the disease. Masturbation-related minor penile skin injury may cause the disease in younger age [110].

Microscopically, massive necrosis of the skin tissue is evident. Mixed bacterial infection, including *Streptococcus* and anaerobic bacteria, is often proven. When streptococci are isolated, it is categorized in fulminant streptococcal infection (**Figure 38**). Secondary surface infection of *Trichosporon* spp. (an opportunistic fungal pathogen) may occur.

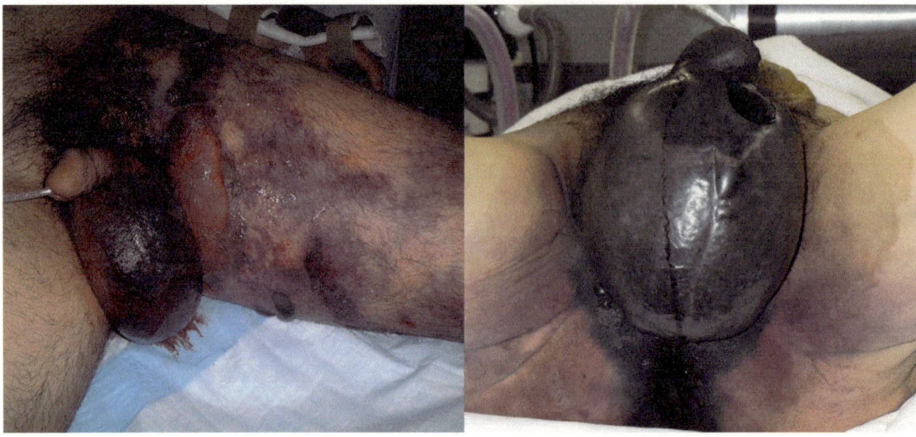

Figure 37.
Fournier's gangrene (gross findings of two male cases). Massive hemorrhagic necrosis started from the scrotum and extended to the left hip and leg (left). Marked black swelling of the scrotum is serious, and necrotizing change extends toward the perianal region (right). The rapidly progressive gangrene caused death in both patients. The penis is spared in the left case, but massively involved in the right case.

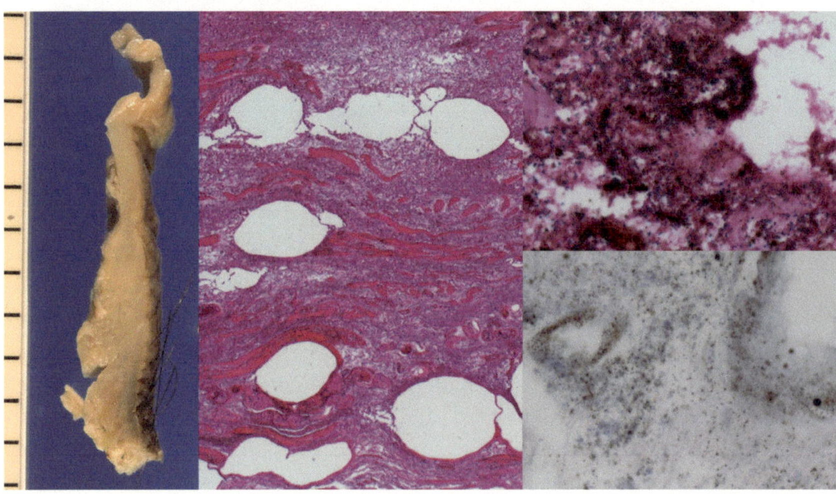

Figure 38.
Fournier's gangrene (gross, H&E, Gram and immunostain). Debridement specimen discloses massive transmural necrosis of the scrotal tissue. Gas bubbles are scattered in the heavily infected necrotic tissue. Gram-positive cocci are immunoreactive for streptococcal antigens. This case represents fulminant streptococcal infection with gas formation (non-clostridial gas gangrene).

As illustrated in **Figure 39**, fulminant necrotizing inflammation involved the lower part of the rectum in a female patient suffering from myelodysplastic syndrome. Emergency surgery disclosed transmural gangrenous necrosis of the rectal wall with massive mixed bacterial infection, including *E. coli*. Occasionally, Fournier's gangrene has been complicated with rectal cancer [111, 112].

12. Necrotizing fasciitis

Necrotizing fasciitis represents clinically severe pyogenic infection (cellulitis) of the skin and underlying soft tissue [113–117]. Deep, painful, and intractable ulceration subacutely progresses predominantly on the extremities (**Figure 40**). Minor trauma may provide the entry for pathogens. The condition uncommonly follows

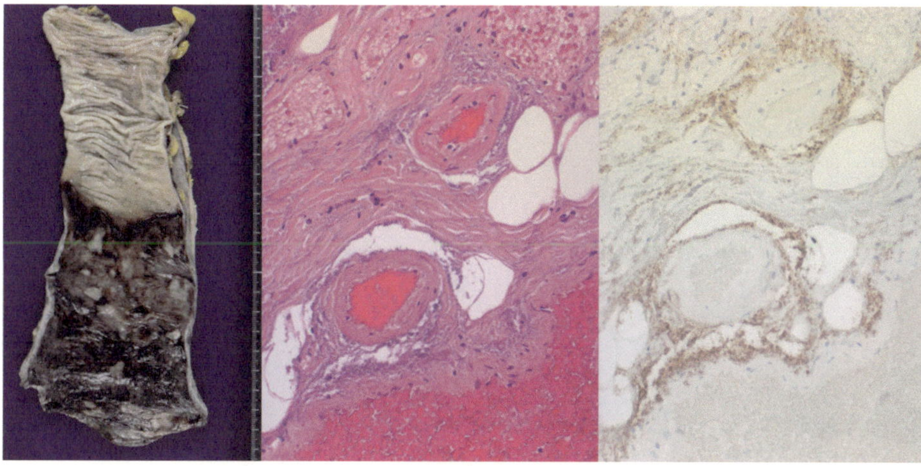

Figure 39.
Lethal Fournier's gangrene of the rectum (gross, H&E and immunostain). Transmural necrotic and gangrenous inflammation is seen in the lower part of the rectum in a female patient aged 60's suffering from myelodysplastic syndrome. Gram-negative rods are immunoreactive for E. coli*-related lipopolysaccharide.*

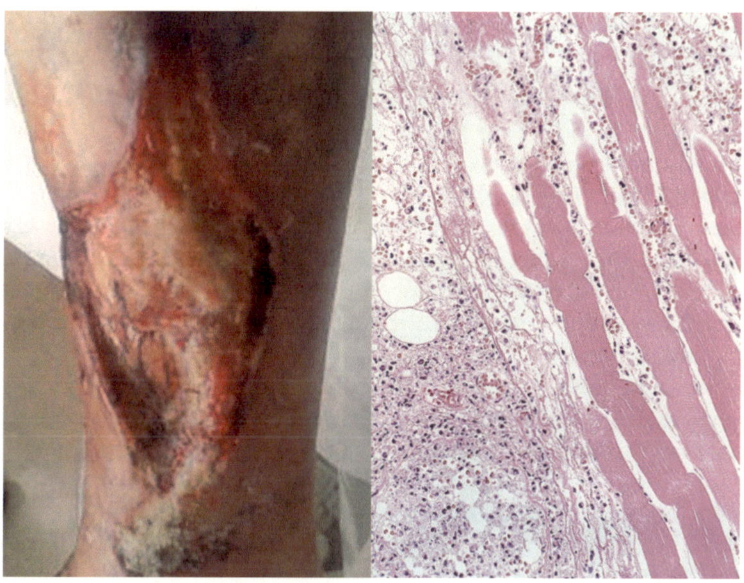

Figure 40.
Necrotizing fasciitis (gross and H&E). Deep and painful ulceration is caused by local and invasive bacterial infection. This aged male diabetic case had a history of arterial replacement therapy for atherosclerosis obliterans. In order to relieve pain and to avoid septicemic spread of infection, amputation surgery was performed. Necrotizing inflammation extends to the striated muscle layer.

surgical procedures. Diabetes mellitus, immunosuppression, alcoholism, drug abuse, atherosclerosis-related ischemia, and malnutrition may be prodromal to this troublesome condition. It may be seen in healthy persons [118]. Necrotizing fasciitis is categorized into two types: type I (polymicrobial infection) and type II (monobacterial infection).

In **Figure 41**, necrotizing fasciitis seen in a poorly controlled diabetic male patient is presented. In the wintertime, a fan heater gave the patient a severe burn on his sole, because he did not feel pain sensation due to diabetic peripheral neuropathy. The doctor-shy patient did not visit a hospital for 1 week, and this allowed the lesion far progressed. Severe atherosclerosis had provoked dry gangrene in his

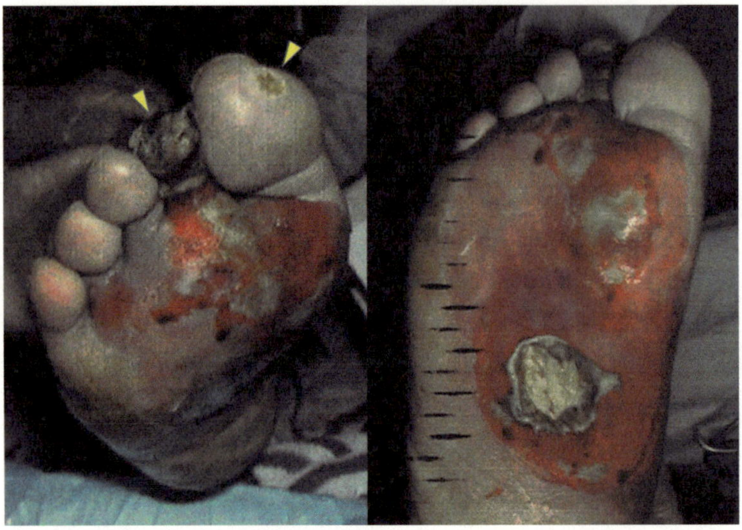

Figure 41.
Localized severe burn on the sole of a diabetic male caused by a fan heater, resulting in necrotizing fasciitis (gross appearance). Because of diabetic neuropathy, deep ulcers occurred on the senseless foot. Dry gangrene on the first and second toes (arrowheads) indicates the association of diabetes-related atherosclerosis obliterans. The importance of foot care in diabetic patients should be emphasized.

toes. Diabetes-related neutrophilic dysfunction provided him with the vulnerability to infection. Polymicrobial (type I) necrotizing fasciitis resulted in septicemia. Emergency amputation saved his life. The importance of foot care for patients with diabetes mellitus should be emphasized.

Infrequently, necrotizing fasciitis is caused by *Pseudomonas aeruginosa* [119, 120]. Reportedly, the mortality rate of this type II lesion is 30%, and the infection often happens in the immunocompromised patients. Clinicians should consider empiric pseudomonal antibiotic coverage for preventing the progression of necrotizing limb infection.

An 18-year-old female patient had suffered from anorexia nervosa for 6 years. She happened to develop phlegmonous inflammation on her left lower leg, rapidly progressing to multifocal ulceration and gangrene. In 3 days, she underwent surgical amputation. *Pseudomonas aeruginosa* was cultured from blood and the leg lesion of necrotizing fasciitis. Immunohistochemical identification of the pseudomonal microbe was achieved by using a commercial monoclonal antibody. Representative features are illustrated in **Figure 42**.

Classic pathogens of cellulitis represent group A β-hemolytic *Streptococcus* and less frequently *Staphylococcus aureus*, but a diverse range of microorganisms, including *Pseudomonas aeruginosa* (as described above), cause cellulitis. Erythematous nodular lesions formed on the leg of neutropenic or leukemic patients were caused by *Stenotrophomonas maltophilia* [121]. Facial cellulitis may result from *Haemophilus influenzae* infection [122].

13. Fulminant coccal infection without gangrene of the extremities

Gram-positive cocci occasionally provoke fulminant, lethal systemic infection without gangrene of the extremities. The pathophysiology resembles that of flesh-eating bacteria infection, accompanying pronounced hypercytokinemia and poor cellular reactions. Streptococcal, pneumococcal, staphylococcal, and enterococcal etiologies are described below.

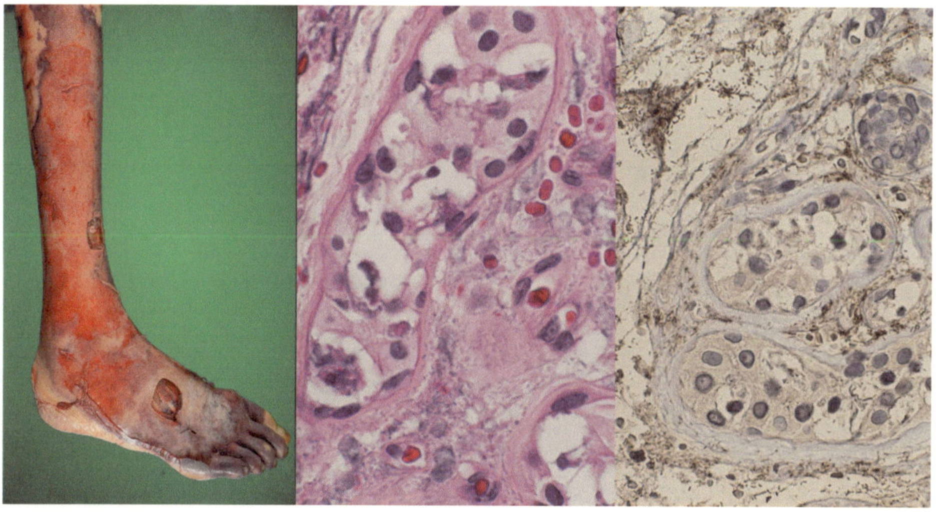

Figure 42.
Pseudomonas-*related necrotizing fasciitis in a young lady suffering from anorexia nervosa (gross, H&E and immunostain). Her leg with massive necrotic/gangrenous lesions was amputated (left, after sampling of histological specimens). Massive bacterial growth provoked little inflammatory reaction. The bacteria are immunoreactive for* Pseudomonas aeruginosa *antigen detected by a monoclonal antibody (the courtesy by Dr. Takashi Tsuchida, a pathologist in Hamamatsu Medical University Hospital, Hamamatsu, Japan).*

13.1 Fulminant streptococcal infection without gangrene of the extremities

Fulminant infection of group A β-hemolytic *Streptococcus* (*Streptococcus pyogenes*) is typically featured by progressive gangrene in the soft tissue of the extremities, as described above in the Section 10.1. Streptococcal toxic shock syndrome provokes an aggressive lethal condition without predisposing diseases [123, 124]. It should be of note that fulminant group A streptococcal infection is also encountered in cases without gangrenous lesions of the extremities [125]. Streptococcal infection in the internal organs may cause the fatal disease.

We experienced five cases of fulminant streptococcal infection without gangrene of the extremities (**Table 1**). Four of five cases were young and immunocompetent, and encountered at forensic autopsy. Infectious foci were seen in internal organs such as the tonsil, bronchus, puerperal endometrium, and urinary bladder. The clinical course was very short ranging from 2 to 4 days. Infective and hemorrhagic cystitis with systemic streptococcal dissemination was encountered in an aged female patient with a history of cerebral infarction and femoral neck fracture (**Figure 43**). Necrotizing endometritis in a puerperal lady was the cause of streptococcal toxic shock-like syndrome, as illustrated in **Figure 44**. It can be categorized in so-called puerperal fever. Pregnancy-associated lethal infection should be of particular notice [126]. Group A *Streptococcus* infection was proven by microbial culture in two cases, and immunoreactivities of streptococcal antigens and Strep A were shown on the Gram-positive cocci in all five cases. Strep A is a carbohydrate antigen specific for group A *Streptococcus* [127].

There are two different pathological mechanisms in fulminant streptococcal infection without gangrene of the extremities [125]. One form with overwhelming bacterial growth is characterized by secondary systemic bacterial dissemination accompanying bacterial emboli with poor neutrophilic reaction. Bacterial embolism in the adrenal gland provokes bilateral adrenal hemorrhage (acute adrenocortical insufficiency), being categorized in Waterhouse-Friderichsen syndrome [128] (**Figure 45**). Another form without bacterial embolism was featured by bacterial

Case	Age/ sex	Clinical course	PD	Primary lesion	BE	MC	Autopsy findings
1	86 F	3 days	+	Hemorrhagic cystitis	+	ND	Bilateral renal cortical necrosis, bilateral adrenal hemorrhage, and DIC
2	30 M	2 days	–	Acute tonsillitis	+	–	Bilateral renal cortical necrosis, bilateral adrenal hemorrhage, and DIC (microthrombosis)
3	38 F	4 days	–	Necrotizing endometritis (Puerperal fever)	–	+*	Hemophagocytic syndrome, bilateral renal cortical necrosis, leukostasis, DIC (microthrombosis), myocardial ischemia, and liver congestion
4	24 F	3 days	+	Necrotizing bronchitis	–[a]	ND	Hemophagocytic syndrome, acute renal tubular necrosis, DIC, myocardial ischemia, pulmonary edema, and tonsillar hyperplasia
5	35 M	3 days	–	Necrotizing bronchitis	–	+	Hemophagocytic syndrome, acute renal tubular necrosis, DIC, myocardial ischemia, liver congestion, and pulmonary edema

PD—preexisting disease (case 1: cerebral infarct and femoral neck fracture; case 4: Graves' disease), BE—bacterial embolus formation in distant organs and tissues, and MC—microbial culture (ND: not done).
**Negative in the blood but positive from the uterine cervix.*
[a]Aspiration of coccal colonies into the alveolar space seen.

Table 1.
Summary of five autopsy cases of fulminant streptococcal infection without gangrene of the extremities [125].

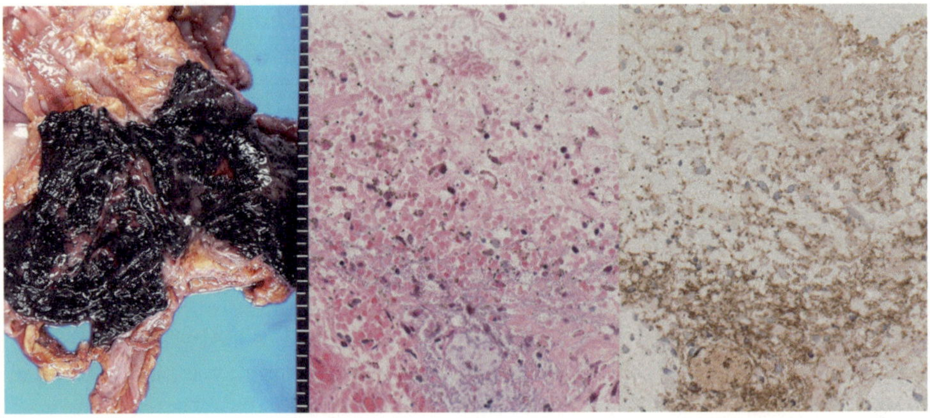

Figure 43.
Fulminant streptococcal infection with hemorrhagic cystitis an 86-year-old female patient (gross, H&E and immunostain). Massive hemorrhagic cystitis is evident. The cocci infected in the eroded bladder wall are immunoreactive for streptococcal antigens.

toxin-induced hemophagocytosis by activated macrophages, reflecting a hypercytokinemic state [129] (**Figure 46**). Hypercytokinemia and disseminated intravascular coagulation (DIC) are common phenomena in both forms, and bilateral renal cortical necrosis may be observed as an extreme manifestation of DIC [130]. Hematopoiesis in the bone marrow appear to be normal, but neutrophilic reactions are limited in the primary and disseminated infective foci. Supposedly, neutrophilic functions are acutely suppressed through two different mechanisms during the process of the fulminant disease. The disease is categorized in streptococcal toxic shock-like syndrome mediated by streptococcal superantigens [94, 95].

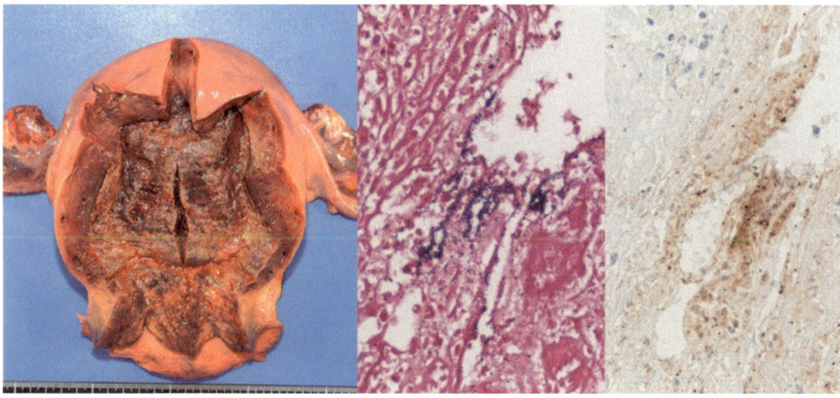

Figure 44.
Fulminant streptococcal infection with necrotizing endometritis in a 38-year-old female patient (gross, Gram, immunostain). The eroded postpartum endometrium 4 days after delivery is colonized by Gram-positive cocci with positive immunoreactivity for Strep A, a carbohydrate antigen of group A Streptococcus. *Neutrophilic reaction is limited in the endometrium. This condition is categorized as puerperal fever.*

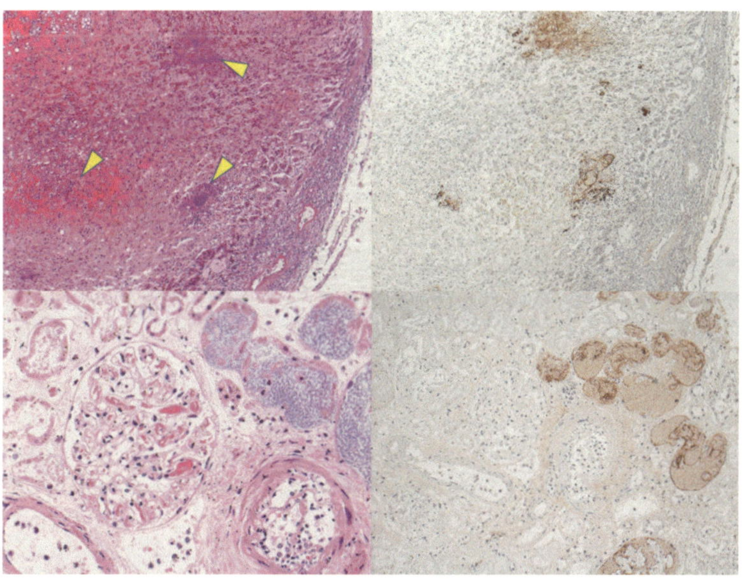

Figure 45.
Fulminant streptococcal infection showing septic embolism, Waterhouse-Friderichsen syndrome, and bilateral renal cortical necrosis in the case demonstrated in ***Figure 43*** *(adrenal and kidneys; H&E and immunostain). The adrenal glands show massive hemorrhagic necrosis. Septic streptococcal emboli (arrowheads) are seen in capillary vessels of the adrenal. The kidneys show bilateral cortical necrosis with marked fibrin thrombosis in the glomeruli and streptococcal colonization in the renal tubules (streptococcal antigens-positive).*

Physicians should keep the possibility of fulminant streptococcal infection in mind, particularly when examining the patient manifesting progressive shock symptoms even without gangrene of the extremities. Autopsy prosecutors (diagnostic and forensic pathologists) must realize the difficulty in making an autopsy diagnosis, particularly when bacterial embolism is not identified under a microscope. The knowledge of these types of fulminant syndrome and the appropriate microscopic recognition of hemophagocytosis in the bone marrow, liver, and spleen are critically important for the autopsy prosecutors. When the association of the hypercytokinemic state was not suspected clinically and microscopically, one can hardly reach the correct autopsy diagnosis.

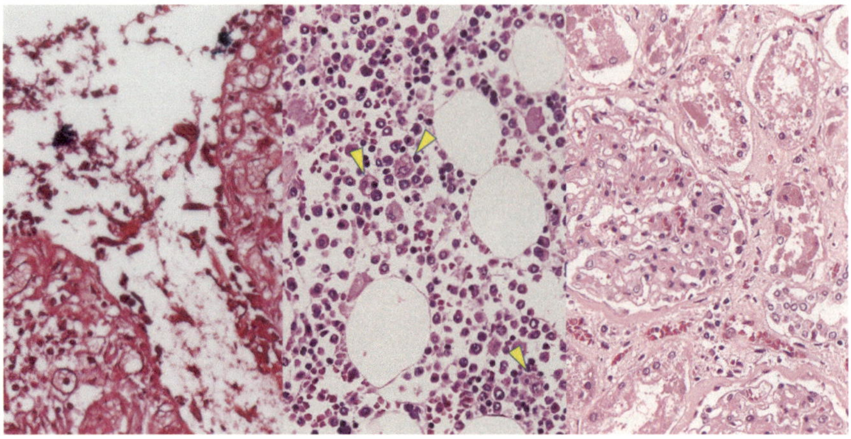

Figure 46.
Fulminant streptococcal infection without septic embolism caused by erosive bronchitis (bronchus: Gram, bone marrow and kidney: H&E). Local infection of Gram-positive cocci on the bronchus provoked hypercytokinemia and disseminated intravascular coagulation. Activated hemophagocytic macrophages (arrowheads) are distributed in the bone marrow. The kidney shows acute tubular necrosis.

13.2 Fulminant pneumococcal infection

Streptococcus pneumoniae (so-called *Pneumococcus*), a capsule-forming Gram-positive coccus, is a leading cause of community-acquired pneumonia. Fulminant pneumococcal infection is a life-threatening disease, resulting in DIC and multiorgan failure [131, 132]. "Purpura fulminans" represents an extreme skin manifestation of DIC and Waterhouse-Friderichsen syndrome (caused by bilateral adrenal hemorrhage). The disease is often seen in splenectomized or immunosuppressed patients [133–135], while it is also observed in healthy patients without a history of splenectomy [136].

A pregnant woman aged 20's manifested high fever and systemic skin rash. She had a history of splenectomy 10 years earlier. The total clinical course was as short as 2 days: septic shock provoked DIC and generalized petechiae. The disease represented puerperal fever. At autopsy, the uterus contained a dead fetus. The placenta contained small abscesses with infection of Gram-positive cocci with immunoreactivity of pneumococcal antigens (**Figure 47**). In the blood, α-hemolytic *Streptococcus* was isolated. Cytokine storm-related hemophagocytosis was observed in the bone marrow and spleen. Neither gangrene of the extremity nor pneumonia was associated. The final diagnosis was fulminant pneumococcal infection as a form of overwhelming postsplenectomy infection.

Another case (a 60-year-old male patient) of fulminant pneumococcal infection is displayed in **Figure 48**. Total clinical course was 3 days. The small-sized spleen was observed. Neither limb gangrene nor pneumonia was observed. The entry of *S. pneumoniae* was unclear. The glomeruli showed bacterial embolism by capsule-forming Gram-positive cocci immunohistochemically expressing pneumolysin (a pneumococcal hemolytic exotoxin). The capsule formation is visualized with the colloidal iron method that stains the acidic substances blue.

13.3 Fulminant staphylococcal infection

Community-acquired methicillin-resistant *Staphylococcus aureus* (CA-MRSA) often infects the skin and soft tissue of healthy young people. Severe invasive CA-MRSA infections include necrotizing pneumonia, necrotizing fasciitis, "purpura

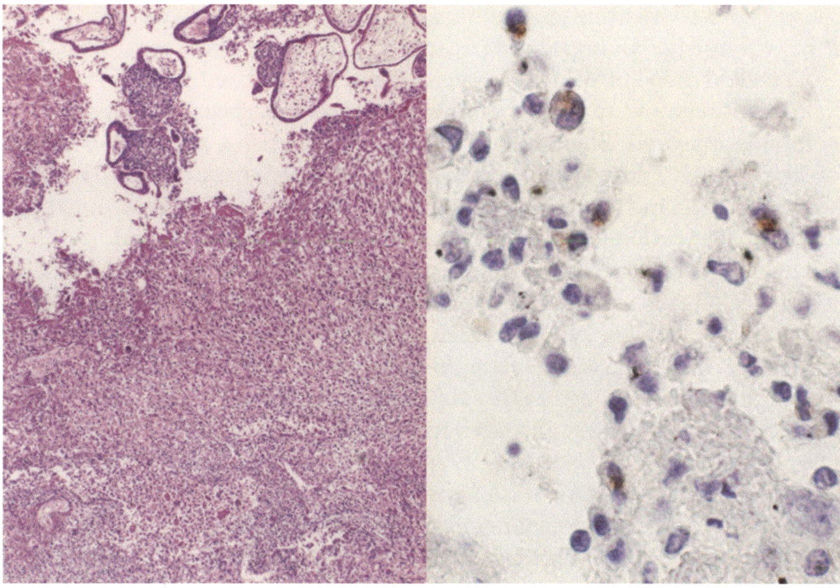

Figure 47.
Fulminant pneumococcal infection (H&E and immunostain). In this young lady with a history of splenectomy, the placenta was the entry of Gram-positive cocci. The bacteria with immunoreactivity of pneumococcal antigens are identified in the cytoplasm of neutrophils in a small abscess among placental villi.

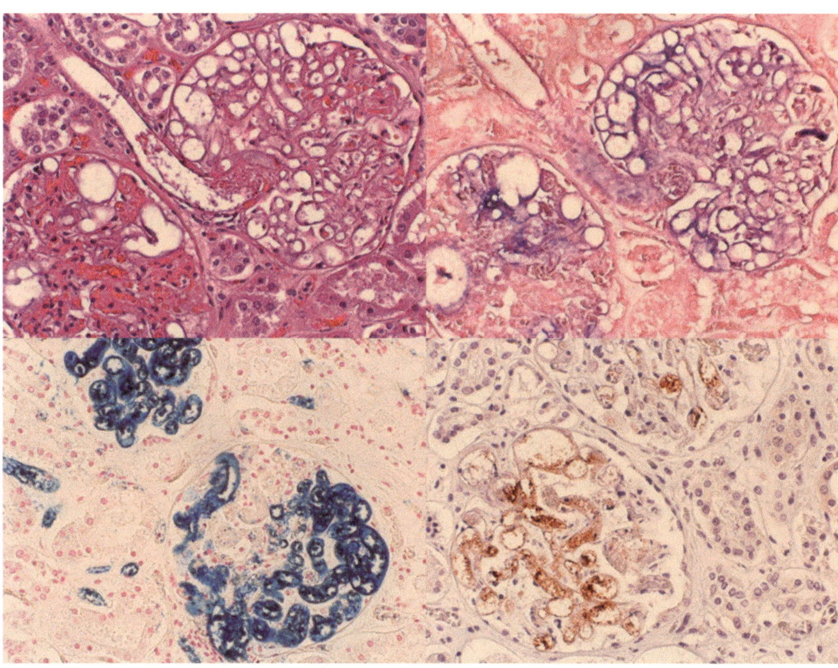

Figure 48.
Fulminant pneumococcal infection (H&E, Gram, colloidal iron and immunostain). Systemic spread of capsule-forming Gram-positive cocci drastically killed the patient. The glomeruli show septic embolism by cocci with colloidal iron-positivity (stained blue) and pneumolysin immunoreactivity (stained brown).

fulminans" (Waterhouse-Friderichsen syndrome) and disseminated infection with septic emboli [137–139]. The severe life-threatening infection may be caused by CA-MRSA, bearing the staphylococcal cassette chromosome mec gene type IV and expressing Panton-Valentine leucocidin, an exotoxin lethal to leukocytes [140]. CA-MRSA has emerged as an important pathogen in the community worldwide.

A 70-year-old man suffering from hepatitis virus C-related liver cirrhosis complained of fever and sudden abdominal pain. He soon became septicemic and skin eruptions appeared. Blood microbial culture identified CA-MRSA. The patient died of septic shock 5 days after onset. Autopsy revealed massive septic emboli of Gram-positive cocci in systemic organs and tissues (**Figure 49**). Disseminated intravascular coagulation was associated. Hypercytokinemia activated hemophagocytosis by macrophages. No gangrene of the extremities was observed. Bacterial entry was unclear. The pathophysiological process resembled that of staphylococcal toxic shock syndrome: the bacteria secrete toxic shock syndrome toxin-1 to activate Vβ2-positive T-lymphocytes secreting cytokines [141].

Another male inpatient aged 60's suffering from liver cirrhosis received endoscopic ligation therapy for esophageal varices. The next day, he manifested high fever and hematemesis. He died of DIC and septic shock in 2 days. The entry of hospital-acquired MRSA (HA-MRSA) was the esophagus, and disseminated septic emboli provoked bilateral adrenal hemorrhage (Waterhouse-Friderichsen syndrome) and hemophagocytosis. **Figure 50** demonstrates glomerular septic emboli of MRSA and massive adrenal hemorrhage.

13.4 Fulminant enterococcal infection

Enterococci may rarely cause a fulminant form of systemic infection [142–144]. Enterococcal gangrenous inflammation in the bile duct was already described in the Section 8.3. Opportunistic, necrotizing, and lethal enterococcal enteritis may be encountered in immunocompromised patients. A diabetic male patient aged 80's with acute thrombosis of the superior mesenteric artery is presented. In the surgical specimen, the transmurally necrotic small bowel wall was heavily colonized by Gram-positive and enterococcal antigens-positive cocci (**Figure 51**), and *Enterococcus faecalis* was identified by microbial culture. Formation of capsules (biofilm), rich in acidic substances, was evident with colloidal iron stain. Septic dissemination of enterococci followed to kill the patient.

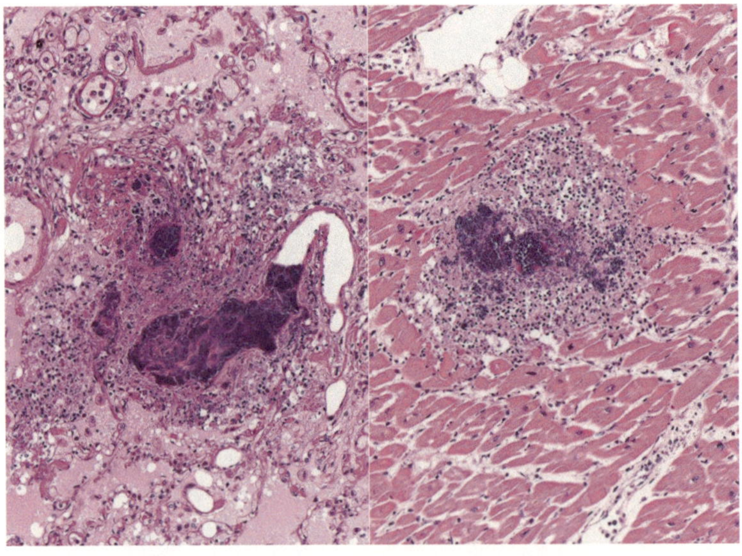

Figure 49.
Fulminant CA-MRSA infection (lung and heart: H&E). Septic emboli of Gram-positive cocci are pronounced in the pulmonary artery branches. Microabscess is formed in the heart muscles.

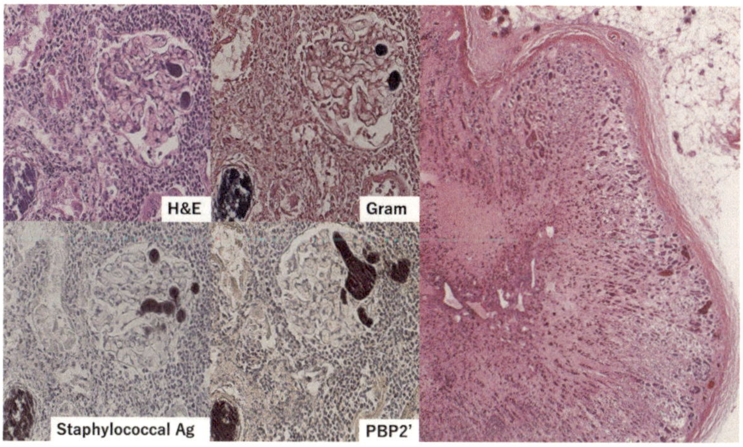

Figure 50.
Fulminant HA-MRSA infection (kidney: H&E, Gram, immunostain for staphylococcal antigen and PBP2'; and adrenal: H&E). Septic emboli in the glomerulus represent Gram-positive cocci with positivity for staphylococcal antigens and penicillin-binding protein 2' (PBP2'), immunohistochemically confirming MRSA septicemia (see also ***Figure 10****). Marked adrenal hemorrhage (right panel) indicated an extreme form of DIC or Waterhouse-Friderichsen syndrome.*

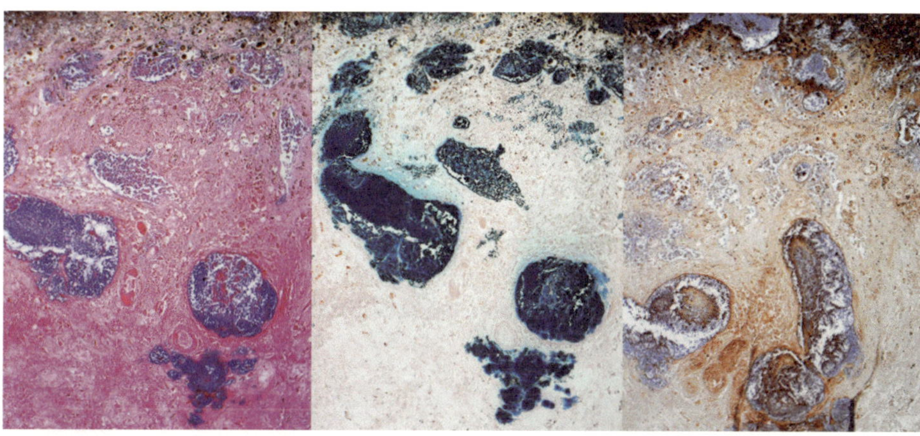

Figure 51.
Fulminant enterococcal infection (H&E, colloidal iron, and immunostain). Enterococcal necrotizing enteritis followed acute thrombosis of superior mesenteric artery in a diabetic male patient aged 80's. In the surgical specimen, the transmurally necrotic small bowel wall is heavily colonized by Gram-positive cocci with colloidal iron-stained thick acidic capsules. Enterococcal antigens are proven. Microbial culture identified Enterococcus faecalis. *Septic systemic dissemination killed the patient.*

14. Gangrenous inflammation associated with uncontrolled diabetes mellitus

As abovementioned repeatedly, diabetes mellitus predisposes gangrenous inflammation, particularly when the disease is poorly controlled. Here, three special disease situations as severe complications of diabetes mellitus are described.

14.1 Malignant otitis externa

The external ear canal guards against infection by producing a protective layer of cerumen that creates an acidic and lysozyme-rich environment. Malignant otitis externa is a type of life-threatening infection in the aged and poorly controlled diabetic patients. Those immunocompromised patients who suffer from acquired immunodeficiency syndrome, undergo chemotherapy, and take immunosuppressant medications such as glucocorticoids may also be vulnerable to this serious disease [145–149]. Once infection becomes established in the external meatus of the susceptible patient, the bacteria invade the underlying structures of the soft tissue and destroy the temporal bone, and finally resulting in septicemia. Malignant otitis externa should be suspected if tenderness, erythema, and/or edema of the external ear and adjacent tissues are noted on physical examination. *Pseudomonas aeruginosa* is the inciting organism in the vast majority of cases. Features of biofilm infection by Gram-negative rods are characteristic. The biopsy histology is illustrated in **Figure 52**. Much less frequently it is caused by *Staphylococcus aureus* and group A β-hemolytic *Streptococcus*. Fungal etiology is also known, and *Aspergillus* and *Candida* can be the causative microbes. When untreated, the mortality rate is around 50%.

14.2 Mucormycosis

Mucormycosis (zygomycosis) is infection by the class *Zygomycetes*, mainly *Mucor ramosissimus, Rhizomucor pusillus* and *Rhizopus oryzae*. Sixteen species of *Zygomycetes* infect the human. *Zygomycetes* (mucoral fungi) are common molds growing in a moist environment. Fungi commonly have chitin as structural polysaccharide, but *Zygomycetes* synthesize chitosan, a deacetylated homopolymer of chitin. Hence, serum β-D-glucan, a laboratory marker of fungal infection, is negative in case of mucormycosis [150].

The main sites of localized mucormycosis are the lung and paranasal cavity. Formation of conidiophores is rarely encountered in case of paranasal cavity infection. The gross features of systemic mucormycosis represent hemorrhagic infarction of the involved tissues and organs [151]. Microscopically, faintly basophilic and wide hyphae, showing the lack of septum formation and wide angle of lamification, are seen in the mycotic thrombus. Stamp smear preparations (**Figure 53**) reveal typical microscopic morphology of mucormycosis. Infection of *Zygomycetes* is microscopically featured by angioinvasiveness and weak reactivity with Grocott

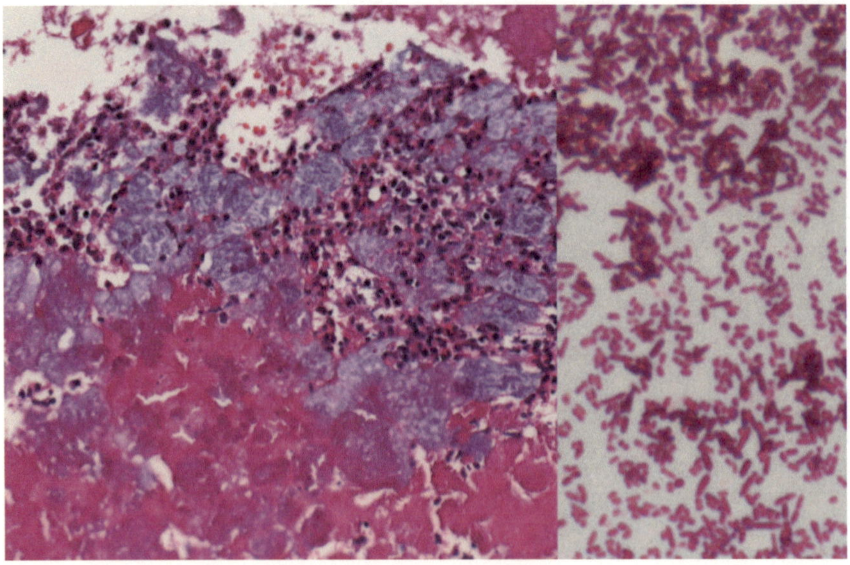

Figure 52.
Malignant otitis externa (H&E and Gram stain on smear preparation). In this lethal diabetic case (a female patient aged 40's) accompanying pseudomonal septicemia, Gram-negative rods densely colonize the necrotic debris in necrotizing petrositis. Myxoid matrix of the colony indicates biofilm infection. Gram-negative rods are demonstrated in the smear preparation.

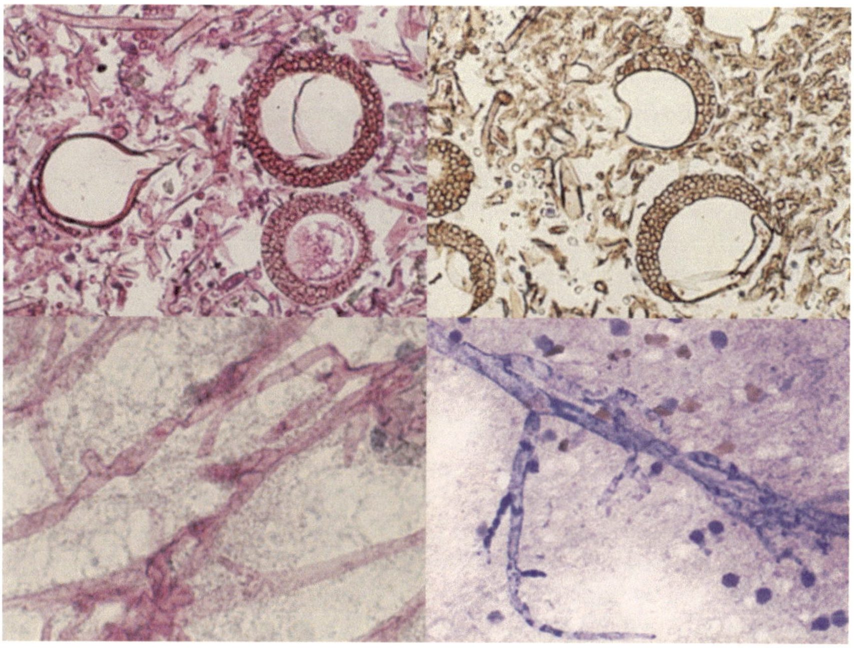

Figure 53.
Mucormycosis. Formation of conidiophores in the paranasal cavity and stamp cytology preparation of cerebral mucormycosis in a pediatric acute leukemia case. Aerated growth condition within the cavity is essential for conidiophore formation (H&E and immunostain for Rhizomucor *antigen). Non-septating hyphae show variable thickness. Wide angle of lamification is distinct from* Aspergillus *hyphae (PAS and Giemsa, the courtesy by Dr. Suzuko Moritani, a pathologist at Shiga Medical University Hospital, Otsu, Japan).*

staining, as illustrated in **Figure 54**. However, some lesions of mucormycosis reveal clear basophilia with strong Grocott reactivity (refer to **Figure 57**, displaying neonatal intestinal mucormycosis).

Cutaneous mucormycosis is infrequently encountered as skin manifestation of systemic mucormycosis [152, 153]. A rare lethal variant is craniofacial

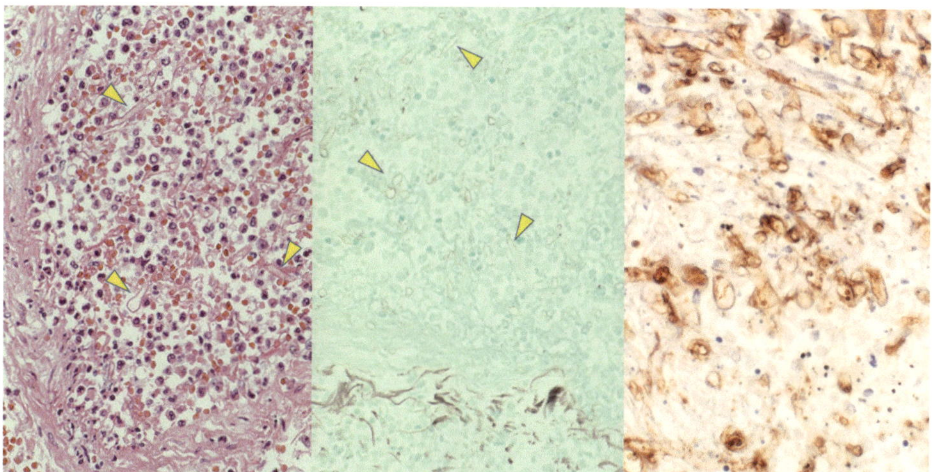

Figure 54.
Angioinvasive mucormycosis (H&E, Grocott, and immunostain). Zygomycetes *frequently shows angioinvasion, resulting in hemorrhagic infarction of the organ and tissue. Weak reactivity with H&E and Grocott stain is characteristic of this opportunistic fungus, as arrowheads indicate. The hyphae are clearly immunoreactive with anti-*Rhizomucor *monoclonal antibody, which is cross-reactive with* Zygomycetes *but not with* Aspergillus *or* Candida.

(rhinocerebral) mucormycosis, which is encountered as a complication of poorly controlled diabetes mellitus [154, 155]. Angioinvasive colonization of *Zygomycetes* aggressively progresses from the paranasal cavity to the overlying facial skin and to the lower part of the frontal lobe of the brain (**Figure 55**).

14.3 *Clostridium butyricum*-induced necrotizing enteritis

Clostridium butyricum is a spore-forming, Gram-positive obligate anaerobic rod with a rugby ball-shaped configuration [156]. It frequently forms spores even in the *in vivo* state, a feature quite different from *C. perfringens*. A male patient aged 30's with severe uncontrolled diabetes mellitus suddenly suffered from mesenteric arterial thrombosis. The surgically resected small bowel accompanied pneumatosis cystoides intestinalis (gas formation in the intestinal wall). Computed tomography scan demonstrated gas embolism filling the portal vein branches in the liver. Microscopically, gas-filled spaces were formed in the submucosa of the small bowel. Spore-forming Gram-positive large rods were discerned in the necrotic gut wall (**Figure 56**). Capsule formation by the spore-forming rods was proven with colloidal iron stain. Microbial culture of the blood identified *C. butyricum*. In contrast to gas gangrene caused by *C. perfringens*, the prognosis of the patient with *C. butyricum*-induced gas gangrene is not so poor. In fact, this patient was alive for years after surgery [157].

Neonatal necrotizing enterocolitis occurs in premature babies. The most likely cause of the disease is infection of *C. butyricum* [158–160]. Symptoms caused by small bowel necrosis include poor feeding, bloating, decreased activity, blood in the stool, or vomiting of bile. Poor blood flow results in ischemic necrosis of the bowel wall. Pneumatosis cystoides intestinalis and perforation with pneumoperitoneum and peritonitis are often associated. Surgery is required in those who have free air in the abdominal cavity. Breastfeeding may prevent the disease. Probiotic studies have reported that peroral administration of *C. butyricum* improves or even prevents clinical manifestation of pseudomembranous colitis due to *C. difficile* infection and hemorrhagic colitis caused by enterohemorrhagic *Escherichia coli* (O-157, H7)

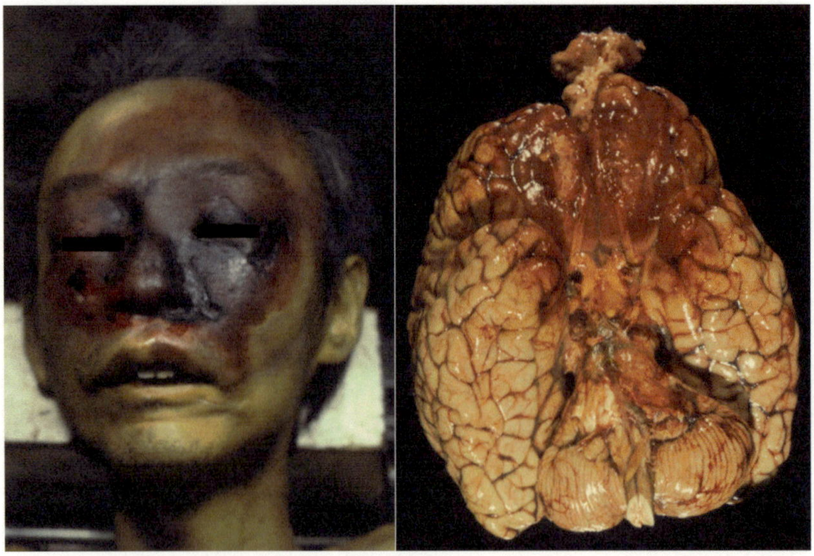

Figure 55.
Lethal mucormycosis of rhinocerebral type in a poorly controlled diabetic male patient aged 60's (gross appearance). Angioinvasive mycosis resulted in hemorrhagic necrosis of the face and anteroinferior part of the brain. Infection had been extended from the paranasal cavity.

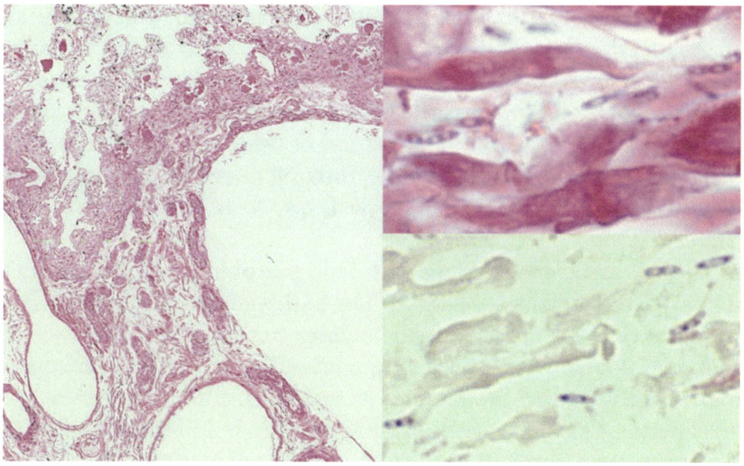

Figure 56.
C. butyricum*-induced gas gangrene (necrotizing enteritis) (H&E and Gram). The small bowel was surgically removed for mesenteric arterial thrombosis in a male patient aged 30's with severe diabetes mellitus. Gas-filled spaces are formed in the submucosa. Spore-forming, Gram-positive, rugby ball-shaped large rods are identified seen in the necrotic gut wall.*

infection [161, 162]. Neonatal intestinal mucormycosis, clinically resembling neonatal necrotizing enterocolitis, is fetal and challenging to make an appropriate diagnosis [163, 164]. Risk factors include premature birth, malnutrition, and asphyxia. The entry of the organism is thought to be the oropharynx or nares. **Figure 57** demonstrates the representative microscopic features of lethal ileal mucormycosis seen in a premature baby.

15. Anthrax and *Bacillus cereus* infection

Anthrax is a zoonotic infection of a large-sized Gram-positive bacillus, *Bacillus anthracis* [165–168]. Formation of spores and capsules is closely related to the pathogenicity of the microbe. Three clinical forms are known, involving the skin, lungs, and intestines. The latter two are often lethal. Skin anthrax, predominantly involving the arm, is an occupation-related infection of veterinarians and those who

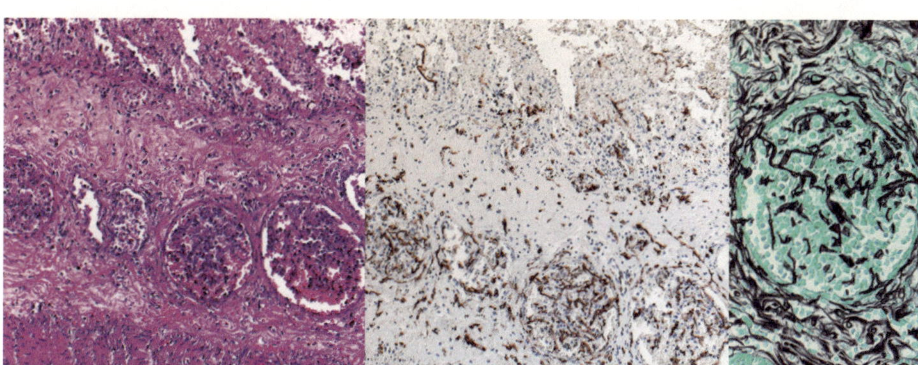

Figure 57.
Neonatal intestinal mucormycosis (H&E, immunostain and Grocott). The premature baby was treated for neonatal necrotizing enterocolitis. Autopsy disclosed necrotic ileal wall with massive transmural infection of Mucor *fungi. Vascular involvement (mycotic embolism) is evident by both immunostaining with a monoclonal antibody against* Rhizomucor *antigen and Grocott silver. Strong Grocott reactivity is noted in this case.*

treat animal hair, skin, or carcass. The latent period is within 4 days. The skin lesion is necrotic and ulcerated to form hemorrhagic crust (eschar or black necrosis) (**Figure 58**). Characteristically, the ulcer is painless. Gram-positive rods are easily found in the exudate. *B. anthracis* is the best-known bioterrorist, because the spores are tolerant to dry conditions for a long period of time, and inhalation of the spored microorganisms provokes lethal necrotizing pneumonia. Ulcer-forming skin infection is also caused by other *Bacillus* species, such as *B. megaterium* and *B. pumilus* [169].

Bacillus cereus is associated mainly with food poisoning, but it may cause potentially fatal non-gastrointestinal infection. The pathogenicity of *B. cereus* is related to the production of tissue-destructive exoenzymes common to *B. anthracis*. *B. cereus* produces a potent β-lactamase, conferring marked resistance to β-lactam antibiotics. Clinically, anthrax-like progressive pneumonia, fulminant sepsis, and devastating central nervous system infections may be seen in immunocompromised individuals, intravenous drug abusers, and neonates. It also occurs in immunocompetent individuals [170]. The primary cutaneous/soft tissue infection of *B. cereus*, mimicking necrotizing fasciitis or non-clostridial gas gangrene induced subsequent to trauma, has been documented [171, 172].

Figure 59 demonstrates primary necrotizing infection of *B. cereus* in the soft tissue of the hip, as a form of necrotizing fasciitis. Gas formation was not associated in this case. Trauma-related soft tissue gangrene, caused by a spore-forming Gram-positive bacillus, *B. cereus*, led this diabetic adult patient to death. Gram-positive rods heavily colonized the necrohemorrhagic muscle tissue.

A 68-year-old housewife received intermittent chemotherapy against lymphoplasmacytic leukemia for 13 years. Her blood contained numbers of indolent small-sized leukemic cells. She happened to take curdled milk, and next day she complained of dyspnea and consciousness disturbance. She expired soon. The

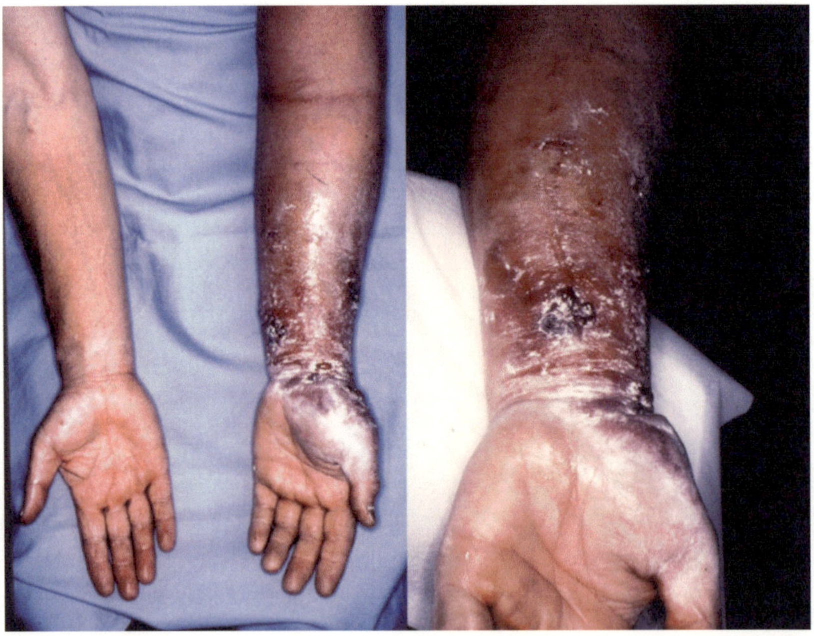

Figure 58.
Cutaneous anthrax (gross appearance). Occupation-related infection in a Japanese veterinarian is shown. The lesion of cutaneous anthrax on the left forearm is necrotic and ulcerated to form hemorrhagic crust (eschar). The courtesy by Dr. Keiko Oka, a dermatologist at Tokyo Hospital of Health Insurance Association of Nippon Express, Tokyo, Japan.

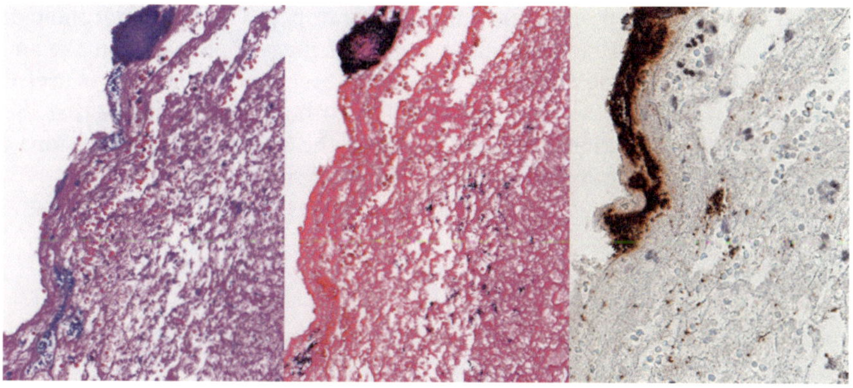

Figure 59.
Bacillus cereus*-induced necrotizing fasciitis (H&E, Gram and immunostain). Trauma-related lethal soft tissue gangrene is formed on the hip of the diabetic patient. Gram-positive rods colonize the necrohemorrhagic soft tissue. Immunostaining for* Bacillus cereus *antigens is strongly positive (the courtesy of Dr. Etsuko Nakamura, a pathologist at Toyohashi Medical Center, Toyohashi, Japan).*

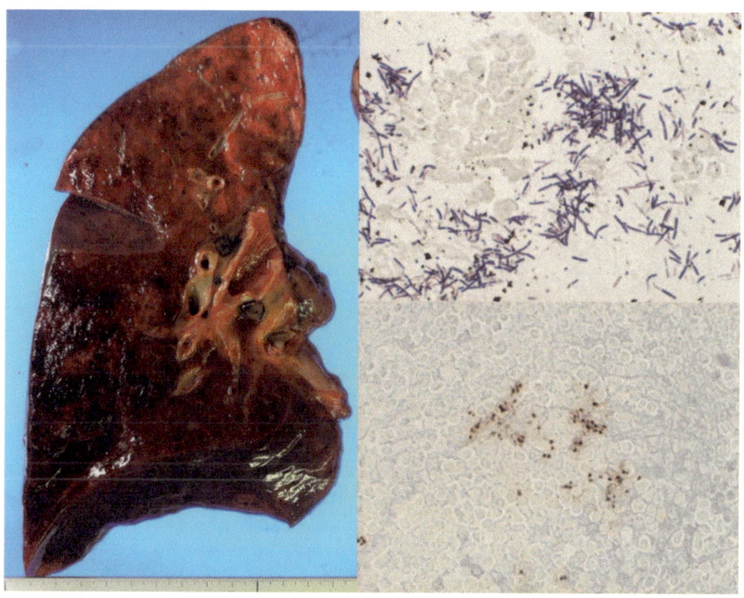

Figure 60.
Lethal Bacillus cereus *pneumonia in a female patient with indolent leukemia (gross, Gram and immunostain). Severe necrotizing hemorrhagic pneumonia was caused by incidental aspiration of sweet-curdled milk. Gram-positive rods grow in the necrotic lesion. The antiserum against* B. cereus *labels spores in the rods.*

growth of *B. cereus* in fluid milk had provoked sweet curdling [173]. Autopsy disclosed massive hemorrhagic and necrotizing pneumonia caused by *B. cereus* in the right lower lobe. Spore-forming Gram-positive rods were identified in the lesion (**Figure 60**). *B. cereus* antiserum clearly labeled spores in the rod-shaped bacteria. It is highly likely that aspiration of the curdled milk resulted in lethal *B. cereus* pneumonia.

16. Conclusive remarks

The author reviewed pathological aspects of a variety of gangrenous lesions. The causative pathogens are commonly anaerobic. Often times, the lesions are clinically severe and fulminant, and often encountered at autopsy. The exact morphological recognition of the respective lesions is essential for the pathologists to make an appropriate histopathologic diagnosis. Immunohistochemical approach is useful for identifying the pathogenic microorganisms. The author sincerely hopes that the present chapter may contribute to brushing up of the knowledge of the lesions with relatively low frequency but with high clinical implications.

Disclosure

There were no funding sources for reporting the present chapter. The author has no conflict of interest for reviewing the present chapter.

Author details

Yutaka Tsutsumi
Diagnostic Pathology Clinic, Pathos Tsutsumi, Nagoya, Japan

*Address all correspondence to: pathos223@kind.ocn.ne.jp

References

[1] Tsutsumi Y. Application of the immunoperoxidase method for histopathological diagnosis of infectious diseases. Acta Histochemica et Cytochemica. 1994;**27**:547-560

[2] Tsutsumi Y. Pathology of Skin Infections. New York: Nova Biomedical; 2013. p. 394. Available from: https://www.pathos223.com/pathology/bookintroduction/pathology_of_skininfections/

[3] Fujii M, Mizutani Y, Sakuma T, et al. *Corynebacterium kroppenstedtii* in granulomatous mastitis: Analysis of formalin-fixed, paraffin-embedded biopsy specimens by immunostaining using low-specificity bacterial antisera and real-time polymerase chain reaction. Pathology International. 2018; **68**:409-418

[4] Tsutsumi Y. Low-Specificity and High-Sensitivity Immunostaining for Demonstrating Pathogens in Formalin-Fixed, Paraffin-Embedded Sections. London, UK: IntechOpen; 2019. pp. 1-46. DOI: 10.5772/intechopen.85055. Available from: https://www.intechopen.com/books/immunohistochemistry-the-ageless-biotechnology/low-specificity-and-high-sensitivity-immunostaining-for-demonstrating-pathogens-in-formalin-fixed-pa

[5] Tsutsumi Y. Pathology of Infectious Diseases. Nagoya, Japan: Pathos Tsutsumi; 2003. Available from: https://pathos223.com/en/

[6] Dafiewhare OE, Agwu E, Ekanem P, et al. A Review of Clinical Manifestations of Gangrene in Western Uganda. London, UK: IntechOpen; 2013: 1-14. DOI: 10.5772/55862. Available from: https://www.intechopen.com/books/gangrene-management-new-advancements-and-current-trends/a-review-of-clinical-manifestations-of-gangrene-in-western-uganda

[7] Jacob J, Gionfriddo RJ. Gangrene. Symptoms, diagnosis and treatment. BMJ Best Practice. 2018. Available from: https://bestpractice.bmj.com/topics/en-us/1015

[8] Al WA. Autoamputation of diabetic toe with dry gangrene: A myth or a fact? Diabetes, Metabolism Syndrome and Obesity: Targets and Therapy. 2018;**11**: 255-264. DOI: 10.2147/DMSO.S164199

[9] Anaya DA, Bulger EM, Kwon YS, et al. Predicting death in necrotizing soft tissue infections: A clinical score. Surgical Infections. 2009;**10**:517-522

[10] Hakkarainen TW, Kopari NM, Pham TN, Evans HL. Necrotising soft tissue infections: Review and current concepts in treatment, systems of care, and outcomes. Current Problems in Surgery. 2014;**51**:344-362

[11] McGovern T, Wright IS. A vascular disease. American Heart Journal. 1941; **22**:583-606

[12] Cribier B, Djeridi N, Peltre B, Grosshans E. A histologic and immunohistochemical study of chilblains. Journal of the American Academy of Dermatology. 2001;**45**: 924-929

[13] Boada A, Bielsa I, Fernándes-Figueras MT, Ferrándiz C. Perniosis: Clinical and histopathological analysis. The American Journal of Dermatopathology. 2010;**32**:19-23

[14] Nixon J, Cranny G, Bond S. Pathology, diagnosis, and classification of pressure ulcers: Comparing clinical and imaging techniques. Wound Repair and Regeneration. 2005;**13**:365-372

[15] Livesley NJ, Chow AW. Infected pressure ulcers in elderly individuals. Clinical Infectious Diseases. 2002;**35**: 1390-1396

[16] Heym B, Rimareix F, Lortat-Jacob A, et al. Bacteriological investigation of infected pressure ulcers in spinal cord-injured patients and impact on antibiotic therapy. Spinal Cord. 2004; **42**:230-234

[17] Kitagawa A, Sanada H, Nakatani T, et al. Histological examination of pressure ulcer tissue in terminally ill cancer patients. Japan Journal of Nursing Science. 2004;**1**:35-46

[18] Shibuya H, Terashi H, Kurata S, et al. Gas gangrene following sacral pressure sores. The Journal of Dermatology. 1994;**21**:518-523

[19] Pirett CCNS, Braga IA, Ribas RM, et al. Pressure ulcer colonized by MRSA as a reservoir and risk for MRSA bacteremia in patients at a Brazilian university hospital. Wounds. 2012;**24**: 67-75

[20] NPUAP. Pressure ulcers in America: Prevalence, incidence, and implications for the future. An executive summary of the National Pressure Ulcer Advisory Panel. Advances in Skin & Wound Care. 2001;**14**:208-215

[21] Shimomura R, Tsutsumi Y. Histochemical identification of methicillin-resistant *Staphylococcus aureus*: Contribution to preventing nosocomial infection. Seminars in Diagnostic Pathology. 2007;**24**:217-226

[22] Darke SG, King AM, Slack WK. Gas gangrene and related infection: Classification, clinical features and aetiology, management and mortality. A report of 88 cases. The British Journal of Surgery. 1977;**64**:104-112

[23] McArthur HL, Dalal BI, Kollmannsberger C. Intravascular hemolysis as a complication of *Clostridium perfringens* sepsis. Journal of Clinical Oncology. 2006;**24**:2387-2388

[24] Stevens DL, Aldape MJ, Bryant AE. Life-threatening clostridial infections. Anaerobe. 2012;**18**:254-259

[25] Sushma M, Vidyadhar TVV, Mohanraj R, Babu M. A review on gas gangrene and its management. PharmaTutor. 2014;**2**:65-74

[26] Valentine EG. Nontraumatic gas gangrene. Annals of Emergency Medicine. 1997;**30**:109-111

[27] Sasaki T, Nanjo H, Takahashi M, Sugiyama T, Ono I, Masuda H. Non-traumatic gas gangrene in the abdomen: Report of six autopsy cases. Journal of Gastroenterology. 2000;**35**:382-390

[28] Birgisson H, Stefánsson T, Andresdóttir A, Möller PH. Emphysematous pancreatitis. The European Journal of Surgery. 2001;**167**: 918-920

[29] Ikegami T, Kido A, Shimokawa H, Ishida T. Primary gas gangrene of the pancreas: Report of a case. Surgery Today. 2004;**34**:80-81

[30] Stockinger ZT, Corsetti RL. Pneumoperitoneum from gas gangrene of the pancreas: Three unusual findings in a single case. Journal of Gastrointestinal Surgery. 2004;**8**: 489-492

[31] Nanjappa S, Shah S, Pabbathi S. *Clostridium septicum* gas gangrene in colon cancer: Importance of early diagnosis. Case Reports in Infectious Diseases. 2015;**2015**:694247

[32] Griffin AS, Crawford MD, Gupta RT. Massive gas gangrene secondary to occult colon carcinoma. Radiology Case Reports. 2016;**11**:67-69

[33] Miller C, Florman S, Kim-Schluger L, et al. Fulminant and fatal gas gangrene of the stomach in a healthy live liver donor. Liver Transplantation. 2004;**10**:1315-1319

[34] Srivastava I, Aldape MJ, Bryant AE, Stevens DL. Spontaneous *C.* septicum gas gangrene: A literature review.

Journal of Clinical Microbiology. 2017; **48**:165-171

[35] Dempsey A. Serious infection associated with induced abortion in the United States. Clinical Obstetrics and Gynecology. 2012;**55**:888-892

[36] Bessman AN, Wagner W. Nonclostridial gas gangrene. Report of 48 cases and review of the literature. JAMA. 1975;**233**:958-963

[37] Hubens G, Carly B, De Boeck H, Vansteenland H, Wylock P. "Spontaneous" non clostridial gas gangrene: Case report and review of the literature. Acta Chirurgica Belgica. 1989; **89**:25-28

[38] Weisenfeld LS, Luzzi A, Picciotti J. Nonclostridial gas gangrene. The Journal of Foot Surgery. 1990;**29**:141-146

[39] Jain AKC, Viswanath S. Non-clostridial gas gangrene in diabetic lower limbs with peripheral vascular disease. OA Case Reports. 2013;**2**(9):83

[40] Takazawa K, Otsuka H, Nakawaga Y, Inokuchi S. Clinical features of non-clostridial gas gangrene and risk factors for in-hospital mortality. The Tokai Journal of Experimental and Clinical Medicine. 2015;**40**:124-129

[41] Shigemoto R, Anno T, Kawasaki F, et al. Non-clostridial gas gangrene in a patient with poorly controlled type 2 diabetes mellitus on hemodialysis. Acta Diabetologica. 2018;**55**:99-101

[42] Li CM, Chen PL, Ho YR. Non-clostridial gas gangrene caused by *Klebsiella pneumoniae*: A case report. Scandinavian Journal of Infectious Diseases. 2001;**33**:629-630

[43] Overcamp M, Pfohl M, Klier D, et al. Spontaneous gas-forming myonecrosis caused by group B streptococci and peptostreptococci. The Clinical Investigator. 1992;**70**:441-443

[44] Carr NJ. The pathology of acute appendicitis. Annals of Diagnostic Pathology. 2000;**4**:46-58

[45] Bhangu A, Søreide K, Di Saverio S, Assarsson JH, Drake FT. Acute appendicitis: Modern understanding of pathogenesis, diagnosis, and management. Lancet. 2015;**386**: 1278-1287

[46] Nordin AB, Diefenbach K, Sales SP, Christensen J, Besner GE, Kenney BD. Gangrenous appendicitis: no longer complicated. Journal of Pediatric Surgery. 2019;**54**:718-722

[47] Marne C, Pallarés R, Martín R, Sitges-Serra A. Gangrenous cholecystitis and acute cholangitis associated with anaerobic bacteria in bile. European Journal of Clinical Microbiology. 1986;**5**: 35-39

[48] Ganapathi AM, Speicher PJ, Englum BR, Perez A, Tyler DS, Zani S. Gangrenous cholecystitis: A contemporary review. The Journal of Surgical Research. 2015;**197**:18-24

[49] Önder A, Kapan M, Ülger BV, Oğuz A, Türkoğlu A, Uslukaya Ö. Gangrenous cholecystitis: Mortality and risk factors. International Surgery. 2015; **100**:254-260

[50] Liao CY, Tsai CC, Kuo WH, et al. Emphysematous cholecystitis presenting as gas-forming liver abscess and pneumoperitoneum in a dialysis patient: A case report and review of the literature. BMC Nephrology. 2016;**17**:23. DOI: 10.1186/s12882-016-0237-3

[51] Lee J. Diagnosis and management of acute cholangitis. Nature Reviews. Gastroenterology & Hepatology. 2009; **6**:533-541

[52] Zimmer V, Lammert F. Acute bacterial cholangitis. Viszeralmedizin. 2015;**31**:166-172

[53] Weber A, Huber W. Spectrum of pathogens in acute cholangitis in

patients with and without biliary endoprosthesis. The Journal of Infection. 2013;**67**:111-121

[54] Nomura T, Shirai Y, Hatakeyama K. *Enterococcal bactibilia* in patients with malignant biliary obstruction. Digestive Diseases and Sciences. 2000;**45**: 2183-2186

[55] Phillips LG, Rao KVS. Gangrene of the lung. The Journal of Thoracic and Cardiovascular Surgery. 1989;**97**: 114-118

[56] Chen CH, Huang WC, Chen TY, Hung TT, Liu HC, Chen CH. Massive necrotizing pneumonia with pulmonary gangrene. The Annals of Thoracic Surgery. 2009;**87**:310-311

[57] Chatha N, Fortin D, Bosma KJ. Management of necrotizing pneumonia and pulmonary gangrene: A case series and review of the literature. Canadian Respiratory Journal. 2014;**21**:239-245

[58] Nayeemuddin M, Wiseman O, Turner A. Emphysematous pyelonephritis. Nature Reviews. Urology. 2005;**2**:108-112

[59] Mahesan T, Reddy UD, Chetwood A, et al. Emphysematous pyelonephritis: A review of a rare condition. Current Bladder Dysfunction Reports. 2015;**10**:207-211

[60] Irfaan AM, Shaikh NA, Jamshaid A, Qureshi AH. Emphysematous pyelonephritis: A single center review. Pakistan Journal of Medical Sciences. 2020;**36**:S83-S86

[61] Kawaguchi Y, Mori H, Izumi Y, Ito M. Renal papillary necrosis with diabetes and urinary tract infection. Internal Medicine. 2018;**57**(22):3343. DOI: 10.2169/internalmedicine.0858-18

[62] Kernt M, Kampik A. Endophthalmitis: Pathogenesis, clinical presentation, management, and perspectives. Clinical Ophthalmology. 2010;**4**:121-135

[63] Durand ML. Endophthalmitis. Clinical Microbiology and Infection. 2013;**19**:227-234

[64] Aulakh S, Nair AG, Gandhi R, et al. Orbital cellulitis with endogenous panophthalmitis caused by methicillin-sensitive *Staphylococcus aureus* in pregnancy. Japanese Journal of Infectious Diseases. 2017;**70**:314-316

[65] Ferrer C, Alio J, Rodriguez A, Andreu M, Colom F. Endophthalmitis caused by *Fusarium* proliferation. Journal of Clinical Microbiology. 2005; **43**:5372-5375

[66] Al-Jundi W, AliShebl A. Emphysematous gastritis: Case report and literature review. International Journal of Surgery. 2008;**6**:e63-e66

[67] López-Maroto DG-L, Cuéllar ER, García CN, Pozuelo AM, Herrero EF. Emphysematous esophagitis with gastric perforation. Revista Española de Enfermedades Digestivas. 2019;**111**: 884-886

[68] Chou YH, Hsu HL, Lee JC, Lin BR, Liu KL. Emphysematous colitis of ascending colon with portal venous air caused by diffuse large B-cell lymphoma. Journal of Clinical Oncology. 2010;**28**:e496-e497

[69] Amano M, Shimizu T. Emphysematous cystitis: A review of the literature. Internal Medicine. 2014; **53**:79-82

[70] Tariq T, Farishta M, Rizvi A, et al. A case of concomitant emphysematous cystitis and *Clostridium difficile* colitis with pneumoperitoneum. Cureus. 2018; **10**:e2897

[71] Liao W-C, Chou J-W. Emphysematous pyelonephritis, ureteritis and cystitis in a diabetic

patient. QJM: An International Journal of Medicine. 2010;**103**:893-894

[72] Chen S, Rahim S, Datta SN. A rare case of emphysematous urethritis. Urology & Nephrology Open Access Journal. 2016;**2**(3):1-4

[73] Vijayan P. Gangrene of the penis in a diabetic male with multiple amputations and follow up. Indian Journal of Urology. 2009;**25**:123-125

[74] Mathur A, Manish A, Maletha M, Luthra NB. Emphysematous epididymo-orchitis: A rare entity. Indian Journal of Urology. 2011;**27**:399-400

[75] Lau HW, Yu CH, Yu SM, Lee LF. Emphysematous epididymo-orchitis: An uncommon but life-threatening cause of scrotal pain. Hong Kong Medical Journal. 2018;**24**:426.e1-426.e2

[76] Chua YJ, Meharry S, Harding S, Stewart CJ. Endometrial pneumatosis (emphysematous endometritis). International Journal of Gynecological Pathology. 2014;**33**:511-514

[77] Lima-Silva J, Vieira-Baptista P, Cavaco-Gomes J, Maia T, Beires J. Emphysematous vaginitis. Journal of Lower Genital Tract Disease. 2015;**19**: e43-e44

[78] Agrawal S, Jayant K, Agarwal R. Breast gangrene: A rare source of severe sepsis. BMJ Case Reports. 2014;**2014**: bcr2013203467

[79] Luey C, Tooley D, Briggs S. Emphysematous osteomyelitis: A case report and review of the literature. International Journal of Infectious Diseases. 2012;**16**:e216-e220

[80] Smith-Singares E, Boachie JA, Iglesias IM, Jaffe L, Goldkind A, Jeng EI. *Fusobacterium* emphysematous pyomyositis with necrotizing fasciitis of the leg presenting as compartment syndrome: A case report. Journal of Medical Case Reports. 2017;**11**:332. DOI: 10.1186/s13256-017-1493-y

[81] Urgiles S, Matos-Casano H, Win KZ, et al. Emphysematous aortitis due to *Clostridium septicum* in an 89-year-old female with ileus. Case Reports in Infectious Diseases. 2019;**2019**:1094837

[82] Ye R, Yang J, Hong H, et al. Descending necrotizing mediastinitis caused by *Streptococcus constellatus* in an immunocompetent patient: Case report and review of the literature. BMC Pulmonary Medicine 2020;**20**:43. DOI: 10.1186/s12890-020-1068-3

[83] Kim CJ, Yi JE, Kim Y, Choi HJ. Emphysematous endocarditis caused by AmpC beta-lactamase-producing *Escherichia coli*: A case report. Medicine (Baltimore). 2018;**97**(6):e9620

[84] American Academy of Periodontology. Consensus report: Necrotizing periodontal diseases. Annals of Periodontology. 1999;**4**(1):78

[85] Corbet EF. Diagnosis of acute periodontal lesions. Periodontology. 2000/2004;**34**:204-216

[86] Paster BJ, Falkler WA Jr, Enwonwu CO, et al. Prevalent bacterial species and novel phylotypes in advanced Noma lesions. Journal of Clinical Microbiology. 2002;**40**: 2187-2191

[87] Enwonwu CO. Noma: The ulcer of extreme poverty. The New England Journal of Medicine. 2006;**354**:221-224

[88] Parikh TB, Nanavati RN, Udani RH. Noma neonatorum. Indian Journal of Pediatrics. 2006;**73**:439-440

[89] Cunningham MW. Pathogenesis of group A streptococcal infections. Clinical Microbiology Reviews. 2000;**13**: 470-511

[90] Fox KL, Born MW, Cohen MA. Fulminant infection and toxic shock

syndrome caused by *Streptococcus pyogenes*. The Journal of Emergency Medicine. 2002;**22**:357-366

[91] Overkamp D, Pfohl M, Klier R, Domres B, Schmülling R-M. Spontaneous gas-forming bacterial myonecrosis caused by group B streptococci and peptostreptococci. The Clinical Investigator. 1992;**70**:441-443

[92] Humar D, Datta V, Bast DJ, et al. Streptolysin S and necrotising infections produced by group G streptococcus. Lancet. 2002;**359**:124-129

[93] Bryant AE, Bayer CR, Huntington JD, et al. Group a streptococcal myonecrosis: Increased vimentin expression after skeletal-muscle injury mediates the binding of *Streptococcus pyogenes*. The Journal of Infectious Diseases. 2006;**193**:1685-1689

[94] Spaulding AR, Salgado-Pabón W, Kohler PL, Horswill AR, Leung DY, Schlievert PM. Staphylococcal and streptococcal superantigen exotoxins. Clinical Microbiology Reviews. 2013;**26**: 422-447

[95] Lin L, Xu L, Lv W, et al. An NLRP3 inflammasome-triggered cytokine storm contributes to streptococcal toxic shock-like syndrome (STSLS). PLoS Pathogens. 2019;**15**(6):e1007795

[96] Chen Y, Satoh T, Tokunaga O. *Vibrio vulnificus* infection in patients with liver disease: Report of five autopsy cases. Virchows Archiv. 2002;**441**:88-92

[97] Chiang SR, Chuang YC. *Vibrio vulnificus* infection: Clinical manifestations, pathogenesis, and antimicrobial therapy. Journal of Microbiology, Immunology, and Infection. 2003;**36**:81-88

[98] Inoue Y, Ono T, Matsui T, et al. Epidemiological survey of *Vibrio vulnificus* infection in Japan between 1999 and 2003. The Journal of Dermatology. 2008;**35**:129-139

[99] Jones MK, Oliver JD. *Vibrio vulnificus*: Disease and pathogenesis. Infection and Immunity. 2009;**77**: 1723-1733

[100] Vukmir RB. *Aeromonas hydrophila*: Myofascial necrosis and sepsis. Intensive Care Medicine. 1992;**18**:172-174

[101] Grant A, Hoddinott C. *Aeromonas hydrophila* infection of a scalp laceration (with synergistic gas gangrene). Archives of Emergency Medicine. 1993; **10**:232-234

[102] Lin SH, Shieh SD, Lin YF, et al. Fatal *Aeromonas hydrophila* bacteremia in a hemodialysis patient treated with deferoxamine. American Journal of Kidney Diseases. 1996;**27**:733-735

[103] Furusu A, Yoshizuka N, Abe K, et al. *Aeromonas hydrophila* necrotizing fasciitis and gas gangrene in a diabetic patient on haemodialysis. Nephrology, Dialysis, Transplantation. 1997;**12**: 1730-1734

[104] van der Burg BL B, Bronkhorst MW, Pahlplatz PV. *Aeromonas hydrophila* necrotizing fasciitis. A case report. The Journal of Bone and Joint Surgery. American Volume. 2006;**8**:1357-1360

[105] Vally H, Whittle A, Cameron S, et al. Outbreak of *Aeromonas hydrophila* wound infections associated with mud football. Clinical Infectious Diseases. 2004;**38**:1084-1089

[106] Lamerton AJ. Fournier's gangrene: Non-clostridial gas gangrene of the perineum and diabetes mellitus. Journal of the Royal Society of Medicine. 1986; **79**:212-215

[107] Eke N. Fournier's gangrene: A review of 1726 cases. The British Journal of Surgery. 2000;**87**:718-728

[108] Thwaini A, Khan A, Malik A, et al. Fournier's gangrene and its emergency management. Postgraduate Medical Journal. 2006;**82**:516-519

[109] Erol B, Tuncel A, Hanci V, et al. Fournier's gangrene: Overview of prognostic factors and definition of new prognostic parameter. Urology. 2010;**75**: 1193-1198

[110] Heiner JD, Eng KD, Bialowas TA, et al. Fournier's gangrene due to masturbation in an otherwise healthy male. Case Reports in Emergency Medicine. 2012;**2012** Article ID 154025

[111] Bruketa T, Majerovic M, Augustin G. Rectal cancer and Fournier's gangrene. Current knowledge and therapeutic options. World Journal of Gastroenterology. 2015;**21**:9002-9020

[112] Yoshino Y, Funahashi K, Okada R, et al. Severe Fournier's gangrene in a patient with rectal cancer: Case report and literature review. World Journal of Surgical Oncology. 2016;**14**:234. DOI: 10.1186/s12957-016-0989-z

[113] Kotrappa KS, Bansal RS, Amin NM. Necrotizing fasciitis. American Family Physician. 1996;**53**:1691-1697

[114] Shimizu T, Tokuda Y. Necrotizing fasciitis. Internal Medicine. 2010;**49**: 1051-1057

[115] Machado NO. Necrotizing fasciitis: The importance of early diagnosis, prompt surgical debridement and adjuvant therapy. North American Journal of Medical Sciences. 2011;**3**: 107-118

[116] Schwartz RA. Dermatologic Manifestations of Necrotizing Fasciitis. New York, NY, USA: Medscape; 2011. Available from: http://emedicine.medscape.com/article/1054438-overview

[117] Lancerotto L, Tocco I, Salmaso R, et al. Necrotizing fasciitis. Classification, diagnosis, and management. Journal of Trauma and Acute Care Surgery. 2012; **72**:560-566

[118] Mahoning County Public Health. Necrotizing fasciitis: A rare disease, especially for the healthy. Youngstown, OH, USA; 2012. Available from: https://www.mahoninghealth.org/health/necrotizing-fasciitis-a-rare-disease-especially-for-the-healthy/

[119] Akamine M, Miyagi K, Uchihara T, et al. Necrotizing fasciitis caused by *Pseudomonas aeruginosa*. Internal Medicine. 2008;**47**:553-556

[120] Reisman JS, Weinberg A, Ponte C, et al. Monomicrobial *Pseudomonas* necrotizing fasciitis: A case of infection by two strains and a review of 37 cases in the literature. Scandinavian Journal of Infectious Diseases. 2012;**44**:216-221

[121] Stens O, Wardi G, Kinney M, Shin S, Papamatheakis D. *Stenotrophomonas maltophilia* necrotizing soft tissue infection in an immunocompromised patient. Case Reports in Critical Care. 2018;**2018**: 1475730

[122] Arnold CJ, Garrigues G, St Geme JW 3rd, Sexton DJ. Necrotizing fasciitis caused by *Haemophilus influenzae* serotype f. Journal of Clinical Microbiology. 2014;**52**:3471-3474

[123] Stevens DL. Streptococcal toxic-shock syndrome: Spectrum of disease, pathogenesis, and new concepts in treatment. Emerging Infectious Diseases. 1995;**1**:69-78

[124] Tajiri T, Tate G, Miura K, et al. Sudden death caused by fulminant bacterial infection: Background and pathogenesis of Japanese adult cases. Internal Medicine. 2008;**47**:1499-1504

[125] Kato S, Yanazaki M, Hayashi K, Satoh F, Isobe I, Tsutsumi Y. Fulminant group A streptococcal infection without

gangrene in the extremities: Analysis of five autopsy cases. Pathology International. 2018;**68**:419-424

[126] Ooe K, Udagawa H. A new type of fulminant group A streptococcal infection in obstetric patients: Report of two cases. Human Pathology. 1997;**28**: 509-512

[127] Onouchi T, Mizutani Y, Shiogama K, et al. Application of an enzyme-labeled antigen method for visualizing plasma cells producing antibodies against Strep A, a carbohydrate antigen, of *Streptococcus pyogenes* in recurrent tonsillitis. Microbiology and Immunology. 2015;**59**: 13-27

[128] Tormos LM, Schandl CA. The significance of adrenal hemorrhage: Undiagnosed Waterhouse-Friderichsen syndrome, a case series. Journal of Forensic Sciences. 2013;**58**:1071-1074

[129] Fujiwara F, Hibi S, Imashuku S. Hypercytokinemia in hemophagocytic syndrome. The American Journal of Pediatric Hematology/Oncology. 1993; **15**:92-98

[130] Fukusato T. Clinicopathological studies of renal cortical necrosis, with special reference to its pathogenesis. The Japanese Journal of Nephrology. 1984;**26**:1461-1478

[131] Naito R, Miyazaki T, Kajino K, et al. Fulminant pneumococcal infection. BMJ Case Reports. 2014;**2014**:bcr201420 5907

[132] Yu VL, Chiou CC, Feldman C, et al. An international prospective study of pneumococcal bacteremia: Correlation with in vitro resistance, antibiotics administration, and clinical outcome. Clinical Infectious Diseases. 2003;**37**: 230-237

[133] Waghorn DJ, Mayon-White RT. A study of 42 episodes of overwhelming post-splenectomy infection: Is current guideline for asplenic individuals being followed? The Journal of Infection. 1997; **35**:289-294

[134] Hale AJ, LaSalvia M, Kirby JE, Kimball A, Baden R. Fatal purpura fulminans and Waterhouse-Friderichsen syndrome from fulminant *Streptococcus pneumoniae* sepsis in an asplenic young adult. IDCases. 2016;**6**:1-4

[135] Theilacker C, Ludewig K, Serr A, et al. Overwhelming postsplenectomy infection: A prospective multicenter cohort study. Clinical Infectious Diseases. 2016;**62**:871-878

[136] Murph RC, Matulis WS, Hernandez JE. Rapidly fatal pneumococcal sepsis in a healthy adult. Clinical Infectious Diseases. 1996;**22**: 375-376

[137] Bukhari E, Al-Otaibi FE. Severe community-acquired infection caused by methicillin-resistant *Staphylococcus aureus* in Saudi Arabian children. Saudi Medical Journal. 2009;**30**:1595-1600

[138] Bukharie HA. Increasing threat of community-acquired methicillin-resistant *Stapylococcus aureus*. The American Journal of the Medical Sciences. 2010;**340**:378-381

[139] Holden R, Yankaskas J. Fulminant community-acquired MRSA infection in a previously healthy young adult. American Journal of Respiratory and Critical Care Medicine. 2020;**201**:A1805

[140] Vandenesch F, Naimi T, Enright MC, et al. Community-acquired methicillin-resistant *Staphylococcus aureus* carrying Panton-Valentine leukocidin genes: Worldwide emergence. Emerging Infectious Diseases. 2003;**9**:978-984

[141] McCormick JK, Tripp TJ, Llera AS, et al. Functional analysis of the TCR binding domain of toxic shock syndrome toxin-1 predicts further diversity in MHC class II/superantigen/

TCR ternary complexes. Journal of Immunology. 2003;**171**:1385-1392

[142] Linden P. Can enterococcal infections initiate sepsis syndrome? Current Infectious Disease Reports. 2003;**5**:372-378

[143] Gilmore MS, Clewell DB, Ike Y, Shankar N. *Enterococci*: From Commensals to Leading Causes of Drug Resistant Infection. Boston: Massachusetts Eye and Ear Infirmary; 2014. p. NBK190424

[144] Pericás JM, Zboromyrska Y, Cervera C, et al. Enterococcal endocarditis revisited. Future Microbiology. 2015;**10**:1215-1240

[145] Amorosa L, Modugno GC, Pirodda A. Malignant external otitis: Review and personal experience. Acta Oto-Laryngologica. Supplementum. 1996;**521**:3-16

[146] Franco-Vidal V, Blanchet H, Bebear C, et al. Necrotizing external otitis: A report of 46 cases. Otology & Neurotology. 2007;**28**:771-773

[147] Soudry E, Joshua BZ, Sulkes J, et al. Characteristics and prognosis of malignant external otitis with facial paralysis. Archives of Otolaryngology – Head & Neck Surgery. 2007;**133**: 1002-1004

[148] Ling SS, Sader C. Fungal malignant otitis externa treated with hyperbaric oxygen. International Journal of Infectious Diseases. 2008;**12**:550-552

[149] Illing E, Olaleye O. Malignant otitis externa: A review of aetiology, presentation, investigations and current management strategies. WebmedCentral Otorhinolaryngology. 2011;**2**(3):WMC001725

[150] Rinaldi MG. Zygomycosis. Infectious Disease Clinics of North America. 1989;**3**:19-41

[151] Roden MM, Zaoutis TE, Buchanan WL, et al. Epidemiology and outcome of mucormycosis: A review of 929 reported cases. Clinical Infectious Diseases. 2005;**41**:634-653

[152] Song KR, Wong MS, Yeung C. Primary cutaneous zygomycosis in an immunodeficient infant: A case report and review of the literature. Annals of Plastic Surgery. 2008;**60**:433-436

[153] Skiada A, Petrikkos G. Cutaneous zygomycosis. Clinical Microbiology and Infection. 2009;**15**(Suppl 5):41-45

[154] Margo CE, Linden C, Strickland-Marmol LB, et al. Rhinocerebral mucormycosis with perineural spread. Ophthalmic Plastic and Reconstructive Surgery. 2007;**23**:326-327

[155] Teixeira CA, Medeiros PB, Leushner P, Almeida F. Rhinocerebral mucormycosis: Literature review apropos of a rare entity. BML Case Reports. 2013;**2013**:bcr2012008552

[156] Cassir N, Benamar S, La Scola B. *Clostridium butyricum*: From beneficial to a new emerging pathogen. Clinical Microbiology and Infection. 2016;**22**: 37-45

[157] Tsutsumi Y. *Clostridium butyricum*-Induced Gas Gangrene Accompanying Pneumatosis Cystoides Intestinalis. Pathology of Infectious Diseases. Nagoya, Japan: Pathos Tsutsumi; 2003. Available from: https://pathos223.com/en/case/case011.htm

[158] Sturm R, Staneck JL, Stauffer LR, et al. Neonatal necrotizing enterocolitis associated with penicillin-resistant, toxigenic *Clostridium butyricum*. Pediatrics. 1980;**66**:928-931

[159] Brook I. Clostridial infection in children. Journal of Medical Microbiology. 1995;**42**:78-82

[160] Rich BS, Dolgin SE. Necrotizing enterocolitis. Pediatrics in Review. 2017; **38**:552-559

[161] Seki H, Shiohara M, Matsumura T, et al. Prevention of antibiotic-associated diarrhea in children by *Clostridium butyricum* MIYAIRI. Pediatrics International. 2003;**45**:86-90

[162] Dahiya DK, Malik R, Dangi AR, et al. Chapter 44. New-generation probiotics. In: Faintuch J, Faintuch S, editors. Microbiome and Metabolome in Diagnosis, Therapy, and Other Strategic Applications. Cambridge, MA, USA: Academic Press; 2019. pp. 417-424. DOI: 10.1016/B978-0-12-815249-2.00044-0

[163] Patra S, Vij M, Chirla DK, Kumar N, Samal SC. Unsuspected invasive neonatal gastrointestinal mucormycosis: A clinicopathological study of six cases from a tertiary care hospital. Journal of Indian Association of Pediatric Surgeons. 2012;**17**:153-156

[164] Vallabhaneni S, Mody RK. Gastrointestinal mucormycosis in neonates: A review. Current Fungal Infection Reports. 2015;**9**:269-274

[165] Dixon TC, Meselson M, Guillemin J, et al. Anthrax. The New England Journal of Medicine. 1999;**341**: 815-826

[166] Tutrone WD, Scheinfeld NS, Weinberg JM. Cutaneous anthrax: A concise review. Cutis. 2002;**69**:27-33

[167] Tena D, Martinez-Torres JA, Perez-Pomata MT, et al. Cutaneous infection due to *Bacillus pumilus*: Report of 3 cases. Clinical Infectious Diseases. 2007; **44**:e40-e42

[168] Doganay M, Metan G, Alp E. A review of cutaneous anthrax and its outcome. Journal of Infection and Public Health. 2010;**3**:98-105

[169] Duncan KO, Smoth TL. Primary cutaneous infection with *Bacillus megaterium* mimicking cutaneous anthrax. Journal of the American Academy of Dermatology. 2011;**65**:e60-e61

[170] Ishida R, Ueda K, Kitano T, et al. Fatal community-acquired Bacillus cereus pneumonia in an immunocompetent adult man: A case report. BMC Infectious Diseases. 2019; **19**:197

[171] Darbar A, Harris IA, Gosbell IB. Necrotizing infection due to *Bacillus cereus* mimicking gas gangrene following penetrating trauma. Journal of Orthopaedic Trauma. 2005;**19**:353-355

[172] Bottone EJ. *Bacillus cereus*, a volatile human pathogen. Clinical Microbiology Reviews. 2010;**23**:382-398

[173] Overcast WW, Atmaram K. The role of *Bacillus cereus* in sweet curdling of fluid milk. Journal of Milk and Food Technology. 1974;**37**:233-236

Chapter 8

Fournier's Gangrene of the Shoulder Girdle

Gyoguevara Patriota, Luiz Marcelo Bastos Leite, Nivaldo Cardozo Filho, Paulo Santoro Belangero and Benno Ejnisman

Abstract

Fournier's gangrene is uncommon, a high-mortality infection that affects the subcutaneous tissue with rapidly progressive necrosis. Reports on cases involving the shoulder girdle are more rare. Similar to the presentation on other regions of the human body, fundamental is early diagnosis and surgical intervention.

Keywords: Fournier's gangrene, necrotizing fasciitis, shoulder girdle infection, acute necrotizing, shoulder

1. Introduction and epidemiology

Fournier's gangrene is known by a variety of other names, necrotizing fasciitis (NF) such as hospital gangrene and hemolytic streptococcal gangrene, among others [1]. Necrotizing fasciitis (NF) is a rare and serious infection, characterized by extensive and rapidly progressive necrosis. Around 500–1500 cases/year are reported in the United States [2] but do not exist studies with profile in Brazil. It has an estimated greater involvement of males (3:1) at a mean age of 50 years. The predominant region of the body is in the perineal area. The mean mortality rate is high (32.2%), and if untreated, it can reach 100%.

Fournier's gangrene is a death-threatening infection caused by aerobic and/or anaerobic microorganisms. It affects the fascia and subcutaneous tissue with micro-circulation thrombosis and rapidly progressive necrosis of the skin in the affected region (evolution reaches 2–3 cm/h) [3–6].

Reports on cases involving the shoulder girdle are uncommon. There are only eight cases of Fournier's gangrene (on the shoulder girdle) described in the literature, beginning after surgical procedures in the shoulder (arthroscopy or osteosynthesis) [7–9], or intra-articular infiltration in the shoulder (with corticosteroids) [10], or after closed trauma [11–13], or even without any trauma reported [14]. The description of two cases of necrotizing fasciitis after closed trauma, without the presence of injuries or immune depression conditions, intrigues and alerts us to the possibility of occurrence in any person. Fundamental is the early diagnosis and intervention in early surgery.

IntechOpen

2. Etiology

The acute necrotizing inflammatory process initially affects the deep subcutaneous tissue and the fascia. The most superficial tissues and the skin are affected secondarily, due to vascular injury, thrombosis, and ischemia—resulting from the action of pro-inflammatory cytokines, proteinases, and endothelin. The destruction of subcutaneous nerves occurs in advanced stages. Initially described as a disease of unknown cause, it is now known that an underlying pathological process can be found in most cases of Fournier's gangrene; nonetheless, in a significant number of patients, the cause cannot be determined [15–17]. Therefore, a careful investigation can indicate the point of entry, which is located primarily in the digestive tract, in cutaneous affections, in the urogenital tract, or in cutaneous affections [18]. Eke et al. [16], in 2000, published a series of 1726 cases published cutaneous conditions accounted for 24% of the cases. The agents most associated with NF are group A beta-hemolytic *Streptococcus* and *Staphylococcus aureus*. However, other pathogens have already been linked to this disease, namely, *Clostridium perfringens*, *Peptostreptococcus*, Enterobacteriaceae, Proteus, *Pseudomonas*, and *Klebsiella* [16].

The most commonly observed comorbidity is diabetes mellitus, with a prevalence of 40–60%. Other common comorbidities include hepatic cirrhosis, immunodeficiencies, heart failure, systemic lupus erythematosus, obesity, alcoholism, hypertension, Addison's disease, and peripheral vascular disease. However, NF can present in healthy individuals, without comorbidities [19].

3. Classification

There are two classifications for necrotizing fasciitis. The US Food and Drug Administration (FDA) classifies NF according to its microbiological characteristics: type I, aerobic/anaerobic polymicrobial pattern (streptococci, staphylococci, enterococci, *Bacteroides*); type II, only one agent (*S. aureus* or more commonly group A beta-hemolytic *Streptococcus*) and less aggressive lesions, accounting for 10–15% of the cases; and type III, gastric myonecrosis and necrotizing fasciitis caused by *Clostridium perfringens* (less than 5% of the cases) [20].

Féres et al. [21] proposed an anatomic classification. This classification considers two relevant criteria: extension of the necrosis area and correlated it with mortality; these authors defined four groups (increasing mortality), in which group I presented a 12.5% rate, while the mortality rate in group IV was 68.75% (**Table 1**).

Groups	Description	Mortality (%)
Group I	Necrosis of the anterior perineum, scrotum, and penis or vulva	12.5
Group II	Group I + posterior perineum, perianal region up to 7 cm in diameter, rectum, and perirectal fat	34
Group III	Group II + sacral region, gluteal, inguinal region, and necrosis of the penis	37
Group IV	Group III + abdominal wall, suprapubic region, flank, thoracic wall, axillary region, and retroperitoneum	68.75

Table 1.
Anatomical classification of the necrosis area and correlation with mortality in Fournier gangrene. Féres et al. [21].

4. Diagnosis

The diagnosis is eminently clinical and corroborated by surgical findings, which include low adherence of the subcutaneous tissue, observed to the surgical manipulation, absence of bleeding, and liquefaction of the subcutaneous fat. Due to its severity and speed of evolution, Fournier's gangrene is an emergency. The clinical diagnosis must be suspected of the classic triad, pain, edema, and erythema with fever/tachycardia as soon as possible so that early treatment can be initiated. Necrotizing fasciitis could evolve rapidly with necrotic tissues and hemorrhagic blisters [19]. Patients may present laboratory abnormalities such as elevated serum urea (more than 18 mg/dl), serum creatinine (more than 1.2 mg/dl), leukocytosis (>20,000 WBC/mm^3), and CPK (more than 600 µ/l) [22].

Furthermore, imaging exams may also be used, such as radiographs (subcutaneous gas formation—low sensitivity and specificity), ultrasonography (has little practical use), computed tomography (lesion extension and gas formation), and magnetic resonance imaging (more accurate but more costly) [23]. CT provides additional information, such as asymmetric thickening of the fascia and changes in subcutaneous fat, as well as the presence of gas and abscesses. MRI, on the other hand, is considered superior to other imaging methods. It has high sensitivity and allows to define the area of necrosis of the fascia and to schedule the surgical procedure. The absence of changes in the deep fascia practically excludes in the diagnosis. It is worth mentioning that the critical condition of the patient often makes transportation impossible to perform the exam, limiting its use.

Culturing the debrided tissue is important in order to guide antibiotic therapy [19]. Biopsy of the fascia is considered the gold standard for diagnosis and should be performed in all patients during debridement, even in those whose macroscopic appearance is normal.

5. Case report

A 42-year-old female patient [12] who had previously been a victim of a motorcycle accident was attended to at a hospital unit in the interior of the state, diagnosed with a fracture of the middle third of the clavicle (Allman's group I) with deviation (>2 cm).

The physical examination showed no neurovascular deficit, no imminence of bone exposure at the fracture site, or apparent deformity, but she presented with right shoulder abrasions (posteriorly). Initially she was medicated with analgesics but has not received orientation about the use of a sling or necessity of follow-up with a specialist.

After the trauma (2 weeks), she presented with fever, local hyperemia, pain, and fever, requiring hospitalization (city of origin). She evolved local abscess (right clavicle region) and followed by spontaneous drainage of a purulent secretion (through a small orifice). Seventeen days after trauma, the patient underwent abscess drainage with 0.9% saline solution (in the ward) but no debridement (cultures/swab wasn't collected) (**Figure 1**). Results of laboratory exams are as follows: white blood cells (WBC), 2000/mm^3; erythrocyte sedimentation rate (ESR), 25 mm/h; and C-reactive protein (CRP), 11 mm/dl. At this stage, intravenous antibiotic therapy was initiated (with clindamycin 600 mg 8/8 h, metronidazole 500 mg 8/8 h, and ceftriaxone 1 g 12/12 h) and was transferred to a referral hospital in orthopedics surgery in the city of Salvador-BA (Brazil).

Despite the use of intravenous antibiotics, she evolved with toxemia, sepsis (heart rate (HR), 110 bpm; respiration frequency (RF), 26 ripm; temperature, 38.5°C), and an extensive lesion of the right hemithorax and base of the neck (with necrosis) and clavicle bone exposure but no neurovascular alterations (**Figure 2**). Their exams evolved worse (21,000 WBC/mm^3; ESR, 44 mm/h; CRP, 20 mm/dl, creatinine, 1.3 mg/dl; urea, 48 mg/dl; and CPK, 900 u/l), and the magnetic resonance imaging (MRI) of the thorax evidenced inflammatory process in the anterior region of the thorax (but no involving deep tissue layers), typical of Fournier's gangrene. Because of this clinical condition, she was admitted in the intensive care unit (ICU).

After clinical stabilization, the patient needed surgical debridement (with collected culture material), with clavicle preservation and modified antibiotic therapy (changed to vancomycin 1 g 12/12 h and meropenem 1 g 8/8 h).

The Fournier's gangrene continued with an increase in the necrotic area and of osteolysis in the clavicle exposure area. The orthopedics surgeons decided to perform a total clavicle resection with debridement (**Figure 3**). Two days after,

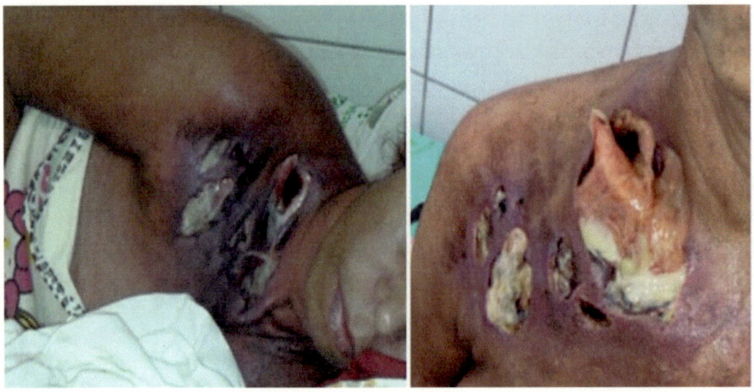

Figure 1.
Initial Fournier's gangrene of the shoulder girdle.

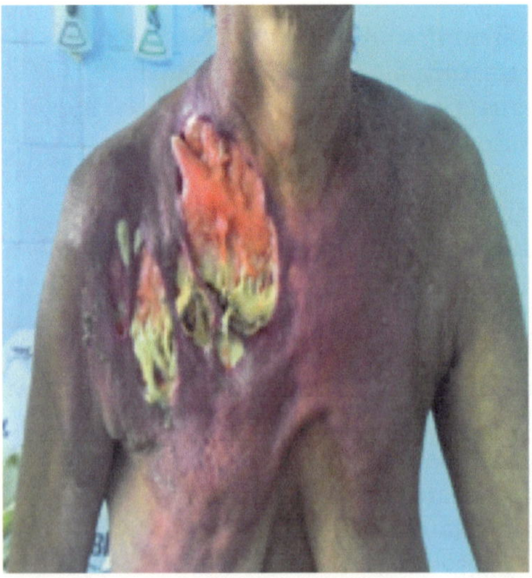

Figure 2.
Evolution of Fournier's gangrene of the shoulder girdle.

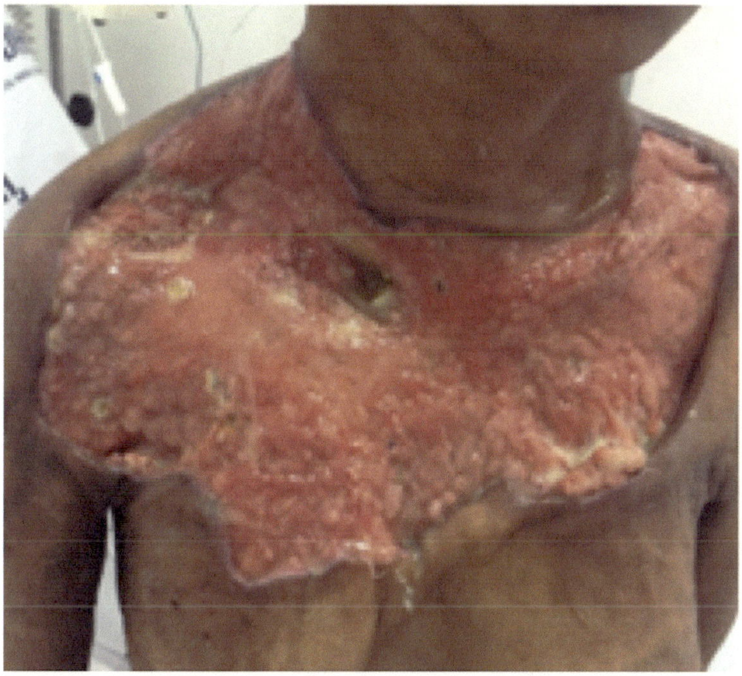

Figure 3.
Lesion before grafting.

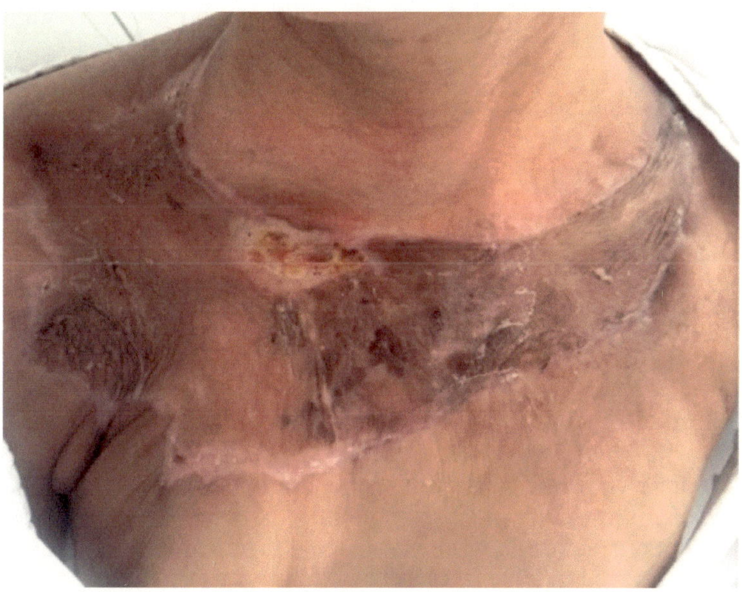

Figure 4.
Three months after grafting.

she evolved with a reduction of the WBC/inflammatory markers and presented an important clinical improvement.

The borders of the lesion ceased to evolve with necrosis and purulent secretion (granulation tissue started). The bone and soft tissue culture results were negative (maybe due to previous use of antibiotics).

Twenty days after the clavicle resection, with normal laboratory tests, a skin graft was performed by the plastic's surgeon. The patient was discharged from

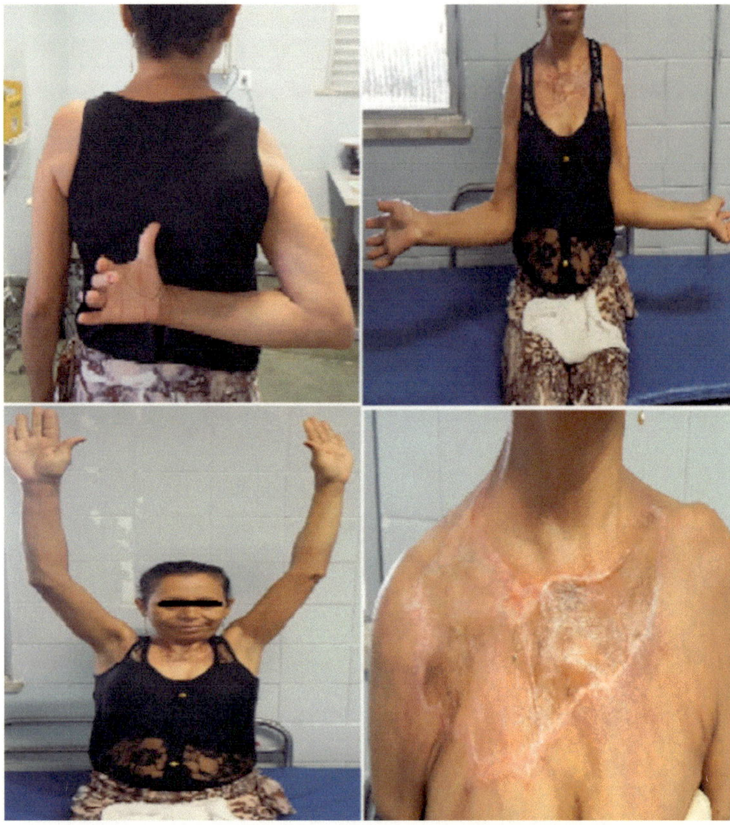

Figure 5.
Six months after the procedure, the patient presented excellent functional results and a completely healed wound.

the hospital after evolving without new signs of infection. The wound presented complete healing of the graft after 60 days **(Figure 4)**.

At the last outpatient visit (after 6 months of trauma), the patient presented excellent upper limb function (33 points on the UCLA score and 93 points on the constant score) **(Figure 5)**.

6. Treatment

Once the diagnosis is established, treatment must be instituted immediately and consists of volume replacement; ample surgical debridement, with removal of all necrotic material, including the fascia; and the use of broad-spectrum antibiotics.

Although didactic, the classification of NF, in types I–IV, has little practical utility and should not be decisive in the choice of antimicrobials. The polymicrobial form is responsible in 80% of cases, which justifies the initial broad-spectrum empirical antibiotic therapy, formed by the association of clindamycin, with aminoglycoside or ciprofloxacin, used in the reported cases. Recently, the American Society of Infectious Diseases indicated the combination of ampicillin-sulbactam, clindamycin, and ciprofloxacin as the scheme of choice for community-acquired infections. In cases of nosocomial infection, the association of carbapenems with anaerobicides is indicated, according to the profile sensitivity of the most prevalent bacteria in the institution [10].

Mallikarjuna et al. [24] described the treatment of Fournier's gangrene consists of drainage, radical debridement of the necrotic tissues, and antibiotic therapy for approximately 4 to 6 weeks (initially with ampicillin/sulbactam or ampicillin combined with clindamycin or metronidazole and de-escalation guided by culture results), plus good hemodynamic stabilization. Hyperbaric oxygen therapy (OH) and the use of immunoglobulins are adjuvant and remain controversial; at the same time, further studies are needed before they can be recommended [24, 25]. The use of a vacuum drain dressing has shown to be beneficial in the follow-up after debridement; this dressing should be changed every 24–72 h [26]. Tetanus prophylaxis should be performed; however, randomized controlled trials are still required to prove the efficacy of the use of immunoglobulins as a neutralizer of *Streptococcus* toxins [27]. After the absence of infectious sign and clinical stabilization, reconstructive surgery would could be performed with grafting (if necessary) [28].

Conflict of interest

The authors do not have any conflicts of interest to declare.

Author details

Gyoguevara Patriota[1*], Luiz Marcelo Bastos Leite[1], Nivaldo Cardozo Filho[1], Paulo Santoro Belangero[2] and Benno Ejnisman[2]

1 Cardiopulmonar Hospital – Salvador-BA, Salvador, Brazil

2 Sports Medicine Division, Orthopaedics Department, Federal University of São Paulo/UNIFESP, São Paulo, Brazil

*Address all correspondence to: ombroecotovelo@drpatriota.com.br

References

[1] Stephens BJ, Lathrop JC, Rice WT, Gruenberg JC. Fournier's gangrene: Historic (1764-1978) versus contemporary (1979-1988) differences in etiology and clinical importance. The American Surgeon. 1993;**59**(3):149-154

[2] World Health Organization. Necrotizing fasciitis [Fasciite nécrosante]. Weekly Epidemiological Record [Relevé épidémiologique hebdomadaire]. 1994;**69**(22):165-166. Available from: https://apps.who.int/iris/handle/10665/229069

[3] Jones J. Investigation upon the nature, causes, and treatment of hospital gangrene as it prevailed in the confederate armies. In: Surgical Memories of the War of Rebellion. New York: United States Sanitary Commission; 1871. pp. 1861-1865

[4] Laucks SS 2nd. Fournier's gangrene. The Surgical Clinics of North America. 1994;**74**(6):1339-1352. DOI: 10.1016/s0039-6109(16)46485-6

[5] Smith GL, Bunker CB, Dinneen MD. Fournier's gangrene. British Journal of Urology. 1998;**81**(3):347-355. DOI: 10.1046/j.1464-410x.1998.00532.x

[6] Yaghan RJ, Al-Jaberi TM, Bani-Hani I. Fournier's gangrene: Changing face of the disease. Diseases of the Colon and Rectum. 2000;**43**(9):1300-1308. DOI: 10.1007/bf02237442

[7] Reid AB, Stanley P, Grinsell D, Daffy JR. Severe, steroid-responsive, myositis mimicking necrotizing fasciitis following orthopedic surgery: A Pyoderma variant with Myonecrosis. PRS Global Open. 2015;**15**(2):1-5. DOI: 10.1097/GOX.0000000000000124

[8] Zani S. Babigian a. necrotizing fasciitis of the shoulder following routine rotator cuff repair. The Journal of Bone and Joint Surgery. American Volume. 2008;**90**:1117-1120. DOI: 10.2106/JBJS.G.00173

[9] Včelák J, Šuman R, Beneš J. Pyoderma gangrenosum mimicking necrotising fasciitis after rotator cuff reconstruction. Acta Chirurgiae Orthopaedicae et Traumatologiae Cechoslovaca. 2016;**83**(2):127-130

[10] Rodrigues JB, Judas F, Rodrigues JP, Oliveira J, Simões P, Lucas F, et al. Necrotizing faciitis after shoulder mobilization and intra-articular infiltration with betametasone. Acta Médica Portuguesa. 2013;**26**(4):456-459

[11] Joshy S, Haidar SG, Iossifidis A. Necrotising fasciitis of the shoulder following muscular strain. International Journal of Clinical Practice. 2006;**60**(7):856-857. DOI: 10.1111/j.1742-1241.2006.00635.x

[12] Cardozo Filho N, Patriota G, Falcão R, Maia R, Daltro G, Alencar D. Case report: Treatment of Fournier's gangrene of the shoulder girdle. Revista Brasileira de Ortopedia. 2018;**53**(4):493-498. DOI: 10.1016/j.rboe.2018.05.008

[13] Kim HJ, Kim DH, Ko DH. Coagulase-positive staphylococcal necrotizing fasciitis subsequent to shoulder sprain in a healthy woman. Clinics in Orthopedic Surgery. 2010;**2**:256-259. DOI: 10.4055/cios.2010.2.4.256

[14] Smyth A, Houlihan DD, Tuite H, Fleming C, O'Gorman TA. Necrotising fasciitis of the shoulder in association with rheumatoid arthritis treated with etanercept: A case report. Journal of Medical Case Reports. 2010;**4**:367. DOI: 10.1186/1752-1947-4-367

[15] Sarani B, Strong M, Pascual J, Schwab CW. Necrotizing fasciitis:

Current concepts and review of the literature. Journal of the American College of Surgeons. 2009;**208**(2):279-288. DOI: 10.1016/j.jamcollsurg.2008.10.032

[16] Eke N. Fournier's gangrene: A review of 1726 cases. The British Journal of Surgery. 2000;**87**(6):718-728. DOI: 10.1046/j.1365-2168.2000.01497.x

[17] Quatan N, Kirby RS. Improving outcomes in Fournier's gangrene. BJU International. 2004;**93**(6):691-692. DOI: 10.1111/j.1464-410X.2003.04753.x

[18] Tang WM, Ho PL, Fung KK, Yuen KY, Leong JC. Necrotising fasciitis of a limb. The Journal of Bone and Joint Surgery. 2001;**83**(5):709-714. DOI: 10.1302/0301-620x.83b5.10987

[19] Martinschek A, Evers B, Lampl L, Gerngroß H, Schmidt R, Sparwasser C. Prognostic aspects, survival rate, and predisposing risk factors in patients with Fournier's gangrene and necrotizing soft tissue infections: Evaluation of clinical outcome of 55 patients. Urologia Internationalis. 2012;**89**(2):173-179. DOI: 10.1159/000339161

[20] US Department of Health and Human Services, Food and Drug Administration, Center for Drug Evaluation and Research. Guidance for Industry. Uncomplicated and Complicated Skin and Skin Structure Infections: Developing Antimicrobial Drugs Treatment; 2010

[21] Féres O, Andrade JI, Rocha JJR, Aprilli F. Fournier's gangrene: A new anatomic classification. In: Reis Neto JA, editor. Proceedings of the 18th Biennial Congress of the International Society of University Colon and Rectal Surgeons. Sao Paulo, Brazil: Monduzzi Editore; 2000. pp. 103-107

[22] Wong CH, Khin LW. Clinical relevance of the LRINEC (laboratory risk indicator for necrotizing fasciitis) score for assessment of early necrotizing fasciitis. Critical Care Medicine. 2005;**33**(7):1677. DOI: 10.1097/01.ccm.0000129486.35458.7d

[23] Ruiz-Tovar J, Córdoba L, Devesa JM. Prognostic factors in Fournier's gangrene. Asian Journal of Surgery. 2012;**35**(1):37-41. DOI: 10.1016/j.asjsur.2012.04.006

[24] Mallikarjuna MN, Vijayakumar A, Patil VS, Shivswamy BS. Fournier's gangrene: Current practices. ISRN Surgery. 2012;**2012**:942437. DOI: 10.5402/2012/942437

[25] Escobar SJ, Slade JB Jr, Hunt TK, Cianci P. Adjuvant hyperbaric oxygen therapy (HBO2)for treatment of necrotizing fasciitis reduces mortality and amputation rate. Undersea & Hyperbaric Medicine. 2005;**32**(6):437-443

[26] Mouës CM, van den Bemd GJ, Heule F, Hovius SE. Comparing conventional gauze therapy to vacuum-assisted closure wound therapy: A prospective randomised trial. Journal of Plastic, Reconstructive & Aesthetic Surgery. 2007;**60**(6):672-681. DOI: 10.1016/j.bjps.2006.01.041

[27] Norrby-Teglund A, Muller MP, Mcgeer A, Gan BS, Guru V, Bohnen J, et al. Successful management of severe group A streptococcal soft tissue infections using an aggressive medical regimen including intravenous polyspecific immunoglobulin together with a conservative surgical approach. Scandinavian Journal of Infectious Diseases. 2005;**37**(3):166-172. DOI: 10.1080/00365540410020866

[28] Butler CE. The role of bioprosthetics in abdominal wall reconstruction. Clinics in Plastic Surgery. 2006;**33**(2):199-211. DOI: 10.1016/j.cps.2005.12.009

Chapter 9

Leprosy in the Modern Era

Syed Manzoor Kadri, Marija Petkovic, Arshi Taj and Ailbhe H. Brady

Abstract

Leprosy is a chronic infective disease that originates from the presence of pathogen agent *Mycobacterium leprae*. *Mycobacterium leprae* was discovered by the Norwegian doctor Gerhard Henrik Armauer Hansen in 1873. For the zoonotic transmission of M. leprae in the US the responsible insects are armadillos (*Dasypus novemcinctus*). M. leprae is an intracellular microorganism leading to loss of sensibility, innervation, intraepidermal impairment and lesions due to the absence of myelin in Schwann cells. *Mycobacterium leprae* has high infectivity and low pathogenicity. Incubation period is from 2 to 7 years. Leprosy is an infectious neurodegenerative disorder of the peripheral nervous system. Leprosy is the major cause of human disability due to neurological damage. Leprosy still represents one of the major causes of disabilities in humans. The most common complications are muscle weakness leading to atrophy, bone loss, amputations and blindness. In the case of chronic cutaneous hyperalgesia, there is a local increase in NGF levels. The application of anti-NGF antibodies may be of benefit in treating hyperalgesia in patients with neuropathy and impaired nerve endings. If combined, NGF, NT-3 and glial cell-line derived neurotrophic factor may be sustainable. In over 90% of human individuals an overall genetic resistance has been noted.

Keywords: *Mycobacterium leprae*, diagnosis, epidemiology, treatment, adult, children

1. Introduction

Leprosy is a chronic infective disease that originates from the presence of pathogen agent *Mycobacterium leprae*. *Mycobacterium leprae* was discovered by the Norwegian doctor Gerhard Henrik Armauer Hansen (**Figure 1**) in 1873 as the first proof that a person's disease is caused by bacteria [1].

For the zoonotic transmission of M. leprae in the US the responsible insects are armadillos (*Dasypus novemcinctus*).

Since ancient times, leprosy has been considered as one of the first major human epidemic diseases.

The first noted leprosy epidemic was in Europe in 1873 due to the detection of *Mycobacterium leprae* in Norway (Bergen).

These regions included the Iberian Peninsula, Sicily, the Balkans, southern Romania, the Baltics, and Scandinavia. In Norway in particular, scientists investigated the disease known as "Spedalsked", the Norwegian name for leprosy [1].

In 2015, the number of newly diagnosed cases of leprosy was 210,758 worldwide (**Figure 2**).

Figure 1.
Gerhard Henrik Armauer Hansen.

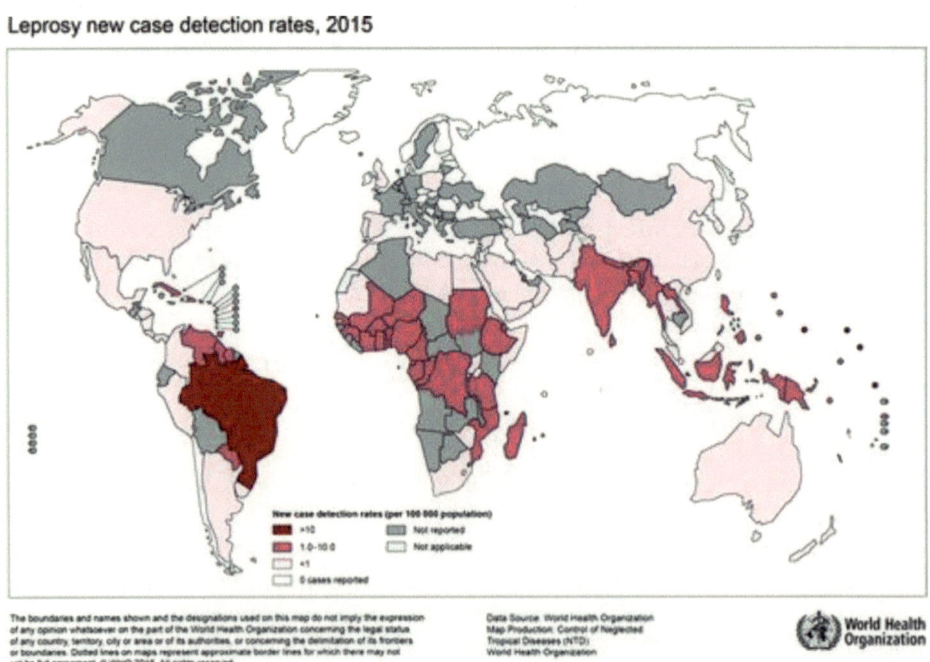

Figure 2.
Newly diagnosed cases of leprosy worldwide [2].

Transmission of leprosy persists despite efforts made by the World Health Organization (WHO). With the highest incidence in India, Brazil and Indonesia.

Mycobacterium leprae is a non-motile, acid-fast pathogen and cannot be cultured on any known medium. The most common route of transmission in humans may be via nasal droplets. Predisposing factor may be a prolonged contact with an infected individual with a high-bacterial load.

Leprosy still represents one of the major causes of disabilities in humans. It has been estimated that approximately 3 million individuals are affected.

Based on 193, 118 cases at the end of 2017, prevalence rate corresponds to 0.3/10,000.

2. History and epidemiology

M. leprae is an intracellular microorganism and an acid-resistant bacilli leading to a loss of sensibility, innervation, intraepidermal impairment and lesions due to the absence of myelin in Schwann cells. If the bacilli are numerous, they can be grouped in parallel or arranged in parallel.

M. leprae has high infectivity and low pathogenicity. The incubation period is from 2 to 7 years.

An individual is infected by inhalation of infectious aerosol or through the skin while contacting with nasal secretions and/or skin changes of the infected individual.

Children are more susceptible to leprosy than adults. Due to the slow proliferation of leprosy (the time of one generation is 14 days), the incubation period of the disease is quite prolonged (2–5 years).

Due to infiltration of the peripheral nerves, neuritis, anesthesia, trophic ulcerations, muscle atrophy and bone resorption occur.

3. Basic scientific considerations and pathology

Leprosy is an infectious neurodegenerative disorder of the peripheral nervous system. Thus, leprosy is the major cause of human disability due to neurological damage.

To this date, M. leprae has not been cultivated on artificial nutrients.

Nerve injury-associated tissue damage is the most prominent clinical consequence of leprosy.

In the process of leprosy-associated neuropathy, the presence of bacilli in nerve endings and Schwann cells induces a response mediated by macrophages and other cells that eventually leads to the appearance of immune-mediated lesions.

The most important cytokines are TNF-α, IL-6, IL-17 that are involved in the progression of neural lesions (**Figure 3**).

According to Antunes et al., it was concluded that in the individuals with neuritic leprosy, NGF-R immunoexpression was lower (nerve fiber, Schwann cells) than in control group (normal individuals). In the leprosy group, hypoesthesia was associated with decreased expression of NGF-R and PGP (protein gene product) 9.5.

TrkA receptors are detected in subepidermal fibers. TrkA receptor messenger RNA is produced in the skin. NGF are also present in keratinocytes and are in correlation with deficient thermal sensation [3].

NGF levels are depleted in nerve and skin lesions in leprosy the loss of NGF-dependent nociceptive fibers in damaged skin. An additional cause of decreased NGF is an impaired interaction between keratinocytes and nerves in affected skin.

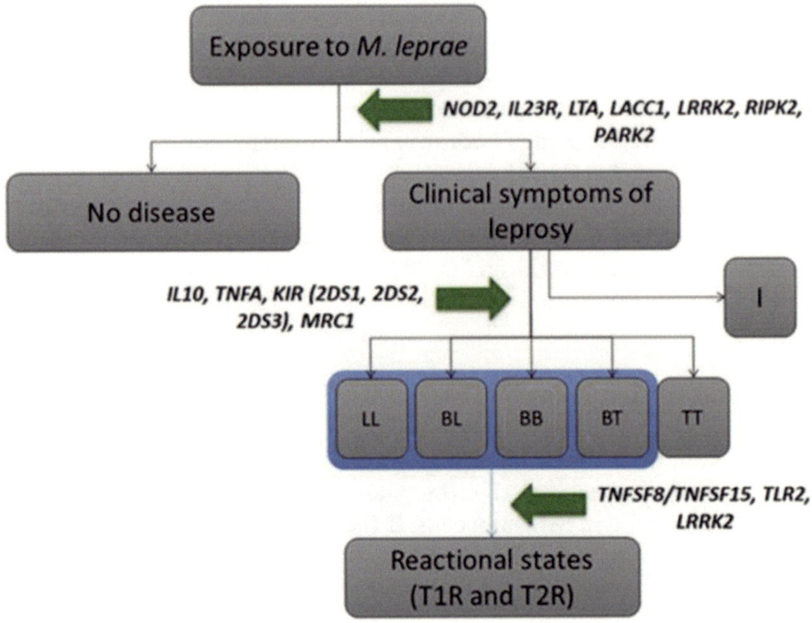

Figure 3.
Pathogenesis of leprosy [4].

Higher levels of NGF are observed in the lepromatous forms, while lower NGF levels are present in tuberculoid forms.

High levels of NGF are associated with lepromatous and decreased levels are associated with tuberculoid forms [5].

4. Classification of leprosy

Classification of leprosy according to immunity by Ridley and Jopling (**Figure 4**) consists of a 5 group-system based upon the clinical, histopathological, immunological and bacilloscopic factors:

1. Borderline-tuberculoid

2. Borderline-lepromatous

3. Borderline-borderline

4. Tuberculoid

    - Granulomatous lesions

    - Th1 (cell-mediated) immune chapter

5. Lepromatous

    - Th2 (an anti-inflammatory cytokine) characteristics

    - Multiplication of pathogen in macrophage phagosomes.

WHO	Ridley-Jopling	ICD-10	MeSH
Paucibacillary	tuberculoid ("TT"), borderline tuberculoid ("BT")	A30.1, A30.2	Tuberculoid
Multibacillary	midborderline or borderline ("BB")	A30.3	Borderline
Multibacillary	borderline lepromatous ("BL"), and lepromatous ("LL")	A30.4, A30.5	Lepromatous

Figure 4.
Leprosy classification.

WHO classification [6] according to the number of lesions consists of three groups:

1. Paucibacillary single-lesion leprosy (single skin lesion)

2. Paucibacillary leprosy (2–5 skin lesions)

3. Multibacillary leprosy (more than 5 skin lesions)

5. Clinical and laboratory diagnosis

Leprosy is transmitted via close and prolonged contact amongst healthy individuals and a bacillus-infected patient through inhalation of the bacilli contained in nasal secretions or Flügge droplets. The main route of transmission is the nasal mucosa. Leprosy, similar to other infectious diseases is a consequence of pathogen invasion of the host organism and transposition of the immunological barrier [7].

Transmission can occur by skin erosions, blood, vertical transmission, breast milk, and insect bites.

In infected individuals, there is a transitional period of nasal release of bacilli. The presence of specific DNA sequences of M. leprae in swabs or nasal biopsies and seropositivity suggest the carrier plays a role in the transmission of leprosy.

In the individuals with the solid immune system, leprosy is presented in tuberculoid form (solitary papules and plaques). Such skin changes may form erythematous plaques with raised borders and an annular appearance.

In the case of a defected immune system, lepromatous leprosy is presented with an impaired T-cell immunity leading to anergy. Clinically, this manifestation is shown as multiple red-brown-nodular infiltrate (lepromas) in the skin and mucous membranes.

"Leonine facies" is defined as the symmetrical centrofacial presentation of the cushion-like lesions, loss of the eyelashes and eyebrows [7].

Involvement of the nasal mucosa leads to destruction of the septum and deformity of the nasal skeleton (saddle nose). Subsequently, this destructive inflammatory process may include the entire nasopharynx, clinically characterized by mucosal ulcerations of the palate and larynx.

Multiple, poorly demarcated, hypopigmented papules, nodules, and/or infiltrated plaques are the hallmark of this form.

In case of chronic cutaneous hyperalgesia, there is a local increase in NGF levels. The application of anti-NGF antibodies may be of benefit in treating hyperalgesia in patients with neuropathy and impaired nerve endings. If combined, NGF, NT-3 and glial cell-line derived neurotrophic factor may be sustainable [8].

NGF has a potential modulatory role in nociception.

NGF may restore pain sensitivity and prevent the ulcer formation caused by nociceptive loss. Anti-NGF treatment may be of benefit in patients with hyperalgesia.

Other cytokine imbalances may be described as the imbalance in the proNGF/NGF ratio, increased TNF-α and p75 neurotrophin [9].

The main targets of M. leprae are Schwann cells. In case of the onset of Schwann cells degradation, peripheral neuropathy is the most common consequence. It has been documented that NGF may act as a protective factor for Schwann cells. Thus, low levels of NGF directly are responsible for the development of neuropathy.

NGF are involved in the reparation of neuronal cells and induction of fibroblast migration. They exhibit proliferative and antiapoptotic effects on keratinocytes and endothelial cells [10].

The incidence of nerve impairment in the individuals with paucibacillary leprosy is present in 10% of cases, whilst in multibacillary leprosy in 40%.

TGF-β is involved in the tissue reparation processes due to its anti-inflammatory attributes.

Leprosy is diagnosed mainly on the basis of a clinical picture. For bacteriological diagnosis, a nose swab, scraping or cutaneous skin changes are taken.

The diagnosis is based on the findings of acidoalcohol resistant bacteria in lepromatous leprosy and histopathological findings. Experimental animals (armadillo, mouse) are used to isolate M. leprae. Skin test with lepromine (Mitsuda test) is positive for tuberculoid leprosy. A positive lepromin test is linked to the ability to develop a granulomatous response [11].

A serological test method with good sensitivity in multibacillary forms (approximately 70%) involves the measurement of antibodies against a phenolic glycolipid (PGL-1; 35 kDa) in the bacterial cell wall.

It is recommended to perform skin biopsies taken from the margins of the lesions and should also include subcutaneous tissue.

Bacterial index (BI): scale for assessing the number of leprosy bacteria in skin smears may be according to the bacterial count per visual field(s)	
a. 1–10/100	1 +
b. 1–10/10	2 +
c. 1–10/1	3 +
d. 10–100/1	4 +
e. 100–1000/1	5 +
f. > 1000/1	6 +

In certain individuals, pseudoabscesses along nerves and nerve thickening may be present.

Thickened nerves may be detected by palpation along the course of the supraorbital, retroauricular, ulnar, median, superficial radial, common peroneal, superficial peroneal, posterior tibial, and sural nerves. The initial functional tests implicate weakness (paresis) or loss (paralysis) of muscle strength [12].

Thermosensitivity is checked using a heated test tube with a lighter. The test tube is then held to the patient's skin lesion and the corresponding skin. The test is performed placing a tube with water at room temperature to the skin lesions. The patient is then asked to detect the difference between hot and cold [13].

The neurological examination conducted by a specialist includes EMG, nerve ultrasound, and magnetic resonance imaging. Nerve biopsies are preferably taken from thickened superficial and thus readily accessible nerves such as the sural nerve, the superficial peroneal nerve, the ulnar nerve, and the saphenous nerve.

6. Clinical presentation

The initial stage of leprosy is non-specific presenting with one or more hypopigmented macules.

There have been four noted immunologic leprosy reactions. Type 1 reaction represents a hypersensitivity reaction to M. leprae antigens clinically characterized by sudden onset of urticarial swelling of the leprous skin lesions. It may also be associated with acute and very painful neuritides with loss of sensory and motor function.

Pathophysiologically, type 2 reaction (Syn. erythema nodosum leprosum) is characterized by the occurrence of painful violaceous- erythematous cutaneous or subcutaneous nodules. Type 3 reaction (Syn. Lucio's phenomenon).

Myalgia, arthralgia, and osseous pain are symptoms associated with a type 2 reaction [13].

The first leprosy classification by WHO was applied in 1966 based upon the histological picture – the Ridely-Jopling classification. It shows two forms of leprosy – its mild and severe defect of cell-mediated immunity: tuberculoid (paucibacillary) and lepromatous (multibacillary) leprosy with the following subgroups: borderline tuberculoid (BT), borderline lepromatous (BL) and borderline lepromatous leprosy (BL) [14]. This classification is detailed earlier in the chapter.

7. Complications

The most common complications are muscle weakness leading to atrophy, bone loss, amputations and blindness.

In the case of chronic cutaneous hyperalgesia, there is a local increase in NGF levels. The application of anti-NGF antibodies may be of benefit in treating hyperalgesia in patients with neuropathy and impaired nerve endings. If combined, NGF, NT-3 and glial cell-line derived neurotrophic factor may be sustainable.

In over 90% of human individuals an overall genetic resistance has been noted.

8. Systemic involvement and special situations

Overall, leprosy is a granulomatous inflammatory process. It causes intraneural pressure induced atrophy, palpable nerve thickening and progressive loss of neural functions with necrosis.

Histologically, there is epi-, peri- and endoneural fibrosis.

The most commonly affected nerve structures are: n. ulnaris located in the ulnar groove, n. medianus in the vicinity of the carpal tunnel, *n. tibialis* posterior, superficial branch of the radial nerve, sural nerve posteriorly of the malleolus, great auricular nerve as well as facial nerve, frontal and cervical branches.

Clinically, the resultant ocular muscle paralysis causes lagophthalmos, subsequently facilitating secondary corneal infections due to incomplete lid closure (Bell's palsy).

Sensory loss of the ophthalmic branch (V1) of the trigeminal nerve, too, results in corneal anesthesia, thus facilitating bacterial corneal ulceration.

9. Therapeutics

In the individuals with type 1 leprosy reactions (reversal reaction), the systemic administration of corticosteroids (initial dose of 40–60 mg prednisolone for 14 days, than gradually increase by 5 mg every 14 days).

Type 2 reaction is treated with the administration of thalidomide (100–400 mg/per day) plus the administration of systemic corticosteroids most commonly prednisolone 40 mg for 5 days [13].

In the case of syn. Lucio's phenomenon (type 3 reaction) the use of systemic corticosteroids are the first choice (**Figure 5**).

In the patient with nerve damage, it is necessary to incorporate active and passive physical therapy, local skin care if acral mutilations are present. Certain individuals require orthopedic prosthesis as well as reconstructive procedure(s) (nose reconstruction, tarsorrhaphy, hand surgery, etc.) [13].

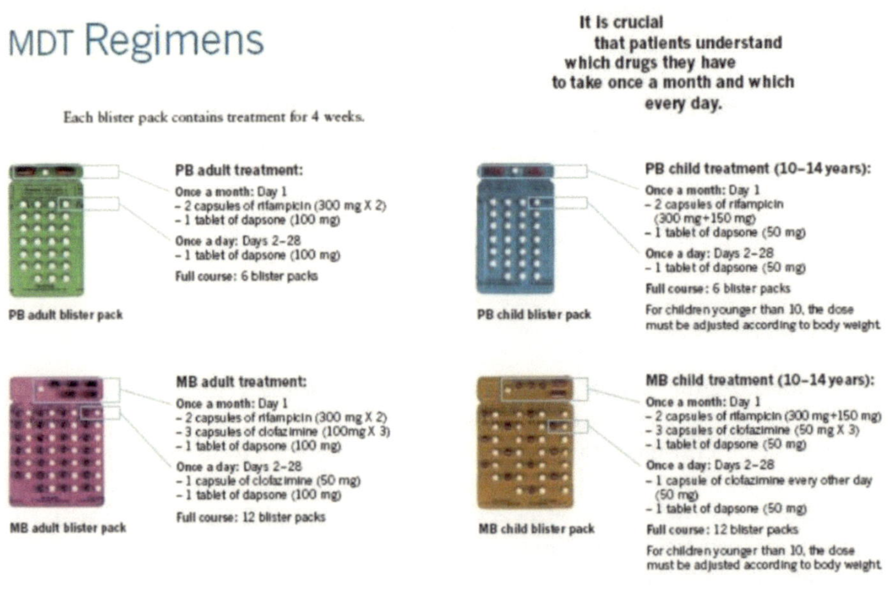

Figure 5.
Leprosy treatment modalities.

10. Prophylaxis and monitoring

The early diagnosis and treatment of leprosy are preventive measures in the initial spreading of this infectious disease. Hemoprophylaxis in children exposed to infection is also required (BCG vaccine). For the treatment of leprosy, sulphonic preparations (dapsone) and rifampicin are used. The duration of treatment is prolonged – from 6 months to 2 years [15].

11. Miscellaneous issues

Recent studies suggested that the presence of the PARK2/PACRG gene (chromosome 6q25-q27) as well as the presence of the NRAMP1 gene (chromosome 2q35) is linked to the higher leprosy susceptibility.

TAP1 and TAP2 (transporter associated with antigen processing) genes (located on chromosome 6p21), TNF-α (tumor necrosis factor alpha) (chromosome 6p21), and the VDR (vitamin D receptor) gene (chromosome 12q12) are associated with the innate and adaptive immunity [15].

In the majority of infected individuals, leprosy takes on an intermediate form, which may – to a variable degree – show clinical features of tuberculoid and lepromatous leprosy. This intermediate form is referred to as borderline leprosy.

In PB forms it has been noted higher cellular response to M.leprae due to the Th1 cytokines such as IFN-y. There are also lower antibody titers to M. leprae – specific antigens. In the individuals with MB form of leprosy there is an absence of the capacity to mount a cell mediated response due to T cell anergy. There is a high antibody titer to M. leprae antigens (PGL-1) [16].

12. Rehabilitation and social issues and future prospects

In untreated individuals with leprosy, there are progressive, destructive and irreversible body impairments. Such deformities and ulcers are the consequence of the skin, muscle and nerve invasion.

Recent studies [15] suggest the role of PARK2 and *LACC1* as one of the major genetic factors involved in the pathogenesis of leprosy, thus implicating the necessity of further investigational studies.

Author details

Syed Manzoor Kadri[1]*, Marija Petkovic[2], Arshi Taj[3] and Ailbhe H. Brady[4]

1 Disease Control, Directorate of Health Services, Kashmir, India

2 University of Medicine, Belgrade, Serbia

3 Department of Anaesthesia, Critical Care and Pain Management, Government Medical College, Srinagar, Kashmir, India

4 Warrington and Halton Teaching Hospitals NHS Foundation Trust, Intensive Care Department, Warrington, United Kingdom

*Address all correspondence to: kadrism@gmail.com

References

[1] Ghosh S, Chaudhuri S. Chronicles of Gerhard-Henrik Armauer Hansen's life and work. Indian Journal of Dermatology. 2015;**60**(3):219-221

[2] World Health Organization. Leprosy: Fact Sheet 101. 2015. Operational Manual – Global Leprosy Strategy 2016-2020

[3] Antunes SL, Chimelli LM, Rabello ET, Valentim VC, Corte-Real S, Sarno EN, et al. An immunohistochemical, clinical and electroneuromyographic correlative study of the neural markers in the neuritic form of leprosy. Brazilian Journal of Medical and Biological Research. 2006;**39**(8):1071-1081

[4] Cambri G, Mira MT. Genetic susceptibility to leprosy—From classic immune-related candidate genes to hypothesis-free, whole genome approaches. Frontiers in Immunology. 2018

[5] Ooi WW, Srinivasan J. Leprosy and the peripheral nervous system: Basic and clinical aspects. Muscle & Nerve. 2004;**30**(4):393-409

[6] Pardillo FE, Fajardo TT, Abalos RM, Scollard D, Gelber RH. Methods for the classification of leprosy for treatment purposes. Clinical Infectious Diseases. 2007;**44**(8):1096-1099

[7] Chapman SJ, Hill AVS. Human genetic susceptibility to infectious disease. Nature Reviews. Genetics. 2012;**13**(3):175-188

[8] Anand P. Neurotrophic factors and their receptors in human sensory neuropathies. Progress in Brain Research. 2004;**146**:477-492

[9] Mohamed R, Coucha M, Elshaer SL, Artham S, Lemtalsi T, El-Remessy AB. Inducible overexpression of endothelial proNGF as a mouse model to study microvascular dysfunction. Biochimica et Biophysica Acta. 2017;**1864**:746-757. DOI: 10.1016/j.bbadis.2017.12.023

[10] de Souza Aarão TL, de Sousa JR, Falcão ASC, Falcão LFM, Quaresma JAS. Nerve growth factor and pathogenesis of leprosy: Review and update. Frontiers in Immunology. 2018;**9**:939

[11] Scollard DM, Adams LB, Gillis TP, Krahenbuhl JL, Truman RW, Williams DL. The continuing challenges of leprosy. Clinical Microbiology Reviews. 2006;**19**(2):338-381

[12] Lastória JC, de Abreu MAMM. Leprosy: A review of laboratory and therapeutic aspects - part 2. Anais Brasileiros de Dermatologia. The Brasilian Society of Dermatology. 2014;**89**(3):389-401

[13] Fischer M. Leprosy – An Overview of Clinical Features, Diagnosis, and Treatment. Wiley Online Libraries; 2017

[14] White C, Franco-Paredes C. Leprosy in the 21st century. Clinical Microbiology Reviews. 2015;**28**(1):80-94

[15] Misch EA, Berrington WR, Vary JC Jr, Hawn TR. Leprosy and the human genome. Microbiology and Molecular Biology Reviews. December 2010;**74**(4):589-620

[16] Spencer JS, Brennan PJ. The role of *mycobacterium leprae* phenolic glycolipid I (PGL-I) in serodiagnosis and in the pathogenesis of leprosy. Leprosy Review. 2011;**82**(4):344-357

Chapter 10

Multidrug-Resistant Bacterial Foodborne Pathogens: Impact on Human Health and Economy

Lilia M. Mancilla-Becerra, Teresa Lías-Macías, Cristina L. Ramírez-Jiménez and Jeannette Barba León

Abstract

The drug abuse known to occur during growth of animals intended for food production, because of their use as either a prophylactic or therapeutic treatment, promotes the emergence of bacterial drug resistance. It has been reported that at least 25% of the foodborne isolates show drug resistance to one or more classes of antimicrobials (FAO 2018). There are diverse mechanisms that promote drug resistance. It is known that the use of sub-therapeutic doses of antibiotics in animals intended for food production promotes mutations of some chromosomal genes such as *gyrA-parC* and *mphA*, which are responsible for quinolone and azithromycin resistance, respectively. Also, the horizontal transfer of resistance genes as groups ("cassettes") or plasmids makes the spread of resistance to different bacterial genera possible, among which there could be pathogens. The World Health Organization considers the emergence of multidrug-resistant pathogenic bacteria as a health problem, since the illnesses caused by them complicate the treatment and increase the morbidity and mortality rates. The complication in the illness treatment caused by a multidrug-resistant pathogen causes economic losses to patients for the payment of long stays in hospitals and also causes economic losses to companies due to the absenteeism of their workers.

Keywords: multidrug-resistant bacteria, MDR, foodborne pathogens, antimicrobial resistance, MDR bacteria human health, MDR microorganism economic impact

1. Introduction

Increasing antimicrobial resistance (AMR) is a global public health threat. The excessive use and abuse of drug therapies in humans and in the animals intended for human consumption, and its bad disposition as a waste, have tightened up the problem in recent years [1]. This phenomenon affects any person regardless of sex, age, origin, or social status and threats the ability to effectively solve the treatments of different diseases and also compromises the food security, economy, and development of the countries [2].

Microorganisms are sensitive to antimicrobials when they do not harbor components involved in degrading them. AMR occurs when bacteria, fungi, viruses, and parasites are exposed for a long time to sub-therapeutic doses of drugs such as antibiotics, antifungals, antivirals, antimalarials, or anthelmintics that modify the ecology of microorganism. In order to contend with the residues present in their environment, microorganisms may acquire genetic elements that allow them to cope with these compounds and survive. In some cases, the use of poor quality drug, counterfeit products, incorrect product, and modified dosage can accelerate the development of microbial resistance. Another relevant factor for development of AMR is the inadequate disposal of waste generated in the agricultural production and pharmaceutical and wastewater treatment plants as they can be spread through the environment [3]. One of the recently described phenomena observed is the association between the emergence of multiresistant microorganisms (MMR) and the increase in the isolates that show the production of extended-spectrum beta-lactamase enzymes (ESBL). Currently, more than 200 varieties of BLEE enzymes are recognized with different substrates, and the frequency of isolates producing these enzymes varies from country to country (from 20 to 48%) [3].

Although there are many factors that favor the spread of antibiotic resistance, it affects different sectors, such as human health, animal health, agriculture, environment, and commercial trade [4]. It is estimated that 700,000 people die each year from infections caused by microorganisms resistant to antimicrobials and a large number of sick animals that do not respond to treatments [5]. Within the agricultural and food industry, resistant microorganisms represent a risk for production that threatens the global economy. For this reason, it is important to implement supervised agricultural regulations and practices that ensure the responsible use of antimicrobials in the production of animals and crops.

2. Mechanisms and propagation of resistance

In microorganisms, drug resistance arises in order to contend against a harmful stimulus that threatens their survival. In bacteria, the mechanisms that confer the resistance against antibiotics could be classified as intrinsic (mutations originating in the organism itself) or acquired by transfer of genetics elements during the replication of DNA (vertical transfer) or from different species or genera (horizontal transfer) [6] (**Figure 1**).

2.1 Intrinsic mechanisms or natural resistance

The intrinsic mechanisms can be found in the cell in a natural manner. They are conditions that are universally found in bacterial species and that are independent of antibiotic selectivity [7]. Some of the intrinsic mechanisms are described below.

2.1.1 Permeability or impermeability of the outer membrane or cell wall

Gram-positive bacteria are more susceptible to various antibiotics since they have a thick outer layer of peptidoglycan with polymers of teichoic acid and covalently bound proteins, which allows the easy penetration of small molecules up to 30–57 kDa [8, 9]. In contrast, Gram-negative bacteria have an outer membrane that surrounds them with a relatively thin peptidoglycan layer. The composition of the outer membrane is based on lipid molecules covalently linked to polysaccharide units [10].

The lipid molecule has a large chain of fatty acids that contribute to reduce the fluidity of the lipopolysaccharide (LPS) membrane [10]. The central region of the LPS plays an important role providing a barrier to hydrophobic antibiotics and other compounds. It has been reported that strains that express full-length LPS have an intrinsic resistance to hydrophobic antibiotic class such as macrolides and aminoglycosides. Another modification observed is the alteration of the anionic nature of the LPS; the most common LPS modifications are the cationic substitution of phosphate groups with 4-amino-4-deoxy-L-arabinose (L-Ara4N) or phosphoethanolamine (PEtN), which decreases the net negative load of lipid A from minus 1.5 to minus 1 or from minus 1.5 to 0, respectively [11]. The net positive charge resulting from the LPS modification reduces the binding of some cationic antibiotics such as polymyxins, leading the resistance of the bacteria such as *Escherichia coli*, *Klebsiella pneumoniae*, and *Salmonella enterica* [12].

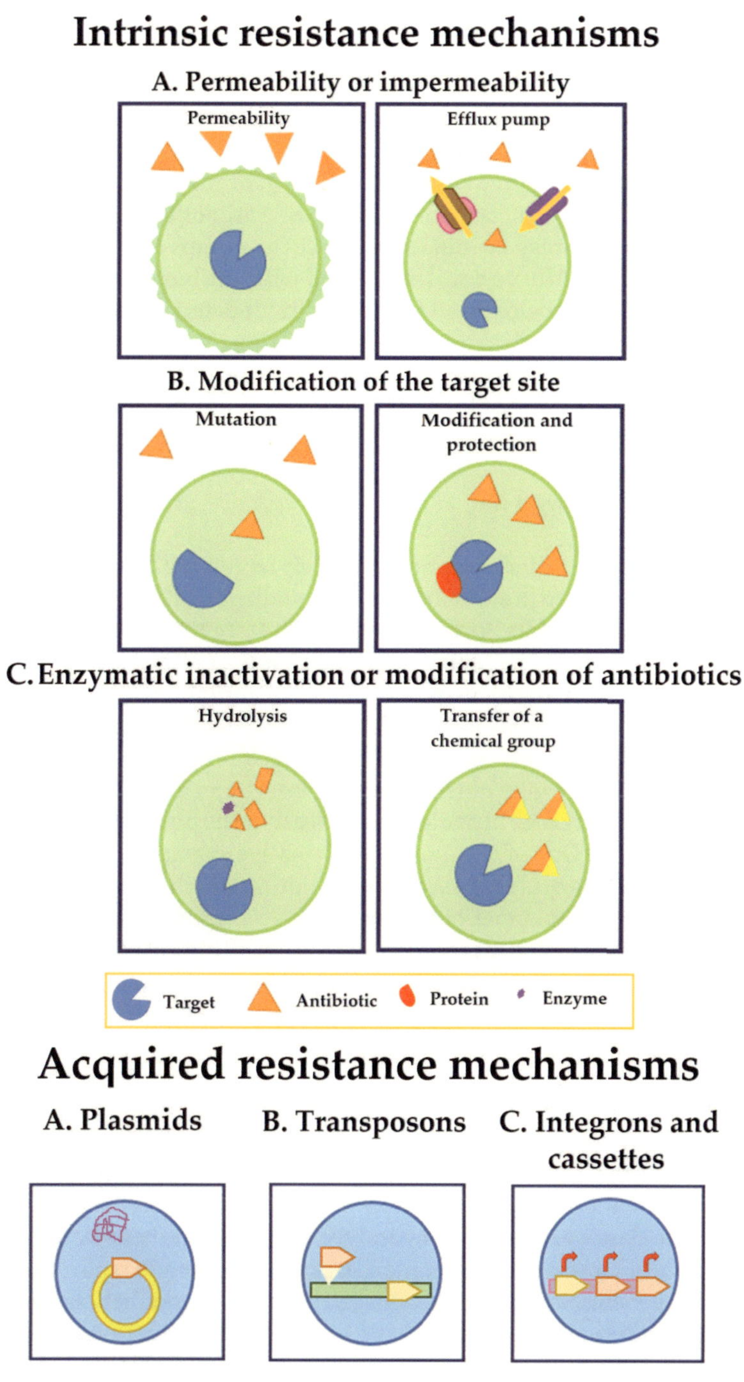

Figure 1.
Schematic representation of the mechanisms of multidrug-resistance acquisition in bacteria.

On the other hand, embedded in the outer layer membrane of Gram-negative bacteria, there are proteins called porins that function as a channel through which the molecules can diffuse. Porins could restrict the influx of numerous antibiotics and contribute to the resistance against them [13]. The mechanism of resistance promoted by the porins consists in changing the hydrophilic composition of some antibiotics such as beta-lactam, chloramphenicol, fluoroquinolones, and tetracyclines. Likewise, the alteration of the amount or modification of the structure of the porins promotes resistance to antibiotics [11].

2.1.2 Expulsion of the antibiotic by active mechanisms

In general those mechanisms are mediated by bacterial flow pumps that actively transport many toxic molecules out of the cell [14]. The outflow pumps can interact with only one molecule (enzyme substrate specific), or they can have a broader spectrum and export distinct classes of molecules. Antibiotic resistance mediated by active outflow pumps may be incidental, since the pumps exhibit a broad-range substrate [9]. However, efflux pumps associated with antibiotic resistance have been described in Gram-positive and Gram-negative bacterial pathogens. The energy of some flow pumps depends of the antibiotic agents in order to extract it from the periplasm to the outside. The overexpression of one or more of these flow pumps prevents the intracellular accumulation of antibiotics at the thresholds necessary to exert their inhibitory activity [15].

2.1.3 Modification of the target site

Most of the antibiotics bind specifically to their targets with high affinity. Changes in target structure prevent an effective binding to antibiotics but still allow the target to carry out its normal function. The target modification could be originated by a mutation in the gene that encodes for the antibiotic target [14]. An example of this mutation mechanism is the linezolid antibiotic, a member of the oxazolidinone class, which inhibits the initiation of bacterial translation by altering multiple copies of the V domain of the 23S rRNA in Gram-positive bacteria. The mutation in one of these copies of the V domain can confer antibiotic resistance [16].

Another mechanism relies in avoiding or releasing the binding of the antibiotic to their target site [17]. One of the most representative examples of this mechanism and of current importance is that used by quinolones, which exert their function by inhibiting important enzymes of bacterial DNA replication such as gyrase and topoisomerases II and IV. The mechanism of evasion of the antibiotic function is by the expression of repeating pentapeptides (PRP), encoded by *qrn* genes, which bind and promote the release of the quinolone from the target enzymes, allowing the normal activity of the topoisomerases [18].

2.1.4 Enzymatic inactivation or modification of antibiotics

In this case, the mechanism of action could be by enzymatic hydrolysis [19] or modification of chemical groups by transfer or addition of different chemical compounds [14]. The classic example of a hydrolytic enzyme is the beta-lactamase, which hydrolyzes the beta-lactam ring, a common structural element in penicillins, cephalosporins, carbapenems, and monobactams [20]. Four classes of beta-lactamases have been described: the classes A, C, and D have a serine hydrolase activity; in contrast, class B has a metalloenzyme activity [21]. Another example of this type of resistance is provided by the enzymes erythromycin esterases EreA and EreB. These enzymes hydrolyze the macrolactone rings of macrolides such as erythromycin. It should be noted that EreB enzyme confers resistance to almost all members of the macrolide class, with the exception of telithromycin, a semisynthetic erythromycin derivative, which belongs to a new class of antibiotics called ketolides [22]. In contrast, the EreA enzyme does not hydrolyze azithromycin and also telithromycin [23].

The modification of antibiotics includes the modification of some element of their structure, which is essential in the union with the bacteria target diminishing its affinity. This mechanism involves the addition or transfer of groups such as N-acetyl, phosphoryl, O-nucleoside, O-ribosyl, and O-glycoside. Unlike hydrolysis, this modification does not destroy the essential structures of the antibiotic but obstructs the interaction of the antimicrobial with its target. An example of this mechanism occurs for polycationic antibiotics such as aminoglycosides, which act between the ionic bonds of the amino and hydroxyl groups of the antibiotics and the 16S rRNA region of the A site of the bacterial ribosome, deteriorating the translation mechanism. The enzymes responsible for the modification of the aminoglycosides are the aminoglycoside phosphotransferases (APH) and nucleotidyltransferases (ANT) that modify the hydroxyl groups and the aminoglycoside acetyltransferases (AAC) which modify the amino groups, changing the size, structure, and electronic properties of the antibiotic [24].

2.2 Genetic mobile elements transfer or acquired resistance mechanisms

Once the bacterial cell acquires some degree of antibiotic resistance by an intrinsic mechanism that implies DNA modification, it can transfer the gene or genes encoding for the resistance marker to the offspring (vertical transfer) or to a different specie or genus (horizontal transfer) [25]. The gene resistance can be acquired by genetic mobile elements such as plasmids, transposons, or integrons [19].

The vertical transfer or vertical evolution occurs when a spontaneous mutation in the bacterial chromosome confers resistance to some members of the bacterial population. Once the resistance genes have arisen, they are transferred to the progeny of the bacteria during DNA replication [6]. When the bacterial genes that confer resistance to antibiotics are mobile, because they are contained within plasmids or are flanked by sequences recognized by some DNA transposition enzymes, they can be transferred between bacteria of a different taxonomic and ecological group. Some genetic mobile elements are plasmids, transposons, integrases, and genetic cassettes; in general this mechanism is called horizontal transfer gene [19, 26].

2.2.1 Plasmids

In bacterial cells, there are circular portions of extrachromosomal DNA that improve the survival characteristics of bacteria. This genetic information could be dispensable when it is no longer necessary to contend with the specific stress to which it imparts protection. Plasmids are self-replicating, given that they do so independently of chromosomal DNA replication. When plasmids contain antibiotic resistance genes, they are called plasmids R, and they can be transferred between bacteria of the same or different genera. Plasmids can be transferred to another bacterial cell by mechanisms called transformation or conjugation [19]. Transformation involves the acquisition of free DNA available in the medium. For this process the recipient bacteria must be in a competitive state; and the translocated DNA must be stabilized, either by integration into the host receptor genome or by recircularization (in the case of plasmid DNA) [27]. In contrast, conjugation involves the transfer of DNA through a multistep process that requires cell-to-cell contact, via cell surface pili or adhesins [28]. The conjugative machinery is encoded by genes in plasmids or by integrative conjugative elements in the chromosome [29].

2.2.2 Transposons

These are already known as jumping genes. These are short chains of DNA that jump from chromosome to plasmid or vice versa. DNA transfer can occur between bacterial chromosome, plasmids, and bacteriophages. The most salient feature of transposons is that DNA acquired is easily integrated into the host chromosome or plasmids. Unlike plasmids, jumping genes are not self-replicating, and they must be kept within a self-replicating structure to replicate them [19].

2.2.3 Integrons and cassette system

Both of them provide a simple mechanism for the acquisition of new genes. The DNA acquisition implies a single event of a site-specific recombination that causes the integration or removal of a single gene or a group of antibiotic resistance genes called cassette [30]. The integrons have certain components which allow a site-specific recombination system that recognizes and captures mobile genes. An integron includes a gene that codes for an integrase (Intl) and a site of specific recombination (*attI*) [31]. The sequences of the integrase enzymes allow integron classification in different classes (I–III) [32].

Genetic cassettes are small mobile elements that include a short sequence of 57 to 141 bp that are a specific recombination site. Cassettes can exist as free circular DNA molecules, and frequently they do not contain a promoter [30]. The lack of the promoter and the recombination sites make the recognition of the cassettes by Intl and Int13 and also by integrases encoded in the integrons possible. The integration of the cassettes to the integron structure allows the cassette genes' transcription from the characteristic integron promoter called Pant [30].

The dangerous feature of the transfer of antibiotic-resistant genes by transposons, integrons, or cassettes is the possibility that the bacterial receptor could acquire several classes of antibiotic resistance genes in a single event. A summary of the resistance to the different antibiotic classes, obtained by the intrinsic and acquired mechanisms, can be found in **Table 1**.

Antibiotic class/ antibiotics	Mode of action	Mechanism of resistance
Beta-lactam [33] *Penicillins** Penicillin G and V Cloxacillin Ampicillin Carbenicillin *Cephalosporins** Cephaloridine Cephalexin Cefuroxime Moxalactam Ceftiofur Cefoperazone Cefepime *Inhibitors* *Beta-lactamase** Clavulanate Sulbactam Tazobactam *Carbapenems** Imipenem/cilastatin Aztreonam	Act as suicide substrates for penicillin-binding proteins (PBP) (transpeptidases) They inhibit cell wall biosynthesis, specifically the peptidoglycan structure	*Acquired* • Plasmids • Transposons • Integrons (*bla*TEM-1, *bla*NDM-1, *bla*KPC, *bla SHV*, *blaCTX-M*, *AmpC*, *bla*VIM, *bla*OXA, and *blaIMI* genes) [34] *Intrinsic* • Bacterial flow pumps (RND, ABC, and 233 transporter)

Antibiotic class/ antibiotics	Mode of action	Mechanism of resistance
Monobactams [33, 35] *Aminoglycosides** Streptomycin Kanamycin Neomycin Gentamicin Spectinomycin	Inhibits the synthesis of proteins by binding to the ribosomal 30S subunit	*Intrinsic* • Bacterial flow pumps (MexXY and ABC transporter) • Enzymatic modification (*bla* KPC gene) • Ribosomal point mutation (*rrs* gene)
Diaminopyrimidines [33] Trimethoprim	They inhibit DNA replication by binding to dihydrofolate reductase, an enzyme involved in the metabolism of folic acid	*Acquired* • Transposon Tn7 (*dhfrI* gene) [36] *Intrinsic* • Competitive inhibition of folic acid synthesis
Phenicols [33] Chloramphenicol Thiamphenicol	They bind to the peptidyl transferase (PTC) center of the 50S ribosomal subunit to inhibit the translation elongation stage	*Intrinsic* • Target modification (*cfr* gen) • Bacterial flow pumps of amphenicols (Cml transporter)
Fluoroquinolones [37] Enrofloxacin Danofloxacin Marbofloxacin	They inhibit DNA synthesis by topoisomerases II and IV	*Intrinsic* • Target modification (*gyrA* and *parC* genes) • Bacterial flow pumps of amphenicols (AcrA transporter)
Glycopeptides [33] Vancomycin Teicoplanin Streptogramins Virginamycin	They inhibit the cell wall biosynthesis in Gram-positive bacteria. They block the binding of the substrate and the transglycosylases	*Intrinsic* • Bacterial flow pumps of amphenicols (AcrF transporter) *Acquired* • Transposon Tn1546 (*van* gene) [38]
Lincosamides [33] Lincomycin Clindamycin Pirlimycin **Macrolides** [33] Erythromycin Oleandomycin Tylosin Spiramycin Tilmicosin	They prevent protein elongation during translation by causing premature dissociation of the tRNA, inhibiting the 50S ribosomal subunit	*Intrinsic* • Ribosomal modification by methylation or mutation (*erm* and *msr* genes) [39] • Bacterial flow pumps (ABC and MFS transporter) • Drug inactivation (*Lnu* and *Mph* genes) [39]
Nitroimidazoles [40] Metronidazole	They inhibit the synthesis of DNA by oxidation. The nitro group is reduced to toxic radical species	*Intrinsic* • Bacterial flow pumps (RND and BME transporter) • Reductive activation by altering • The metabolism of pyruvate (PFOR) *Acquired* • Chromosomal mutations or plasmids acquired (*nim* gen)

Antibiotic class/ antibiotics	Mode of action	Mechanism of resistance
Peptides [41] Polymyxin B Colistin	They displace the Mg^{+2} and Ca^{+2} ions and interact electrostatically with the lipopolysaccharides (LPS) of the external Gram-negative cell membranes	*Intrinsic* • Reduction of specific proteins of the membrane and LPS • Lipid modifications
Rifamycins [42] Rifampicin	They stop transcription by interacting with the β subunit of RNA polymerase (RNAP)	*Intrinsic* • Point mutations in the rifampicin-binding region of the β subunit of RNAP (*rpoB* gene) • Bacterial flow pumps (VceB and Acr transporter) [33]
Sulfonamides [36] Sulfanilamide Sulfadiazine Sulfatiazole	They act as competitive inhibitors of DHPS; they block the folate biosynthesis in the bacterial cell	*Acquired* • Integrons (*sul1* gene) • Plasmids (*IncQ* class: *sul2* gene)
Tetracyclines [43] Doxycycline Minocycline Oxytetracycline	They block the access of the tRNA to the ribosome by binding to the 30S ribosomal subunit	*Intrinsic* • Bacterial flow pumps (SMR, RND, or ABC transporter) • Ribosomal modification • Enzyme inhibition (coded by different classes of *tet*, *otr*, *and tcr* genes)

**The subclasses of the class of beta-lactam*

Table 1.
Modes of action to different classes of antibiotics and their mechanisms of resistance.

3. Overview of resistant pathogens isolated from food

Fruits, vegetables, and foods from animal origin can be contaminated with antibiotic-resistant bacteria at any time in the food chain FAO [2]. There has been an increase in drug resistance in pathogens isolated from food for human consumption since 2000. *Salmonella enterica* and *Escherichia coli* isolates have been considered among the most important pathogens, because they can make zoonotic transfer of resistant genes [44]. However other pathogens, such as *Vibrio* spp., some species of *Aeromonas*, spores of *Clostridium botulinum* type F, or enteric bacteria such as *Campylobacter*, have been linked to gastrointestinal diseases in humans who have consumed foods of animal and marine origin. It has been reported that multidrug-resistant plasmids are easily transferred to *Aeromonas salmonicida* by *E. coli* [45, 46].

Salmonella is a pathogenic bacterium that cause a gastrointestinal disease called salmonellosis. In Latin America, Asia, and Africa, 200–500 cases of salmonellosis per 100,000 inhabitants per year have been documented, where the 95% of the infections come from the consumption of contaminated foods [47]. Worldwide, it was estimated that the infections caused by *Salmonella enterica* are above 93.8 million cases with 155,000 deaths per year [48].

Salmonella enterica is one of the most frequently isolated foodborne pathogens from different kinds of food. In the United States, between 11 and 20% of strains isolated from animals destined to human consumption were resistant to more than five different antibiotics [49]. Other studies mention that 82% of the isolates in strains from food are resistant to at least one antibiotic, associated with high resistance levels to tetracycline, streptomycin, sulfamethoxazole, and ampicillin [49]. In Latin American countries, the average of *Salmonella* resistant isolates is dependent on the region and analyzed food. In Brazil in a study conducted in a salami processing line, a 3.7% resistance to 1 antibiotic and 11.1% resistance to 3 or more antibiotics out of a total of 54 isolates have been reported [50]. In contrast, in a study conducted in pork carcasses, 147 out of 155 *Salmonella* strains isolated (94.85%) were resistant to at least one or more antibiotics [51].

Escherichia coli is one of the most widespread microorganisms in nature, and it is a member of the normal intestinal flora of many organisms, including humans. In a study conducted in Havana, Cuba, 74 *E. coli*-resistant strains were isolated from foods involved in foodborne diseases (ETA). Among foods with the highest CFU of *E. coli* serogroups identified, there were soy yogurt (14.3% of isolates), pork steak (11.9%), chicken hash (11.9%), cheese (9.5%), ham (9.5%), and beef hash (7.1%). Resistance to ampicillin was present in 36.4% of the isolates, and some isolates were also resistant to streptomycin, sulfamethoxazole, and tetracycline [52]. In America, Eastern Mediterranean, Africa, Southeast Asia, and the Western Pacific regions, an increased resistance to third-generation cephalosporins and fluoroquinolones into *E. coli* isolates has been reported [53].

On the other hand, studies conducted in dairy products have shown that 73.3% (33/45) of the *E. coli* strains isolated were susceptible to all antibiotics tested and 24.4% (11/45) showed resistance to ampicillin. The phylogenetic analysis of the *E. coli* isolates resulted in grouping into two phylogroups, A and B1, which have a higher frequency of resistance genes than those that were grouped in B2 and D. It is worth to notice that *E. coli* isolates in this study that belonged to phylogroup A and B1 were commensal strains with few or no virulence factors [54]. These results suggest that the food chain is the vehicle for the transfer of resistant genes, and it has been suggested that *E. coli* strains present in food are the original carrier of many mechanisms of antibiotic resistance in the intestinal microbiota of humans.

Studies conducted in different food classes have isolated other bacterial genera different to *Salmonella enterica* and *E. coli*. One of the most studied is *Staphylococcus aureus*, which causes staphylococcal poisoning. Strains of *S. aureus* have been studied in the last decades because they show resistance to methicillin. The analyses done in 282 *S. aureus* strains isolated from food and manipulators showed that 56.1% of the strains were resistant to one or more antimicrobials [55].

Another bacterial genus of health importance is *Mycobacterium bovis*. In the United States, outbreaks by *M. bovis* have been associated with the consumption of contaminated food. In 2007, 203 samples of cheese imported from Mexico were collected at the California customs office. Of the samples collected, 4.9% tested positive for *Mycobacterium* genus, with drug susceptibility test to streptomycin, isoniazid, rifampicin, ethambutol, and pyrazinamide, showing that they were susceptible to all the antibiotics tested except pyrazinamide [56]. In contrast, in Japan, 58 *M. bovis* strain isolates from dairy cattle reported 7 strains resistant to the fluoroquinolones enrofloxacin, orbifloxacin, and danofloxacin. The fluoroquinolone resistance was associated with the mutation to quinolone resistance-determining regions of *gyrA* and *parC* genes (QRDR). The strains that showed no fluoroquinolone resistance phenotype did not present mutations [57].

Listeria, *Shigella*, and *Campylobacter* are other bacterial genera that have been isolated from foods and have shown antibiotic resistance. *Listeria monocytogenes* strains isolated from cheese have shown resistance to streptomycin, kanamycin, cephalothin, and tetracycline [58]. The analysis of 152 *Shigella* strains isolated from various foods that caused outbreaks of shigellosis in Brazil showed that several strains were resistant to streptomycin (88.6%), followed by ampicillin (84.6%) and sulfamethoxazole/trimethoprim (80.5%). The resistant strains were grouped into 73 patterns, where pattern A (resistance to ampicillin, sulfamethoxazole/trimethoprim, tetracycline, streptomycin, and chloramphenicol and intermediate resistance to kanamycin) grouped the highest number of isolates ($n = 36$) [59]. In Malaysia, *Campylobacter* spp. was reported with a prevalence of 17.4%, from a total of 340 cattle samples. *Campylobacter* isolates showed resistant to tetracycline (76.9%) and ampicillin (69.2%), while resistance to chloramphenicol was low (7.6%) [60].

Even in farms of goldfish (*Carassius auratus*), 70 strains of bacterial genera such as *Aeromonas hydrophila*, *Vibrio fluvialis*, and *V. furnissii* have been identified, with 45% of the isolates being resistant to 6 of the 14 antibiotics tested; 100% of the strains were resistant to cephalothin, 94% to ampicillin, 89% to chloramphenicol, 88% to tetracycline, 85.3% to nitrofurantoin, 61.3% to carbenicillin, and 65.3% to kanamycin. Twenty three percent of the isolates presented sensitivity to amikacin, trimethoprim, cefotaxime, netilmicin, pefloxacin, and gentamicin. Only one strain, *A. hydrophila*, showed resistance to all antibiotics tested. Twenty strains generated resistance to 7 different antibiotics, and 67 of the 70 strains generated resistance to more than 1 antibiotic [46].

As we can see in this overview, the resistance of the different bacterial genera isolated in a great diversity of foods is alarming, since many of these bacterial genera are the cause of many foodborne diseases. The diseases produced by these resistant pathogenic bacteria are difficult to treat, being able to provoke death in some patients.

4. Economic implications/economic impact of the resistant pathogens in food

As we previously mentioned, the resistance occurs when the antibiotics used for the control of bacterial diseases are no longer optimal for their elimination. The most common routes to get infected with pathogenic bacteria are air, direct contact with sick people, or consumption of contaminated water or food. Pathogenic bacteria can be spread through sick people and contaminated fruits, vegetables, or animals that are intended for consumption. Antimicrobial resistance is a risk factor and complication of the disease, being difficult to treat infections, and it could be eventually lead to death [61]. In 2019, 700,000 deaths worldwide can be attributed to antimicrobial resistance, and the figure would rise to 35 million in 35 years, due to the lack of treatments to cure diseases caused by resistant pathogens; the estimated cost for the treatment of these persons will be 100 billion dollars [62, 63].

Morbidity and mortality increase when the administration of effective treatments to counteract infections caused by resistant pathogens is delayed. The duration of the disease and hospitalization of patients with infections by resistant pathogens have an economic impact, since there are extra procedures for the treatment of the disease, the antibiotics that could be administered usually are more expensive than the ones used as first line, and also there are long hospitalization stays. The economic impact for the patient is due to the loss of productivity for taking care of themselves or a family member [61]. It is highlighted that 63.5% of infections are acquired in hospitals and that the groups with the highest incidence are under 1 year or over 65 years old [62, 63].

In Europe and the United States, more than 50,000 people die every year from infections with drug-resistant pathogens, while in India it is estimated that close to

60,000 newborns die due to resistant infections. There are at least 700,000 deaths every year caused by resistant microorganism that generate diseases such as bacterial infections, malaria, HIV/AIDS, or tuberculosis [64]. The Centers for Disease Control and Prevention estimated that 2 million patients will be treated each year for resistant bacteria, of which 23,000 die [65]. In United States, it was estimated that there are an average of 1400 sick people with infection caused by resistant microorganisms, with a medical cost per patient estimated in $18,588 to $29,069 US dollars with a mortality rate of 6.5%. In the European Union, the cost for loss of productivity due to an illness originated by resistant bacteria is estimated in 1.5 billion € per year [66].

It is evident that the problem of bacterial resistance has reached great impact not only in the health of the population but also on its economy. For this reason, it is necessary to undertake actions that help in stopping the acquisition of the genetic elements that promote antibiotic resistance. Without doubt, the implementation of government laws that avoid the excessive use of antibiotics in livestock and fish farming could help to hinder the problem. Also the use of alternative molecules to antibiotics for the prevention of diseases in animals and improvement of the hygiene and vaccination measures in the farming collection and food processes would help to stop the problem of bacterial resistance. It has been proven that countries that have implemented control measures in the use of antibiotics in animals for human consumption and their products reduce up to 39% resistant bacteria. Not less important is the implementation of control measures in hospitals and clinics as well as generation of awareness in the population to avoid the overprescription of antibiotics, elements that all together can make a difference.

Conflict of interest

The authors do not have any conflict of interest.

Author details

Lilia M. Mancilla-Becerra, Teresa Lías-Macías, Cristina L. Ramírez-Jiménez and Jeannette Barba León*
Public Health Department, Biological and Agricultural Sciences Center, University of Guadalajara, Zapopan, Jalisco, Mexico

*Address all correspondence to: jeannette.barba@academicos.udg.mx

References

[1] Cires Pujol M. La resistencia a los antimicrobianos, un problema mundial. Revista Cubana de Medicina General Integral. 2002;**18**(2):165-168

[2] FAO. Food and Agriculture Organization of the United Nations and WHO World Health Organization. FAO/WHO expert meeting on foodborne antimicrobial resistance: Role of environment, crops and biocides [Internet]. 2018. Available from: https://www.who.int/foodsafety/areas_work/antimicrobial-resistance/FAO_WHO_AMR_Summary_Report_June2018.pdf [Accessed: 04 June 2019]

[3] FAO Food and Agriculture Organization of the United Nations. Antimicrobial resistence: What you need to know [Internet]. 2017. Available from: http://www.fao.org/fao-stories/article/en/c/1056781/ [Accessed: 04 June 2019]

[4] Liliam CJ, Andrés MRL, Liliam HCM. Principios generales de la terapéutica antimicrobiana. Acta Medica. 1998;**8**(1):13-27

[5] Pérez RD. Resistencia bacteriana a antimicrobianos: su importancia en la toma de decisiones en la práctica diaria. Información terapéutica del sistema nacional de salud. 1998;**22**(3):57-67

[6] Van Wyk H. Antibiotic resistance. A Pharmaceutical Journal. 2015;**82**(3):20-23

[7] Cox G, Wright GD. Intrinsic antibiotic resistance: Mechanisms, origins, challenges and solutions. International Journal of Medical Microbiology. 2013;**303**(6-7):287-292. DOI: 10.1016/j.ijmm.2013.02.009

[8] Schaffer C, Messner P. The structure of secondary cell wall polymers: How Gram-positive bacteria stick their cell walls together. Microbiology-Sgm. 2005;**151**:643-651. DOI: 10.1099/mic.0.27749-0

[9] Scherrer R, Gerhardt P. Molecular sieving by the *Bacillus megaterium* cell wall and protoplast. Journal of Bacteriology. 1971;**107**(3):718-735

[10] Obst S, Kastowsky M, Bradaczek H. Molecular dynamics simulations of six different fully hydrated monomeric conformers of *Escherichia coli* re-lipopolysaccharide in the presence and absence of Ca^{2+}. Biophysical Journal. 1997;**72**(3):1031-1046. DOI: 10.1016/s0006-3495(97)78755-1

[11] Delcour AH. Outer membrane permeability and antibiotic resistance. Biochimica et Biophysica Acta. 2009;**1794**(5):808-816. DOI: 10.1016/j.bbapap.2008.11.005

[12] Olaitan AO, Morand S, Rolain JM. Mechanisms of polymyxin resistance: Acquired and intrinsic resistance in bacteria. Frontiers in Microbiology. 2014;**5**:643. DOI: 10.3389/fmicb.2014.00643

[13] Ruiz N, Montero T, Hernandez-Borrell J, Viñas M. The role of *Serratia marcescens* porins in antibiotic resistance. Microbial Drug Resistance-Mechanisms Epidemiology and Disease. 2003;**9**(3):257-264. DOI: 10.1089/107662903322286463

[14] Blair JMA, Webber MA, Baylay AJ, Ogbolu DO, Piddock LJV. Molecular mechanisms of antibiotic resistance. Nature Reviews Microbiology. 2015;**13**(1):42-51. DOI: 10.1038/nrmicro3380

[15] Mahamoud A, Chevalier J, Alibert-Franco S, Kern WV, Pages JM. Antibiotic efflux pumps in Gram-negative bacteria: The

inhibitor response strategy. Journal of Antimicrobial Chemotherapy. 2007;**59**(6):1223-1229. DOI: 10.1093/jac/dkl493

[16] Billal DS, Feng J, Leprohon P, Legare D, Ouellette M. Whole genome analysis of linezolid resistance in *Streptococcus pneumoniae* reveals resistance and compensatory mutations. BMC Genomics. 2011;**12**:512. DOI: 10.1186/1471-2164-12-512

[17] Kumar N, Radhakrishnan A, Wright CC, Chou TH, Lei HT, Bolla JR, et al. Crystal structure of the transcriptional regulator Rv1219c of *Mycobacterium tuberculosis*. Protein Science. 2014;**23**(4):423-432. DOI: 10.1002/pro.2424

[18] Hooper DC, Jacoby GA. Mechanisms of drug resistance: Quinolone resistance. Annals of the New York Academy of Sciences. 2015;**1354**:12-31. DOI: 10.1111/nyas.12830

[19] Errecalde JO. Uso de antimicrobianos en animales de consumo [Internet]. 2004. Available from: http://www.fao.org/3/a-y5468s.pdf [Accessed: 05 June 2019]

[20] Sauvage E, Terrak M. Glycosyltransferases and transpeptidases/penicillin-binding proteins: Valuable targets for new antibacterials. Antibiotics (Basel). 2016;**5**(1):1-27. DOI: 10.3390/antibiotics5010012

[21] Bush K, Bradford PA. Beta-lactams and beta-lactamase inhibitors: An overview. Cold Spring Harbor Perspectives in Medicine. 2016;**6**(8). DOI: 10.1101/cshperspect.a025247

[22] DrugBank database. Telithromycin [Internet]. 2019. Available from: https://www.drugbank.ca/drugs/DB00976 [Accessed: 15 July 2019]

[23] Gomes C, Martinez-Puchol S, Palma N, Horna G, Ruiz-Roldan L, Pons MJ, et al. Macrolide resistance mechanisms in Enterobacteriaceae: Focus on azithromycin. Critical Reviews in Microbiology. 2017;**43**(1):1-30. DOI: 10.3109/1040841X.2015.1136261

[24] D'Costa V, Wright GD. Biochemical logic of antibiotic inactivation and modification. In: Mayers DL, editor. Antimicrobial Drug Resistance. Infectious Disease. New York, USA: Humana Press; 2009. pp. 81-95. DOI: 10.1007/978-1-59745-180-2_8

[25] Hermsen R, Deris JB, Hwa T. On the rapidity of antibiotic resistance evolution facilitated by a concentration gradient. Proceedings of the National Academy of Sciences of the United States of America. 2012;**109**(27):10775-10780. DOI: 10.1073/pnas.1117716109

[26] Jain R, Rivera MC, Lake JA. Horizontal gene transfer among genomes: The complexity hypothesis. Proceedings of the National Academy of Sciences of the United States of America. 1999;**96**(7):3801-3806. DOI: 10.1073/pnas.96.7.3801

[27] Thomas CM, Nielsen KM. Mechanisms of, and barriers to, horizontal gene transfer between bacteria. Nature Reviews Microbiology. 2005;**3**(9):711-721. DOI: 10.1038/nrmicro1234

[28] Wozniak RAF, Waldor MK. Integrative and conjugative elements: Mosaic mobile genetic elements enabling dynamic lateral gene flow. Nature Reviews Microbiology. 2010;**8**(8):552-563. DOI: 10.1038/nrmicro2382

[29] Smillie C, Garcillán-Barcia MP, Francia MV, Rocha EPC, de la Cruz F. Mobility of plasmids. Microbiology and Molecular Biology Reviews. 2010;**74**(3):434-452. DOI: 10.1128/mmbr.00020-10

[30] Hall RM, Collis CM. Antibiotic resistance in gram-negative bacteria: The role of gene cassettes and integrons. Drug Resistance Updates. 1998;**1**(2):109-119. DOI: 10.1016/S1368-7646(98)80026-5

[31] Hall RM, Brookes DE, Stokes HW. Site-specific insertion of genes into integrons role of the 59-base element and determination of the recombination cross-over point. Molecular Microbiology. 1991;**5**(8):1941-1959. DOI: 10.1111/j.1365-2958.1991.tb00817.x

[32] Kaushik M, Kumar S, Kapoor RK, Virdi JS, Gulati P. Integrons in Enterobacteriaceae: Diversity, distribution and epidemiology. International Journal of Antimicrobial Agents. 2018;**51**(2):167-176. DOI: 10.1016/j.ijantimicag.2017.10.004

[33] Fair RJ, Tor Y. Antibiotics and bacterial resistance in the 21st century. Perspectives in Medicinal Chemistry. 2014;**6**:25-64. DOI: 10.4137/PMC.S14459

[34] López-Velandia DP, Torres-Caycedo MI, Prada-Quiroga CF. Genes de resistencia en bacilos Gram negativos: Impacto en la salud pública en Colombia. La Revista Universidad y Salud. 2016;**18**(1):190-202

[35] Palomino J, Pachon J. Aminoglycosides. Enfermedades Infecciosas y Microbiologia Clinica. 2003;**21**(2):105-115. DOI: 10.1157/13042869

[36] Huovinen P, Sundström L, Swedberg G, Sköld O. Trimethoprim and sulfonamide resistance. Antimicrobial Agents and Chemotherapy. 1995;**39**(2):279-289. DOI: 10.1128/aac.39.2.279

[37] Drlica K, Malik M. Fluoroquinolones: Action and resistance. Current Topics in Medicinal Chemistry. 2003;**3**(3):249-282. DOI: 10.2174/1568026033452537

[38] Alekshun MN, Levy SB. Molecular mechanisms of antibacterial multidrug resistance. Cell. 2007;**128**(6):1037-1050. DOI: 10.1016/j.cell.2007.03.004

[39] Leclercq R. Mechanisms of resistance to macrolides and lincosamides: Nature of the resistance elements and their clinical implications. Clinical Infectious Diseases. 2002;**34**(4):482-492. DOI: 10.1086/324626

[40] Dingsdag SA, Hunter N. Metronidazole: An update on metabolism, structure-cytotoxicity and resistance mechanisms. Journal of Antimicrobial Chemotherapy. 2018;**73**(2):265-279. DOI: 10.1093/jac/dkx351

[41] Falagas ME, Kasiakou SK. Colistin: the revival of polymyxins for the management of multidrug-resistant gram-negative bacterial infections (2005;**40**:1333). Clinical Infectious Diseases. 2006;**42**(12):1819-1819

[42] Goldstein BP. Resistance to rifampicin: A review. Journal of Antibiotics. 2014;**67**(9):625-630. DOI: 10.1038/ja.2014.107

[43] Roberts MC. Update on acquired tetracycline resistance genes. FEMS Microbiology Letters. 2005;**245**(2):195-203. DOI: 10.1016/j.femsle.2005.02.034

[44] Cantón R, Novais A, Valverde A, Machado E, Peixe L, Baquero F, et al. Prevalence and spread of extended-spectrum β-lactamase-producing Enterobacteriaceae in Europe. European Society of Clinical Microbiology and Infectious Diseases. 2007;**14**(1):144-153. DOI: 10.1111/j.1469-0691.2007.01850.x

[45] Kumar D, Taneja N, Singh Gill H, Kumar R. *Vibrio cholerae* O1 Ogawa serotype outbreak in a village of Ambala district in Haryana, India. Indian Journal of Community Medicine. 2011;**36**(1):66-68

[46] Negrete Redondo P, Romero Jarero J, FJL A. Anatibiotic resistance and presence of plasmids in: *Aeromonas hydrophila*, *Vibrio fluvialis*, and *Vibrio furnissii* isolated from *Carassius auratus auratus*. Veterinaria México. 2004;**35**(1):21-30

[47] Campioni F, Moratto Bergamini AM, Falcao JP. Genetic diversity, virulence genes and antimicrobial resistance of *Salmonella* Enteritidis isolated from food and humans over a 24-year period in Brazil. Food Microbiology. 2012;**32**(2):254-264. DOI: 10.1016/j.fm.2012.06.008

[48] Hendriksen RS, Vieira AR, Karlsmose S, Lo Fo Wong DM, Jensen AB, Wegener HC, et al. Global monitoring of Salmonella serovar distribution from the World Health Organization Global Foodborne Infections Network Country Data Bank: Results of quality assured laboratories from 2001 to 2007. Foodborne Pathogens and Disease. 2011;**8**(8): 887-900. DOI: 10.1089 / fpd.2010.0787

[49] Puig-Peña Y, Leyva-Castillo V, Martino-Zagovalov TK. Estudio de susceptibilidad antimicrobiana en cepas de *Salmonella* sp aisladas de alimentos. Revista Habanera de Ciencias Médicas. 2008;**7**(2):1-9

[50] Ribeiro VB, AndrighetoI C, BersotII LS, BarcellosII V, ReisIII EF, Destro MT. Serological and genetic diversity amongst *Salmonella* strains isolated in a salami processing line. Brazilian Journal of Microbiology. 2007;**38**(1):178-182. DOI: 10.1590/S1517-83822007000100036

[51] Bermúdez-D PM, Rincón-G SM, Suárez-A MC. Evaluation of antimicrobial susceptibility of *Salmonella* spp. strains isolated from pork carcasses on Colombia. Revista Facultad Nacional de Salud Pública. 2013;**32**(1):88-94

[52] Puig-Peña Y, Leyva-Castillo V, Apórtela-López N, Campos-González N, Frerer-Marquez Y, Soto-Rodríguez P. Serogrupos y resistencia antimicrobiana de cepas de *Escherichia coli* aisladas en alimentos procedentes de brotes de enfermedades diarreicas. Revista Cubana de Alimentación y Nutrición. 2014;**24**(2):161-171

[53] WHO. World Health Organizarion. WHO's first global report on antibiotic resistance reveals serious, worldwide threat to public health [Internet]. 2014. Available from: https://www.who.int/mediacentre/news/releases/2014/amr-report/en/ [Accessed: 06 June 2019]

[54] Guillén L, Millán B, Araque M. Caracterización molecular de cepas de *Escherichia coli* aisladas de productos lácteos artesanales elaborados en Mérida, Venezuela. Infection. 2014;**18**(3):100-108. DOI: 10.1016/j.infect.2014.04.004

[55] Puig-Peña Y, Espino-Hernández M, Leyva-Castillo V, Apórtela-López N, Pérez-Muñoz Y, Soto-Rodríguez P. Resistencia antimicrobiana en cepas de estafilococos coagulasa positiva aisladas en alimentos y manipuladores. Revista Cubana de Alimentación y Nutrición. 2015;**25**(2):245-260

[56] Harris NB, Payeur J, Bravo D, Osorio R, Stuber T, Farrell D, et al. Recovery of *Mycobacterium bovis* from soft fresh cheese originating in Mexico. Applied and Environmental Microbiology. 2007;**73**(3):1025-1028. DOI: 10.1128/AEM.01956-06

[57] Sato T, Okubo T, Usui M, Higuchi H, Tamura Y. Amino acid substitutions in GyrA and ParC are associated with fluoroquinolone resistance in *Mycoplasma bovis* isolates from Japanese dairy calves. Journal of Veterinary Medical Science. 2013;**75**(8):1063-1065. DOI: 10.1292/jvms.12-0508

[58] Carolina C, Laura AM. Caracterización de cepas de *Listeria monocytogenes* realizados a partir de queso fresco proveniente de diferentes zonas productoras costarricenses. Archivos Latinoamericanos de Nutrición. 2009;**59**(1):66-70

[59] Daniel de Paula CM, Passos-Geimba M, Heidrich do Amaral P, Cesar Tondo E. Antimicrobial resistance and PCR-ribotyping of *Shigella* responsible for foodborne outbreaks occurred in southern Brazil. Brazilian Journal of Microbiology. 2010;**41**(4):966-977. DOI: 10.1590/S1517-83822010000400015

[60] Premarathne J, Anuar AS, Thung TY, Satharasinghe DA, Jambari NN, Abdul-Mutalib NA, et al. Prevalence and antibiotic resistance against tetracycline in *Campylobacter jejuni* and *C. coli* in cattle and beef meat from Selangor, Malaysia. Frontiers in Microbiology. 2017;**8**:2254. DOI: 10.3389/fmicb.2017.02254

[61] WHO. World Health Organization. Antibiotic resistance [Internet]. 2018. Available from: https://www.who.int/news-room/fact-sheets/detail/antibiotic-resistance [Accessed: 05 June 2019]

[62] Londoño-Restrepo J. Factores de riesgo asociados a infecciones por bacterias multirresistentes derivadas de la atención en salud en una institución hospitalaria de la ciudad de Medellín. Infection. 2015;**20**(2):77-83. DOI: 10.1016/j.infect.2015.09.002

[63] Red Nacional de Vigilancia Epidemiológica (RENAVE). Protocolo de vigilancia y control de microorganismos multirresistentes o de especial relevancia clínico-epidemiológica (ProtocoloMMR). Madrid, España: RENAVE; [Internet]. 2016. Available from: http://www.comunidad.madrid/sites/default/files/doc/sanidad/epid/iras_protocolo-mmr.pdf. [Accessed: 16 August 2019]

[64] O'Neill J. Tackling drug-resistant infections globally: Final report and recommendations [Internet] United Kingdom: Wellcome Trust. 2006. Available from: https://amr-review.org/sites/default/files/160518_Finalpaper_with cover.pdf. [Accessed: 06 June 2019]

[65] CDC. Centers for Disease Control and Prevention. Antibiotic resistance threats in the united states [Internet]. 2013. Available from: https://www.cdc.gov/drugresistance/pdf/ar-threats-2013-508.pdf [Accessed: 06 June 2019]

[66] IS Global Instituto de Salud Global Barcelona. Resistencia a los antibióticos: cuando el problema va más allá de las patentes [Internet]. 2017. Available from: https://www.isglobal.org/documents/10179/5808947/Informe+Resistencia+Antimicrobiana+ES/a74ac65e-7d4b-4f18-8c3b-ec86778034ee [Accessed: 06 June 2019]

Section 4

Diagnosis and Antimicrobial Control

Chapter 11

Urine Tests for Diagnosis of Infectious Diseases and Antibiotic-Resistant Pathogens

Nahla O. Eltai, Hashim Alhussain, Sanjay Doiphode, Asma Al Thani and Hadi Yassine

Abstract

The relation between disease and urine was recognized by physicians since the earliest civilization BC. Urine is considered an ideal diagnostic specimen for its noninvasive and easy method of collection. Urinalysis encompasses a wide range of tests, which includes a variety of chemical tests, urine microscopy, bacterial cultures, and molecular tests. Importantly, urine tests can diagnose patients with antibiotic-resistant urinary tract infections (UTI), directly from urine and/or bacterial culture. This chapter summarizes the most common urine tests in the infectious disease field, with a special focus on diagnosing UTI and characterizing their antibiotic resistant. In addition to describing the advantages and limitation of these tests, the chapter explores the promising emerging technologies and methods in this field. This chapter is beneficial for scientists and healthcare workers in the field.

Keywords: urine, infectious diseases, urinalysis, bacteria, antibiotic resistance

1. Introduction

Urinalysis has been a useful diagnostic tool since thousands of years. Although urine was the first body fluid to be examined by mankind for the diagnosis of diseases [1], it is still one of the most common specimens used in clinical and diagnostic laboratories. Urine samples have been used for the diagnosis of a wide and diverse range of disorders, including but not limited to renal diseases [2, 3], metabolic disorders [4], cancer [5], infectious diseases [6], and others [7–10].

In infectious disease field, urine tests are applied in diagnosing urinary tract infections (UTI) [11–13]. Further, several other infections can be diagnosed by urine tests at different levels [14] including community-acquired pneumonia (CAP) [15], legionellosis [16], tuberculosis [17, 18], congenital cytomegalovirus (CMV) infection [19], and dengue virus [20, 21], and recently, several papers suggest the high value of urinalysis in the detection of Zika virus [22, 23]. Parasites can also be diagnosed from the urine by detection of urinary egg, for example, diagnosis of *Schistosoma haematobium* (*S. haematobium*) [24]. Furthermore, urine has been used for screening of different sexually transmitted diseases (STD) such as *Neisseria gonorrhoeae* and *Chlamydia* sp. [25, 26]. The sensitivity and accuracy of urine test vary according to the agent being detected [14], and in some cases, urinalysis is only performed to exclude other diseases [27].

Many of the aforementioned infections are treated with antimicrobial drugs [28–30], a discovery of the past century that completely changed the medical field and saved millions of lives [31]. Unfortunately, this discovery did not last unchallenged for long; soon after the discovery of penicillin by Sir Alexander Fleming in 1928 [32], the problem of penicillin resistance first emerged in 1947—19 years after its discovery and 4 years after the dug started being mass-produced and was used heavily to treat allied troops fighting in Europe during the World War II. Ever since, antimicrobial resistance (AMR) has become a fierce challenge endangering the existence of many antimicrobial agents [33].

With the emergence of pathogenic strains resistant to almost all available antimicrobial drugs [34, 35] and with only few new drugs in the development and production pipeline [36], AMR is now one of the most urgent global health threats [33]. This emphasizes on the importance of urine analysis and detection of antimicrobial resistance in the diagnostic laboratories.

Treatment of UTI is a good example of AMR impact on the medical field. Many of the antibiotics prescribed traditionally for the treatment of UTIs are now compromised and to a large extent are ineffective [37]. More alarming, recent years have recorded the emergence of bacterial strains that are resistant to even last resort antibiotics such as colistin, making the treatment of UTIs a global challenge [38].

2. Urine tests for infectious disease diagnostics and treatment

A urine specimen is one of the most frequent specimens examined in many of the clinical- and hospital-based laboratories. Urine cultures account for up to 40% of those laboratories' cultures, making it the most common type of culture in such laboratories [27]. Urine is considered an ideal diagnostic specimen for its noninvasive, easy method of collection and the sufficient amount in which it is excreted [14].

The most common infections diagnosed by urinalysis are UTIs, which are one of the most common bacterial infections that require medical intervention. Several other infections such as community-acquired pneumonia and viremia infections can also be diagnosed with the help of urinalysis.

2.1 Urinalysis for diagnosis of urinary tract infection

UTI affects about 150 million around the globe every year [28]. Urinary tract infections are common in women and to less extent in children. Many women experience multiple infections during their lifetimes. Risk factors specific to women for UTIs include female anatomy. A woman has a shorter urethra than a man does, which shortens the distance that bacteria must travel to reach the bladder [39]. UTIs are generally categorized clinically as complicated and uncomplicated based on the presence of risk factors that comprise the urinary tract or the host defense [40]. Both Gram-negative and Gram-positive bacteria can cause UTIs, with *Escherichia coli* representing about 40–70% of the cases [27].

The diagnosis of UTIs is mainly based on urinalysis and the medical history of the patient, with latter being the most essential element [27].

2.1.1 Sample collection

Avoiding contamination is a key element when collecting samples for diagnostic purposes. Bacterial contamination of the urine during collection is always a concern, and midstream urine is recommended for UTI. Clear instruction to the patient is indicated to reduce the risk of contamination with the use of clean containers.

Although methods such as suprapubic aspiration and straight catheter technique can reduce or even eliminate the chances of contamination, they are rarely performed as they are not practical for routine cases [27]. Importantly, these methods are invasive, costly, take a lot of time and effort, and are not risk free. Contamination and infection caused by the catheter used to collect the sample is one example [41]. One of the most common method used for the collection of urine is the clean-catch midstream technique [42], which is simple, noninvasive, and risk free and most importantly gives accurate enough results to be used for routine testing. Nonetheless, contamination of the sample is the main disadvantage of this method [43].

2.1.2 Sample processing

The clinical information obtained from a urine specimen is influenced by the collection method, timing, and handling. An enormous variety of collection and transport containers for urine specimens are available, depending on the type of laboratory test ordered. National Committee for Clinical Laboratory Standards (NCCLS) recommended testing urine sample within 2 hours of collection to avoid false-positive results; however refrigeration or chemical preservation of urine specimens may be utilized if testing or refrigeration within a two-hour window is not possible. A variety of urine preservatives (tartaric and boric acids being the most common) are available that allow urine to be kept at room temperature while still providing results comparable to those of refrigerated urine. Generally, the length of preservation capacity ranges from 24 to 72 hours. Metabolites which can be significantly influenced by the interaction of exposure time and temperature include arginine, glutamine, methionine, phenylalanine, and others, while metabolites which can be significantly influenced by freeze and thaw cycles are the C3 family and histones H1 [44].

2.1.3 Urine microscopy

Bacteria can be simply observed in urine specimens under the microscope, especially after Gram staining. After centrifugation of urine samples, a small amount of the pellet is applied to a glass microscopic slide and stained with the usual Gram-staining protocol. Gram staining can also be done to uncentrifuged specimen; however, there are no definitive criteria to determine positive results with this method, and it is not sensitive for detection of low number of bacteria.

Although Gram-staining test can give relatively fast results about the nature of the causative agent; however it is not practical for routine use, it is labor intensive, insensitive test for concentration of bacteria lower than 10^5 cfu/mL, and time-consuming, making it unsuitable for the patient with uncomplicated UTIs [45].

2.1.4 Urine nitrite test

Enterobacteriaceae are the main causative agent of UTI. They typically produce nitrite, thus making this bacterium chemically detectable. The urine sample for this test should be taken from the first urine produced in the morning, as a minimum of 4 hours are required for the bacteria to produce a detectable amount nitrite. Unfortunately, other bacteria such *as Staphylococcus saprophyticus* cannot produce nitrite, introducing limitation for this test [46].

2.1.5 Pyuria

The presence of pus in the urine (i.e., pyuria) can be detected by various methods. The best and most accurate method is to microscopically measure the urinary leukocyte excretion rate. Other microscopic methods may also include counting the leukocyte in the urine. However, as the microscopic method is unpractical for routine use, other easier methods such as leukocyte esterase tests for detection of pyuria can be performed. The leukocyte esterase tests have many disadvantages as it produces false-positive results due the presence of eosinophils in the urine. Decreased positive results or false-negative results of this test are referred to other reasons including elevated level of glucose and protein in the urine or if the patient is treated with certain drugs such as cephalexin or tetracycline [47]. The commercial products are believed to be more efficient, although they have low sensitivity, but they are highly specific, and they provide information about both pyuria and bacteriuria [27].

2.1.6 Urine culture

Urine culture is the gold standard in diagnosing UTI. It is crucial not only for the diagnosis but also to guide appropriate antimicrobial prescription and treatment. Patients with complicated UTIs, those who have suffered from recurrent UTIs, or those who are not responding to the empirical treatment are the ones usually subjected for urine culture. In this regard, the most common used culture media are the blood agar and MacConkey's agar. This is especially true for specimen from outpatients, knowing that almost all UTIs in outpatients are caused by aerobic and facultative Gram-negative bacteria, making it unnecessary to use a medium that is selective for Gram-positive bacteria. However, for the hospitalized patients, inoculation of Gram-positive bacteria especially cocci should be considered as enterococci is one of the most common causative agents of UTIs in inpatients. Thus, media routinely used should support growth of both Gram-negative and Gram-positive bacteria.

On the other hand, anaerobic bacteria are rarely a cause of UTIs, and cultures of anaerobic bacteria are usually indicated only for the patient with increased risk of infection with anaerobic bacteria, and those are usually patients with anatomical abnormalities.

For the diagnosis of Candiduria, blood agar which is used for routine bacterial culture can be used perfectly for its detection and other funguria.

2.2 Urinalysis for diagnosis of other infectious diseases

Many systemic infections other than UTIs can be diagnosed utilizing urine samples. This is applied for viral and bacterial infections. Some of the viruses are directly shed in urine such as human polyomaviruses and congenital cytomegalovirus. Other infections can be detected by markers and antigen secretion in the urine. With the rapid development in diagnostic technologies, urine can be utilized to diagnose even a larger number of infectious agents.

2.2.1 Streptococcus pneumoniae

Streptococcus pneumoniae is the number one causative agent of community-acquired pneumonia both in adult and children. In addition, it is underdiagnosed because of the lack of reliable and sensitive diagnostic method. CAP can be diagnosed using various samples including blood, sputum, and urine. Recently, multiple publications [48–50] provided evidence showing that urinalysis and urine specimen can be a very helpful in the diagnosis of CAP with relatively highly sensitive results. Urine immunoassay was used by reference laboratories to determine the course of a complicated outbreak of *S. pneumoniae* complicated by influenza A; this clearly indicates the importance of urine as a diagnostic specimen for the detection of S. *pneumonia* [51].

2.2.2 Legionellosis

Legionella pneumophila is the most common cause of the life-threatening atypical pneumonia known as legionellosis or Legionnaires' disease. Rapid urinary antigen detection kits are the primary choice for the diagnosis of legionellosis. It is considered to be a reliable diagnostic method for the detection of legionellosis with acceptable sensitivity. New tests and assays such as *Legionella* fluorescence immunoassay have been developed. And they seem promising with papers showing a higher sensitivity results [52].

2.2.3 Tuberculosis

Tuberculosis is a worldwide health issue. Many factors make tuberculosis hard to control, and one of them is the lack of fast and accurate diagnostic tools. With lipo-arabinomannan (cell wall glycolipid of *Mycobacterium tuberculosis*) being secreted in the urine, several assays and tools have been developed to detect this marker of infection. However, no urine test until now is sensitive enough to be adopted for routine use [17, 18].

2.2.4 Human polyomaviruses

Infections with polyomaviruses with clinical significance occur generally only in immunocompromised patients; the virus is shed in urine in large quantities. The best way to detect the virus is by electronic microscopy, which is highly sensitive, although it is less sensitive than PCR; however, it might be more reliable clinically. That is because a large portion of the adult population are exposed to the virus, and PCR can give positive results to clinically insignificant cases [53].

2.2.5 Congenital cytomegalovirus

Congenital cytomegalovirus is the leading cause of neurological impairment and nongenetic sensorineural hearing loss. The virus can be cultured from urine and diagnosis can be made from various types of specimen which include urine, blood, and saliva [54]. PCR both quantitative and qualitative is widely used for diagnosis of CMV infection. Qualitative PCR test is intended to detect CMV DNA in urine, whereas quantitative PCR test is performed to detect quantitatively CMV DNA in urine specimens as an aid in identifying or management of CMV infections.

2.2.6 Dengue virus

Dengue virus is a mosquito-borne disease affecting more than 50 million people worldwide yearly. Urine specimen can be used for the early detection of this virus, although RT-PCR and ELISA along with other new methods all of which utilizing blood specimen are usually the way to detect dengue virus.

2.2.7 Zika virus

Zika virus is another mosquito-borne pathogen, and it is endemic to Africa and Southeast Asia. The detection of Zika virus can be achieved by ELISA, but it is usually detected by reverse transcription PCR (RT-PCR) from a serum sample. Some evidence shows that the virus can be detected from mother urine sample even after 10 days of the onset of the disease, which is not feasible with serum samples.

This suggests that detection of Zika virus by real-time RT-PCR from urine specimen can be a valuable diagnostic tool [22].

2.2.8 Sexually transmitted disease

Urine specimen is of valuable importance in the diagnosis of sexually transmitted diseases. Urinalysis can help in diagnosis of *Mycoplasma genitalium*, *Chlamydia*, *Neisseria gonorrhoeae*, *Trichomonas vaginalis*, and urethritis to name but a few. For example, leukocyte esterase dipstick test as a point-of-care diagnostic tool for urogenital *Chlamydia*, can be a valuable tool to accurately exclude *Chlamydia*. However, it shows low positive predictive value. Furthermore *Chlamydia trachomatis*, *Mycoplasma genitalium*, and *Neisseria gonorrhea* can be detected with PCR and real-time duplex PCR from urine samples [55–58]. Using urine as a specimen for diagnosis of Gonococcal infections by molecular tests has eased out uncomfortable sample collection by urethral swab and hence increased the number of patients volunteering to give specimen for this disease of public health importance.

2.2.9 Parasite detection in urine

Urine microscopy and sediment test analysis help in the diagnosis of urinary parasites like *Schistosoma urinary egg detection.* However, egg detection method is below optimum and requires multiple samples; other methods for the detection of parasite in urine are being investigated. PCR shows promising results in this regards, and many papers are advocating for the clinical establishment of this method for detection of parasites such as *S. haematobium*, *Leishmania infantum*, *Trypanosoma* sp., and others [59–64].

2.3 New technologies and urinalysis

New technological advances have paved the way for significant progress in automated urinalysis [65]. Time and accuracy are the two key factors for diagnosis. In UTIs, urine dipsticks are very fast and easy to use, but it lacks the accuracy, whereas in the other hand, urine culture for antimicrobial susceptibility testing shows clinically reliable and accurate results, but it takes up to 3 days to give results. Many novel and improved diagnostic technologies and tools are introduced in the market, and some of them are already approved for clinical use and helped significantly in increasing the accuracy and decreasing the time of the test; a good example would be nucleic acid tests and mass spectrometry. Some other technologies show promising future such as the utilization of smartphone for urinalysis [65–68].

2.3.1 Flow cytometry

Recently flow cytometry is being introduced as a reliable method for fast diagnosis of UTIs by counting the bacteria in the urine specimen. With the improved counting precision over visual counting methods, highly accurate positive results can be obtained by this method. Detection of bacteriuria can be achieved with clinical standards using flow cytometry technology [69, 70].

2.3.2 Test strip technology

Major improvement in the test strip technology has been made in recent years. Not only highly sensitive test strips are being introduced, but also now, one can find strips, which give quantitative results for urinary proteins. The financial aspect is also of great importance, especially in the third world and developing countries; inexpensive test strips for various diagnostic reasons such as the diagnosis of diabetes form urine sample are available [71, 72]. Test strip method also shows promising result in antibiotic susceptibility tests; if optimum diagnostic requirement is reached, it can reduce the test time significantly from 2 to 3 days to few hours [73–75].

2.3.3 Automated microscopy

Urine microscopy is one of the most important diagnostic methods for UTIs and other kidney diseases. Manual microscopy is time-consuming and can be labor extensive. Furthermore, with centrifugation decantation and re-suspension always lead to cell loss and cellular lysis. With the current available digital microscopy technologies, a significant time reduction can be archived with much more sample being processed in significantly short time in comparison to manual microscopy. In addition, with the ability to process uncentrifuged urine sample, issues like cell loss and lysis are of no more concern. Many automated analyzers are now available in the market with different kinds of technologies such as laminar flow digital imaging technology and pattern recognition technology [65, 76, 77].

2.3.4 MALDI-TOF

The proteomic method "Matrix-assisted laser desorption ionization–time-of-flight mass spectrometry (MALDI-TOF MS)" for identification of microorganisms directly from culture coupled with Gram stain has given new direction, saved considerable amount of time in diagnosis of UTI, and contributed greatly in the field of clinical microbiology in general. It can identify different pathogens accurately and significantly in short time. The utilization of this technology for the diagnosis of UTIs and furthermore in preforming antibiotic susceptibility tests to decrease the testing time from days to as fast as 2 hours can open wide doors [78, 79].

2.3.5 Urinalysis and smartphones

Smartphone technologies improved the quality of life in countless fronts, and it has large potential for applications in the medical field. With point-of-care testing attracting much attention in recent years, smartphone solutions can be a valuable tool in this regard. It can, for example, increase the compliance of populations with screening programs by offering easy and fast screening method [80]. Studies exploring the possibility of establishing a smartphone-based diagnostic platform for rapid detection of Zika, chikungunya, and dengue viruses showed valuable prospective [20]. Several other smartphone applications utilizing urinalysis for various diagnostic reasons had been tested, and it shows promising future prospective which can greatly help both medical practitioners and patients alike [67, 81, 82].

3. Urine specimen and antibiotic-resistant pathogens

The emergence of the antimicrobial resistant pathogen is worldwide issue threatening thousands if not millions of lives every year and with more and more strains developing not only a single drug resistance but also multidrug resistance (MDR) making the treatment of the disease much more complicated. Furthermore, many pathogens have developed resistance against a second-line or even last resort drugs. Recently the emergence of colistin-resistant strains attracted a lot of attention. Some of these resistant pathogens can cause serious illness or even death.

Some *Mycobacterium tuberculosis* strains developed what is called extensively drug-resistant (XDR), a rare form of MDR which shows resistance to at least one of the second-line drugs, isoniazid, rifampin, and fluoroquinolone [83, 84].

Enterobacteriaceae, the leading cause of UTIs, developed resistance to β-lactam antibiotics by producing β- Lactamases, rendering this class of antibiotic to a large extended ineffective [37]. As mentioned above, with the lack of new drug

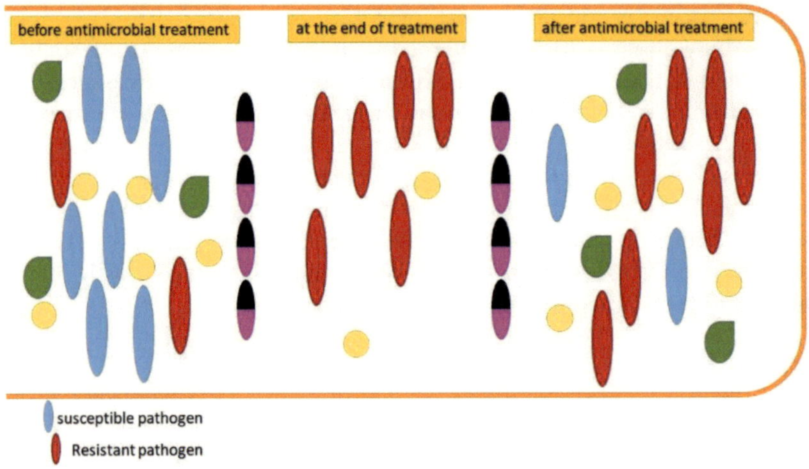

Figure 1.
Schematic demonstrating how antibiotics can contribute to the AMR crisis and how it can change the microbiome of the patient.

	Susceptible		Resistant	
	n	**Estimated proportion [% (95% CI)]***	**n**	**Estimated proportion [% (95% CI)]***
Amoxicillin	215	62.0 (55.5–68.9)	116	38.0 (31.1–44.5)
Amoxicillin/clavulanate	307	91.3 (87.9–94.6)	12	3.5 (1.5–5.5)
Cefuroxime	323	98.0 (96.4–99.7)	8	2.0 (0.3–3.6)
Cefotaxime	323	98.1 (96.5–99.7)	6	1.5 (0.1–3.0)
Ceftazidime	323	98.1 (96.5–99.7)	3	0.9 (0.0–2.1)
Carbapenems	331	100.0	0	0.0
Fosfomycin	331	100.0	0	0.0
Nitrofurantoin	328	99.6 (99.0–99.9)	3	0.4 (0.0–1.0)
Nalidixic acid	311	94.6 (92.1–97.1)	17	4.6 (0.2–7.1)
Ofloxacin	312	94.9 (92.6–97.3)	11	2.8 (1.1–4.4)
Ciprofloxacin	323	98.1 (96.5–99.7)	6	1.5 (0.1–3.0)
Aminoglycoside	327	98.7 (97.0–99.9)	4	1.3 (0.0–3.0)
Trimethoprim/sulfamethoxazole	278	81.9 (75.9–88.0)	51	17.8 (11.7–24.0)

**Estimated proportion with the sampling design and 95% CI.*
n size in the study population.

Table 1.
Resistance rates among 331 Escherichia coli *from urinary tract infection of women over 18 visiting a French GP in 2012–2013 [85].*

development and only few new drugs being in the production pipeline, UTIs with AMR are a major concern, and it is a leading cause of morbidity and a cause of significant financial loss in many countries. It is estimated that 50% of all women will suffer from UTI at some point in their lives.

The abuse and misuse of antimicrobial drugs are the leading causes of this worldwide issue. Controlling the prescription of antimicrobial drugs by practicing judicial drug prescription based on susceptibility testing is of paramount importance not only to control this fast-growing issue of AMR against currently used drugs in the market but to ensure lasting effectiveness of future treatment options and drugs. This cannot be achieved by the effort of the medical practitioner only, but it needs the active effort of policymakers, scientist, and the large community (**Figure 1**).

Urine specimen can play a major role in fighting against this crisis; urine cultures for the diagnosis of UTIs can be used for susceptibility testing, thus following antimicrobial stewardship program recommendation. Urine specimen has the potential to be used for the same reason in other infectious diseases where the pathogen can be found in urine (**Table 1**).

4. Conclusion

Over the ages, urine proved to be an extremely valuable diagnostic specimen; today, it constitutes one of the most common samples processed in clinical and diagnostic laboratories. Its role in the diagnosis of a wide and diverse range of disorders cannot be argued against, ranging from drugs test and metabolic diseases identification to the diagnosis of STDs and lethal infectious disease. Its importance in antimicrobial resistance tests is also of great value, contributing to achieving the antimicrobial stewardship program recommendations.

With the advances in today's technologies, urinalysis now has great potentials, and the merging of test strip technologies with smartphone technologies can lead to tremendous changes in healthcare system and can deeply integrate point-of-care testing into the health system. Furthermore, advances in mass spectrometry can lead to great achievement not only in the diagnostic field by providing a much faster and accurate results, but it can also contribute greatly in the medical and biomedical research fields.

Author details

Nahla O. Eltai[1*], Hashim Alhussain[1], Sanjay Doiphode[2], Asma Al Thani[1] and Hadi Yassine[1]

1 Biomedical Research Center, Qatar University, Doha, Qatar

2 Department of Laboratory Medicine and Pathology, Hamad Medical Corporation, Doha, Qatar

*Address all correspondence to: nahla.eltai@qu.edu.qa

References

[1] Eknoyan G. Looking at the urine: The renaissance of an unbroken tradition. American Journal of Kidney Diseases. 2007:**49**:865-872. DOI: 10.1053/j.ajkd.2007.04.003

[2] Cao W, Jin L, Zhou Z, et al. Overexpression of intrarenal renin-angiotensin system in human acute tubular necrosis. Kidney & Blood Pressure Research. 2016;**41**:746-765. DOI: 10.1159/000450564

[3] Puthumana J, Ariza X, Belcher JM, et al. Urine interleukin 18 and lipocalin 2 are biomarkers of acute tubular necrosis in patients with cirrhosis: A systematic review and meta-analysis. Clinical Gastroenterology and Hepatology. 2017;**15**:1003-1013.e3. DOI: 10.1016/j.cgh.2016.11.035

[4] Blau N, Burgard P. Disorders of phenylalanine and tetrahydrobiopterin metabolism. In: *Physician's Guide to the Treatment and Follow-Up of Metabolic Diseases*. 2006:3-21. DOI: 10.1007/3-540-28962-3_3

[5] Matulewicz RS, DeLancey JO, Pavey E, et al. Dipstick urinalysis as a test for microhematuria and occult bladder cancer. Bladder Cancer. 2017;**3**:45-49. DOI: 10.3233/BLC-160068

[6] Bonaldo MC, Ribeiro IP, Lima NS, et al. Isolation of infective Zika virus from urine and saliva of patients in Brazil. PLoS Neglected Tropical Diseases. 2016;**10**:e0004816. DOI: 10.1371/journal.pntd.0004816

[7] Ricci G, Majori S, Mantovani W, et al. Prevalence of alcohol and drugs in urine of patients involved in road accidents. Journal of Preventive Medicine and Hygiene. 2008;**49**:89-95

[8] Martin GW, Wilkinson DA, Kapur BM. Validation of self-reported cannabis use by urine analysis. Addictive Behaviors. 1988;**13**:147-150. DOI: 10.1016/0306-4603(88)90004-4

[9] Turner JA, Saunders K, Shortreed SM, et al. Chronic opioid therapy urine drug testing in primary care: Prevalence and predictors of aberrant results. Journal of General Internal Medicine. 2014;**29**:1663-1671. DOI: 10.1007/s11606-014-3010-y

[10] Ismail M, Baumert M, Stevenson D, et al. A diagnostic test for cocaine and benzoylecgonine in urine and oral fluid using portable mass spectrometry. Analytical Methods. 2017;**9**:1827-1966. DOI: 10.1039/c6ay02006b

[11] Van Der Zee A, Roorda L, Bosman G, et al. Molecular diagnosis of urinary tract infections by semi-quantitative detection of uropathogens in a routine clinical hospital setting. PLoS One. 2016. DOI: 10.1371/journal.pone.0150755

[12] McKibben MJ, Seed P, Ross SS, et al. Urinary tract infection and neurogenic bladder. Urologic Clinics of North America. 2015;**42**:527-536. DOI: 10.1016/j.ucl.2015.05.006

[13] Stein R, Dogan HS, Hoebeke P, et al. Urinary tract infections in children: EAU/ESPU guidelines. European Urology. 2015;**67**:546-558. DOI: 10.1016/j.eururo.2014.11.007

[14] Tuuminen T. Urine as a specimen to diagnose infections in twenty-first century: Focus on analytical accuracy. Frontiers in Immunology. 2012;**3**:45. DOI: 10.3389/fimmu.2012.00045

[15] Molinos L, Zalacain R, Menéndez R, et al. Sensitivity, specificity, and positivity predictors of the pneumococcal urinary antigen test in community-acquired pneumonia. Annals of the American Thoracic Society. 2015;**12**:1482-1489. DOI: 10.1513/AnnalsATS.201505-304OC

[16] Badoux P, Euser SM, Bruin JP, et al. Evaluation of the bioNexia legionella test, including impact of incubation time extension, for detection of *Legionella pneumophila* serogroup 1 antigen in urine. Journal of Clinical Microbiology. 2017;**55**:1733-1737. DOI: 10.1128/JCM.02448-16

[17] Paris L, Magni R, Zaidi F, et al. Urine lipoarabinomannan glycan in HIV-negative patients with pulmonary tuberculosis correlates with disease severity. Science Translational Medicine. 2017;**9**:e9912807. DOI: 10.1126/scitranslmed.aal2807

[18] Shah M, Hanrahan C, Wang ZY, et al. Lateral flow urine lipoarabinomannan assay for detecting active tuberculosis in HIV-positive adults. Cochrane Database of Systematic Reviews. 2016;**10**:CD011420. DOI: 10.1002/14651858.CD011420.pub2

[19] Ross SA, Novak Z, Pati S, et al. Overview of the diagnosis of cytomegalovirus infection. Infectious Disorders Drug Targets. 2011;**11**:466-474. DOI: 10.2174/187152611797636703

[20] Priye A, Bird SW, Light YK, et al. A smartphone-based diagnostic platform for rapid detection of Zika, chikungunya, and dengue viruses. Scientific Reports. 2017;**7**:447-478. DOI: 10.1038/srep44778

[21] Poloni TR, Oliveira AS, Alfonso HL, et al. Detection of dengue virus in saliva and urine by real time RT-PCR. Virology Journal. 2010;**7**:22. DOI: 10.1186/1743-422X-7-22

[22] Gourinat AC, O'Connor O, Calvez E, et al. Detection of Zika virus in urine. Emerging Infectious Diseases. 2015;**21**:84-86. DOI: 10.3201/eid2101.140894

[23] Bandeira AC, Campos GS, Rocha VFD, et al. Prolonged shedding of Chikungunya virus in semen and urine: A new perspective for diagnosis and implications for transmission. IDCases. 2016;**6**:100-103. DOI: 10.1016/j.idcr.2016.10.007

[24] De Dood CJ, Hoekstra PT, Mngara J, et al. Refining diagnosis of *Schistosoma haematobium* infections: Antigen and antibody detection in urine. Frontiers in Immunology. 2018;**9**:2635. DOI: 10.3389/fimmu.2018.02635

[25] Cook RL, Hutchison SL, Østergaard L, et al. Systematic review: Noninvasive testing for *Chlamydia trachomatis* and *Neisseria gonorrhoeae*. Annals of Internal Medicine. 2005;**142**:914-925. DOI: 10.7326/0003-4819-142-11-200506070-00010

[26] Van der Pol B, Ferrero DV, Buck-Barrington L, et al. Multicenter evaluation of the BDProbeTec ET system for detection of *Chlamydia trachomatis* and *Neisseria gonorrhoeae* in urine specimens, female endocervical swabs, and male urethral swabs. Journal of Clinical Microbiology. 2001;**39**:1008-1016. DOI: 10.1128/JCM.39.3.1008-1016.2001

[27] Bartlett JG. Laboratory diagnosis of urinary tract infections in adult patients. Infectious Diseases in Clinical Practice. 2004;**12**:360-361. DOI: 10.1097/01.idc.0000144910.19687.1f

[28] Flores-Mireles AL, Walker JN, Caparon M, et al. Urinary tract infections: Epidemiology, mechanisms of infection and treatment options. Nature Reviews Microbiology. 2015;**13**:269-284. DOI: 10.1038/nrmicro3432

[29] Diel R. Treatment of tuberculosis. Der Pneumologe. 2019;**16**:117-130. DOI: 10.1007/s10405-019-0234-x

[30] Uranga EA. Duration of antibiotic treatment in community-acquired pneumonia. Archivos de Bronconeumología. 2015;**51**:613-614. DOI: 10.1016/j.arbres.2015.01.003

[31] Tamma PD, Miller MA, Cosgrove SE. Rethinking how antibiotics are prescribed. JAMA. 2018;**321**:139-140. DOI: 10.1001/jama.2018.19509

[32] Tan SY, Tatsumura Y. Alexander Fleming (1881-1955): Discoverer of penicillin. Singapore Medical Journal. 2015;**51**:366-367. DOI: 10.11622/smedj.2015105

[33] Levy SB, Bonnie M. Antibacterial resistance worldwide: Causes, challenges and responses. Nature Medicine. 2004;**10**:s122-s129. DOI: 10.1038/nm1145

[34] Falagas ME, Karageorgopoulos DE. Pandrug resistance (PDR), extensive drug resistance (XDR), and multidrug resistance (MDR) among gram-negative bacilli: Need for international harmonization in terminology. Clinical Infectious Diseases. 2008;**46**:1121-1122. DOI: 10.1086/528867

[35] Cegielski JP, Dalton T, Yagui M, et al. Extensive drug resistance acquired during treatment of multidrug-resistant tuberculosis. Clinical Infectious Diseases. 2014;**59**:1049-1063. DOI: 10.1093/cid/ciu572

[36] Kola I. The state of innovation in drug development. Clinical Pharmacology and Therapeutics. 2008;**83**:227-230. DOI: 10.1038/sj.clpt.6100479

[37] Eltai NO, Al Thani AA, Al-Ansari K, et al. Molecular characterization of extended spectrum β -lactamases enterobacteriaceae causing lower urinary tract infection among pediatric population. Antimicrobial Resistance and Infection Control. 2018;7:90. DOI: 10.1186/s13756-018-0381-6

[38] Schwarz S, Johnson AP. Transferable resistance to colistin: A new but old threat. The Journal of Antimicrobial Chemotherapy. 2016;**71**:2066-2070. DOI: 10.1093/jac/dkw274

[39] Hooton TM, Scholes D, Hughes JP, et al. A prospective study of risk factors for symptomatic urinary tract infection in young women. The New England Journal of Medicine. 2002;**335**:468-474. DOI: 10.1056/nejm199608153350703

[40] Schmiemann G, Kniehl E, Gebhardt K, et al. The diagnosis of urinary tract infection: A systematic review. Deutsches Ärzteblatt International. 2010;**107**:361-367. DOI: 10.3238/arztebl.2010.0361

[41] Tenke P, Mezei T, Bőde I, et al. Catheter-associated urinary tract infections. European Urology Supplements. 2017;**16**:138-143. DOI: 10.1016/j.eursup.2016.10.001

[42] Maher PJ, Brown AEC, Gatewood MOK. The effect of written posted instructions on collection of clean-catch urine specimens in the emergency department. The Journal of Emergency Medicine. 2017;**52**:639-644. DOI: 10.1016/j.jemermed.2016.10.010

[43] Selek MB, Bektöre B, Sezer O, et al. Genital region cleansing wipes: Effects on urine culture contamination. Journal of Infection in Developing Countries. 2017;**11**:102-105. DOI: 10.3855/jidc.8910

[44] Rotter M, Brandmaier S, Prehn C, et al. Stability of targeted metabolite profiles of urine samples under different storage conditions. Metabolomics. 2017;**13**:4. DOI: 10.1007/s11306-016-1137-z

[45] Jenkins RD, Fenn JP, Matsen JM. Review of urine microscopy for Bacteriuria. The Journal of the American Medical Association. 1986;**255**:3397-403. DOI: 10.1001/jama.1986.03370240067039

[46] Mambatta A, Rashme V, Menon S, et al. Reliability of dipstick assay in predicting urinary tract infection. Journal of Family Medicine and Primary Care. 2015;**4**:265-268. DOI: 10.4103/2249-4863.154672

[47] Roberts KB. The diagnosis of UTI: Concentrating on Pyuria. Pediatrics. 2016;**138**:e20162877. DOI: 10.1542/peds.2016-2877

[48] Elberse K, van Mens S, Cremers AJ, et al. Detection and serotyping of pneumococci in community acquired pneumonia patients without culture using blood and urine samples. BMC Infectious Diseases. 2015;**15**:56. DOI: 10.1186/s12879-015-0788-0

[49] Saukkoriipi A, Pascal T, Palmu AA. Evaluation of the BinaxNOW® *Streptococcus pneumoniae* antigen test on fresh, frozen and concentrated urine samples in elderly patients with and without community-acquired pneumonia. Journal of Microbiological Methods. 2016;**121**:24-26. DOI: 10.1016/j.mimet.2015.12.007

[50] Marimuthu S, Summersgill JT, Ghosh K, et al. Preliminary evaluation of an IytA PCR assay for detection of *Streptococcus pneumoniae* in urine specimens from hospitalized patients with community-acquired pneumonia. Journal of Respiratory Infections. 2017;**1**:5. DOI: 10.18297/jri/vol1/iss4/5

[51] Harrison TG, Clark J, Fry NK, et al. Use of a serotype-specific urine immunoassay to determine the course of a hospital outbreak of *Streptococcus pneumoniae* complicated by influenza a. JMM Case Reports. 2016;**3**:e005002. DOI: 10.1099/jmmcr.0.005002

[52] Avni T, Bieber A, Green H, et al. Diagnostic accuracy of PCR alone and compared to urinary antigen testing for detection of *Legionella* spp.: A systematic review. Journal of Clinical Microbiology. 2016;**54**:401-411. DOI: 10.1128/JCM.02675-15

[53] Goetsch HE, Zhao L, Gnegy M, et al. Fate of the urinary tract virus BK human polyomavirus in source-separated urine. Applied and Environmental Microbiology. 2018;**84**:e02374-17. DOI: 10.1128/AEM.02374-17

[54] Ross SA, Ahmed A, Palmer AL, et al. Urine collection method for the diagnosis of congenital cytomegalovirus infection. The Pediatric Infectious Disease Journal. 2015;**34**:903-905. DOI: 10.1097/INF.0000000000000757

[55] Rahimkhani M, Mordadi A, Gilanpour M. Detection of urinary *Chlamydia trachomatis*, *Mycoplasma genitalium* and human papilloma virus in the first trimester of pregnancy by PCR method. Annals of Clinical Microbiology and Antimicrobials. 2018;**17**:25. DOI: 10.1186/s12941-018-0276-7

[56] Venter JME, Mahlangu PM, Müller EE, et al. Comparison of an in-house real-time duplex PCR assay with commercial HOLOGIC® APTIMA assays for the detection of *Neisseria gonorrhoeae* and *Chlamydia trachomatis* in urine and extra-genital specimens. BMC Infectious Diseases. 2019;**19**:6. DOI: 10.1186/s12879-018-3629-0

[57] Bartelsman M, de Vries HJC, Schim van der Loeff MF, et al. Leucocyte esterase dip-stick test as a point-of-care diagnostic for urogenital chlamydia in male patients: A multi-center evaluation in two STI outpatient clinics in Paramaribo and Amsterdam. BMC Infectious Diseases. 2016;**16**:625. DOI: 10.1186/s12879-016-1946-8

[58] Hakre S, Casimier RO, Danboise BA, et al. Enhanced sexually transmitted infection screening for *Mycoplasma genitalium* in human immunodeficiency virus-infected US air force personnel. Clinical Infectious Diseases. 2017;**65**:1585-1588. DOI: 10.1093/cid/cix555

[59] Pessoa-e-Silva R, Mendonça Trajano-Silva LA, Lopes da Silva MA, et al. Evaluation of urine for *Leishmania infantum* DNA detection by

real-time quantitative PCR. Journal of Microbiological Methods. 2016;**131**:34-41. DOI: 10.1016/j.mimet.2016.10.002

[60] Weerakoon KG, McManus DP. Cell-free DNA as a diagnostic tool for human parasitic infections. Trends in Parasitology. 2016;**32**:378-391. DOI: 10.1016/j.pt.2016.01.006

[61] Knopp S, Corstjens PLAM, Koukounari A, et al. Sensitivity and specificity of a urine circulating anodic antigen test for the diagnosis of *Schistosoma haematobium* in low endemic settings. PLoS Neglected Tropical Diseases. 2015;**9**:e0003752. DOI: 10.1371/journal.pntd.0003752

[62] Lodh N, Mikita K, Bosompem KM, et al. Point of care diagnosis of multiple schistosome parasites: Species-specific DNA detection in urine by loop-mediated isothermal amplification (LAMP). Acta Tropica. 2017;**173**:125-129. DOI: 10.1016/j.actatropica.2017.06.015

[63] Lodh N, Caro R, Sofer S, et al. Diagnosis of *Strongyloides stercoralis*: Detection of parasite-derived DNA in urine. Acta Tropica. 2016;**136**:9-13. DOI: 10.1016/j.actatropica.2016.07.014

[64] Nwakanma DC, Gomez-Escobar N, Walther M, et al. Quantitative detection of plasmodium falciparum DNA in saliva, blood, and urine. The Journal of Infectious Diseases. 2009;**199**:1567-1574. DOI: 10.1086/598856

[65] Oyaert M, Delanghe J. Progress in automated urinalysis. Annals of Laboratory Medicine. 2018;**39**:15-22. DOI: 10.3343/alm.2019.39.1.15

[66] Katayama M, Mori T, Hasegawa S, et al. New technology meets clinical knowledge: Diagnosing *Streptococcus suis* meningitis in a 67-year-old man. IDCases. 2018;**12**:119-120. DOI: 10.1016/j.idcr.2018.04.001

[67] Wirth M, Biswas N, Ahmad S, et al. A prospective observational pilot study to test the feasibility of a smartphone enabled uChek© urinalysis device to detect biomarkers in urine indicative of preeclampsia/eclampsia. Health Technology. 2019;**9**:31-36. DOI: 10.1007/s12553-018-0248-0

[68] Bakan E, Bayraktutan Z, Baygutalp NK, et al. Evaluation of the analytical performances of cobas 6500 and sysmex UN series automated urinalysis systems with manual microscopic particle counting. Biochemia Medica. 2018;**28**:020712. DOI: 10.11613/BM.2018.020712

[69] Moshaver B, de Boer F, van Egmond-Kreileman H, et al. Fast and accurate prediction of positive and negative urine cultures by flow cytometry. BMC Infectious Diseases. 2016;**16**:211. DOI: 10.1186/s12879-016-1557-4

[70] Cho SY, Hur M. Advances in automated urinalysis systems, flow cytometry and digitized microscopy. Annals of Laboratory Medicine. 2018;**39**:1-2. DOI: 10.3343/alm.2019.39.1.1

[71] Arnett N, Vergani A, Winkler A, et al. Inexpensive urinalysis test strips to screen for diabetes in developing countries. In: *GHTC 2016 - IEEE Global Humanitarian Technology Conference: Technology for the Benefit of Humanity, Conference Proceedings*. 2016;**13**:549-556. DOI: 10.1109/GHTC.2016.7857339

[72] Delanghe JR, Himpe J, De Cock N, et al. Sensitive albuminuria analysis using dye-binding based test strips. Clinica Chimica Acta. 2017;**471**:107-112. DOI: 10.1016/j.cca.2017.05.032

[73] Kapur S, Gupta S. Indigenous rapid diagnostic technology for antibiotic susceptibility testing in urinary tract infection: From bench side to bedside. BMJ Innovations. 2017;**93**:519-521. DOI: 10.1136/bmjinnov-2015-000111

[74] Van Den Bijllaardt W, Schijffelen MJ, Bosboom RW, et al. Susceptibility of ESBL *Escherichia coli* and *Klebsiella pneumoniae* to fosfomycin in the Netherlands and comparison of several testing methods including Etest, MIC test strip, Vitek2, Phoenix and disc diffusion. The Journal of Antimicrobial Chemotherapy. 2018;**73**:2380-2387. DOI: 10.1093/jac/dky214

[75] Shah PJ, Hanson C, Larson P, et al. Carbapenem-resistant *Escherichia coli* resistant to ceftazidime-avibactam: The need for a commercially available testing product. American Journal of Health-System Pharmacy. 2016;**73**:1029-1030. DOI: 10.2146/ajhp160041

[76] Ince FD, Ellidağ HY, Koseoğlu M, et al. The comparison of automated urine analyzers with manual microscopic examination for urinalysis automated urine analyzers and manual urinalysis. Practical Laboratory Medicine. 2016;**5**:14-20. DOI: 10.1016/j.plabm.2016.03.002

[77] Becker GJ, Garigali G, Fogazzi GB. Advances in urine microscopy. American Journal of Kidney Diseases. 2016;**67**:954-964. DOI: 10.1053/j.ajkd.2015.11.011

[78] Veron L, Mailler S, Girard V, et al. Rapid urine preparation prior to identification of uropathogens by MALDI-TOF MS. European Journal of Clinical Microbiology & Infectious Diseases. 2015;**34**:1787-1795. DOI: 10.1007/s10096-015-2413-y

[79] Cobo F. Use of MALDI-TOF techniques in the diagnosis of urinary tract pathogens. In: The Use of Mass Spectrometry Technology (MALDI-TOF) in Clinical Microbiology. 1st edition. 2018. 298 p. DOI: 10.1016/b978-0-12-814451-0.00010-1

[80] Leddy J, Green JA, Yule C, et al. Improving proteinuria screening with mailed smartphone urinalysis testing in previously unscreened patients with hypertension: A randomized controlled trial. BMC Nephrology. 2019;**20**:132. DOI: 10.1186/s12882-019-1324-z

[81] Ra M, Muhammad MS, Lim C, et al. Smartphone-based point-of-care urinalysis under variable illumination. IEEE Journal of Translational Engineering in Health and Medicine. 2018;**6**:2800111. DOI: 10.1109/JTEHM.2017.2765631

[82] Akraa S, Guo F, Shen H, et al. On the Feasibility of a Smartphone-Based Solution to Rapid Quantitative Urinalysis Using Nanomaterial Bioprobes. In: 14th EAI International Conference on Mobile and Ubiquitous Systems: Computing, Networking and Services. 2017:**10**:523-524. DOI: 10.1145/3144457.3144508

[83] Coll F, Phelan J, Hill-Cawthorne GA, et al. Genome-wide analysis of multi- and extensively drug-resistant *Mycobacterium tuberculosis*. Nature Genetics. 2018;**50**:307-316. DOI: 10.1038/s41588-017-0029-0

[84] Dheda K, Gumbo T, Maartens G, et al. The epidemiology, pathogenesis, transmission, diagnosis, and management of multidrug-resistant, extensively drug-resistant, and incurable tuberculosis. The Lancet Respiratory Medicine. 2017;**5**:291-360. DOI: 10.1016/S2213-2600(17)30079-6

[85] Rossignol L, Vaux S, Maugat S, et al. Incidence of urinary tract infections and antibiotic resistance in the outpatient setting: A cross-sectional study. Infection. 2017;**45**:33-40. DOI: 10.1007/s15010-016-0910-2

Chapter 12

Silver Nanoparticles Offer Effective Control of Pathogenic Bacteria in a Wide Range of Food Products

Graciela Dolores Avila-Quezada and Gerardo Pavel Espino-Solis

Abstract

According to the Food and Agriculture Organization (FAO), food wastage still causes massive economic loss. A major role in this loss is played by the activities of microbial organisms. Treatments such as heat and irradiation can reduce microorganisms in fruits and vegetables and hence reduce postharvest loss. However, some of these treatments can injure the fruit. Effective chemical treatments against bacterial infestations can result in resistance. A more recent method is the use of silver nanoparticles. These can act in a number of ways including at cellular level by inhibiting the cell wall synthesis, by binding to the surface of the cell membrane and by interposing between the DNA base pairs, and by inhibiting biofilm formation, affecting the thiol group of enzymes, affecting bacterial peptides and hence interfering with cell signaling and attaching to the 30S ribosome subunit. A ground-breaking way to survey the effects of the silver nanoparticles on bacterial populations is by flow cytometry. It allows measurement of many characteristics of single cells, including their functional characteristics such as viability and cell cycle. Bacterial viability assays are used with great efficiency to evaluate antibacterial activity by evaluating the physical rupture of the membrane of the bacteria.

Keywords: prevention of postharvest food losses, FAO, fruit pathogens

1. Introduction

1.1 Postharvest pathogens of fruit

Postharvest spoilage of fruits can be caused by a large number of bacterial species. Some of the most important are *Enterobacter cloacae, Erwinia herbicola, Lelliottia amnigena, Pantoea ananatis, Pantoea agglomerans, Pantoea allii, Enterobacter aerogenes, Pseudomonas fluorescens* and *Streptomyces* sp. [1–6]. A wide range of fungal species is similarly involved [2, 7–9].

If adequate postharvest handling and storage practices are not employed, postharvest decays of fruit and vegetables can cause losses of 50% or more [7].

The main triggers for invasion by microorganisms are physiological changes that activate ethylene synthesis or that cause changes to the cuticle or cell walls (loosening), or declines in natural antifungal compounds or high contents of carbohydrates and other nutrients and water. These changes usually occur naturally during ripening [10–12].

Postharvest contamination of fruit by human pathogens can be another key issue in the supply chain. The most commonly reported human pathogen contaminants causing disease outbreaks are bacteria such as *Escherichia coli* (*E. coli*), *Salmonella* spp., *Mycobacterium* spp., *Brucella* spp. and *Pseudomonas aeruginosa* (*P. aeruginosa*). However, good manufacturing and handling practices can significantly reduce these contaminations [13, 14].

Because of the behavior of microbial populations, including fungi and bacteria, an initial infection may originate new infection foci that appear near the primary one, so increasing disease incidence and/or severity [15, 16]. Quality deterioration and loss of fresh fruit and vegetables during storage have an exceptionally high economic impact because by this stage high costs have been incurred in harvesting, grading, packaging, freighting and storage. All these reasons emphasize the importance of defining new practices to reduce populations of the postharvest microorganisms.

2. Silver nanoparticles for pathogen control

Silver nanoparticles (AgNPs) offer oligodynamic action which is also of low toxicity and broad spectrum [17–19]. Moreover, compared with synthetic biocides, there is also only a low chance that microbial resistance might develop. These AgNPs have been exploited against Gram-negative bacteria, such as *Acinetobacter, Escherichia, Pseudomonas, Salmonella* and *Vibrio,* and against Gram-positive bacteria including *Bacillus, Clostridium, Enterococcus, Listeria, Staphylococcus* and *Streptococcus* [20]. A number of research reports have demonstrated that their antimicrobial nature depends on the surface-capping agent and the size and shape of the nanoparticle [21, 22].

The effectiveness of AgNPs also depends on bacterial dose [23]. Silver nanoparticles affect the growth of bacteria in a dose-dependent manner. In a study conducted by Agnihotri et al. [23], concentrations of 10 and 20 μg/ml Ag (10 nm) caused reductions of ~18 and ~53% in *E. coli*, respectively. Meanwhile, AgNP concentrations at 30 and 40 μg/ml eliminated all bacterial growth.

Silver nanoparticles smaller than 100 nm, and containing between 10,000 and 15,000 silver atoms, are effective as antibacterial agents [20]. The AgNPs' antibacterial potential increases as size decreases. This effect is more pronounced for AgNPs of size <10 nm, because contact with the bacterial cell is direct [24].

Research into the antimicrobial activity of AgNPs against Gram-positive and Gram-negative bacteria shows Gram-negative bacteria are more sensitive to AgNPs than Gram-positive ones [23, 25], although their relative sensitivity cannot be explained based only on a difference in the composition of the cell membrane.

In studies using discs impregnated with AgNP in culture media with bacteria, the formation of a clear zone of inhibition around the impregnated discs is an indicator of bactericidal potential of AgNP > 15 nm [21]. Bacteria are unable to survive in this area, possibly because of the release of silver in the form of nanoparticles or of silver ions.

In addition, nanoparticle silver can be released by the mobility of small size AgNPs through the semisolid agar, whereby a zone of inhibition is observed.

In a previous study conducted by Biao et al. [21], chitosan was combined with silver nanoparticles to form composites. They found that chitosan-silver colloid has a high inhibition ratio against the prokaryotes *E. coli* and *Staphylococcus aureus* (*S. aureus*) and the eukaryote *Candida albicans* (*C. albicans*). They concluded that the chitosan-silver colloid had a broad spectrum of antimicrobial activity.

3. Some mechanisms of bactericidal action of silver nanoparticles (AgNPs)

3.1 Electrostatic attraction

A way to transport active silver cations to the bacteria can occur on the cell membrane or within the cell. When combined with protonated chitosan, the positively charged AgNPs bind well to the negatively charged bacterial membrane proteins through electrostatic attraction [23].

3.2 Alterations in the bacterial membrane

The first bacterial contact with AgNP can trigger an antibacterial mechanism by facilitating the entry of AgNPs into the bacterial cells. This is followed by an explosive release of silver ions inside the bacterial cells causing the bactericidal effect.

The nature of the AgNP, bacteria interaction and its antibacterial effect have been analyzed by a number of methods. Bacteria exposed to AgNPs show high protein leakage and morphological changes [26]. As an example, *E. coli* treated with AgNPs (~10 nm) appeared to shrink and develop an irregular shape. Micrographs show AgNPs on the cell membrane attached to the lipopolysaccharide layer of the cell wall, and a proportion of AgNPs were found inside the bacterial cell [23].

Biao et al. [21] noticed that bacterial strains have intact membranes and smooth surfaces in the absence of silver colloid, whereas after exposure to chitosan-silver colloid, the cell membrane and surface become shriveled, invaginated and disrupted. This cell membrane damage indicates the mode of action of chitosan-silver colloid. Its bactericidal effect is attributed to the release of silver cation from AgNPs and to alteration of the bacterial cell wall structure and associated physicochemical changes.

Osmoregulation of the bacterial cell can also be affected causing extrusion of intracellular material and hence cell death. The deformed or wrinkled cell wall can also cause leakage of cytoplasmic contents.

In addition, AgNPs can penetrate bacterial membranes, facilitating internalization. The rupture of perforation of the cell wall is an evidence of internalization of AgNP and of uncontrolled transport through the cytoplasm resulting in cell death [27] (**Figure 1**).

3.3 Silver nanoparticles internalization: effects on DNA

Multiple pathways of AgNP can occur after internalization. Silver atoms in nanoparticles are characterized by a high affinity with sulfur and phosphorus-containing compounds such as DNA. In this way, they readily combine with cell constituents and so destroy the cell.

Silver ions can also inhibit bacterial replication by binding and denaturing bacterial DNA. Silver ions react with the thiol groups of enzymes, followed by DNA condensation resulting in cell death [28–29].

Blocking of respiration is also a result of the interaction with cell membranes [30].

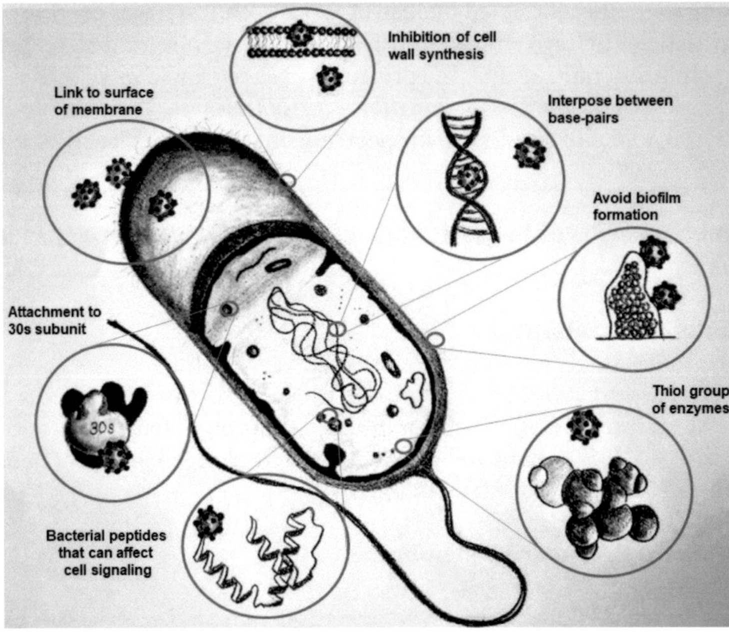

Figure 1.
Mode of action of silver nanoparticles in the bacterial cell.

Disruption of biofilms is another effect of AgNPs. The anti-biofilm action of ~8 nm AgNPs on Gram-negative bacteria has been demonstrated [31]. The outer membrane of Gram-negative contains aquaporins (water-filled channels) which are involved in the transport of Ag ions into the cell where they exert their antibacterial effects [32].

4. Cell status by flow cytometry

Flow cytometry (FCM) is a well-established and powerful analytical tool that has led to many revolutionary discoveries in cell biology and cellular-molecular disease diagnosis and, more recently, has been used to analyze physiological responses of bacteria [33, 34]. In FCM, cells are first introduced to a high-speed (up to 5–20 m/s) laminar flow stream, and after being focused into single file, they are subjected to laser-induced fluorescence, and/or forward and sideways scattered light is detected using photodetector arrays with spectral filters. More recently, FCM has been used to characterize distinct physiological conditions in bacteria including their responses to antibiotics and other cytotoxic chemicals [33]. Once the control of bacterial cells or fungal conidia has been applied, an accurate technique is required to measure the effectiveness of the silver nanoparticles. Flow cytometry is one of the most reliable techniques for detecting and counting living cells and to measure their viability.

When studying response to antibacterial agents such as silver nanoparticles, viability can be evaluated as an indicator of antibiotic susceptibility. There are now reagents available that allow assays of membrane permeability and potential by measuring the production of a fluorescent metabolite from a nonfluorescent precursor [33, 34].

Besides monitoring susceptibility to antibacterial activity, information can be obtained using FCM that can establish mechanisms of antibacterial drug activity [35–40]. Traditional culture-based techniques cannot do this [41]. The use of fluorescent probes to detect specific cell changes provides a unique tool for interrogating bacteria permeability and changes in membrane potential [42] (**Figure 2**). DNA content and metabolic activity [42] are useful indicators of cell viability and thus of antibiotic susceptibility.

The accuracy of cell counting depends on fluorescent staining. The choice of a fluorescent dye should take into account factors such as membrane permeability, photostability, pH and sensitivity to temperature [43, 44]. The total bacterial count is a key quality criterion for food or beverages [45] and a useful tool for detecting the presence of microbes within matrices. Williams et al. [46] used this technique to detect *E. coli* O157:H7 in raw spinach. The presence of plant pathogens during crop growth has been investigated by several authors. Day et al. [47] used FCM to detect and quantify *Phytophthora infestans* sporangia. A study of colonization of root-associated bacteria in rice was carried out by Valdameri et al. [48]. Otherwise, Golan et al. [49] counted *Pectobacterium carotovorum* subsp. *carotovorum* cells tagged with green fluorescent protein (GFP) in *Ornithogalum dubium* seedlings to detect resistant cultivars. The application of FCM is useful to create the bases for predictive models of spore germination, infection and disease development.

Cell viability assays can distinguish between live and dead cell populations and so correlate with other cell functions or treatments. Many companies offer a wide range of viability dyes, including fixable and non-fixable types and ones specific to bacterial or yeast viability tests. FCM can be applied to monitor the efficacy of treatments to reduce contamination of water [43] and foods and beverages [45, 50] by determining the viability of residual microorganisms. In agriculture FCM can be used to test the effectiveness of antibiotics and antifungals against plant pathogens. The advantage of live FCM cell counts compared to plate counts is that FCM allows the determination of several different morbidity stages between living and dead cells. Some of these are membrane integrity, esterase activity, membrane potential, electron transport, total cells, GFP expression, active/dead, mitochondrial activity, intracellular pH and carotenoid content [51–53].

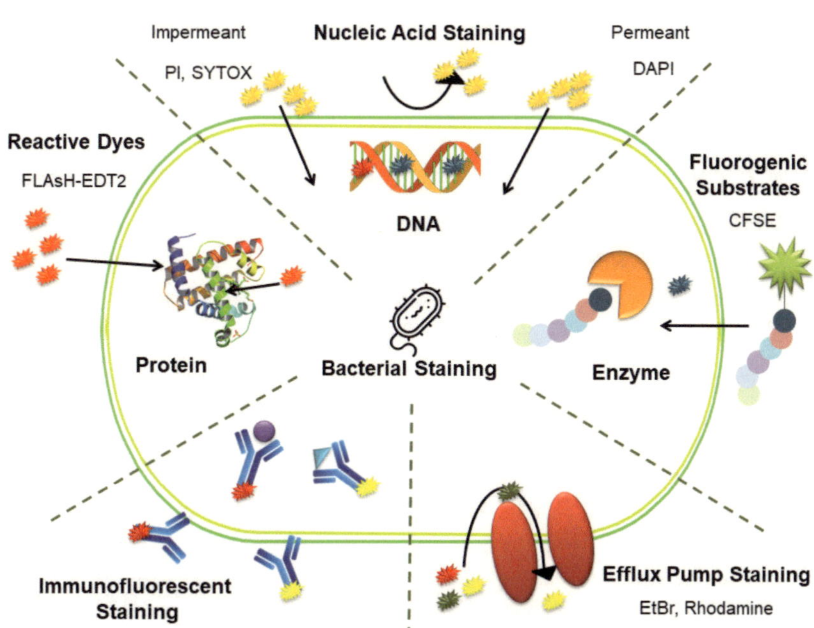

Figure 2.
Fluorescent probes to detect specific bacterial cell changes as an indicator of cell viability.

5. Conclusions

The Food and Agriculture Organization of the United Nations predicts that, globally, about 1.3 billion tons of food is lost per year. A large proportion of this loss is caused by postharvest microbial action. Much of this loss could be averted if more effective procedures and protocols were developed and adopted. Nanotechnology offers a range of novel tools with application in the fight against microbial food spoilage. Silver nanoparticles can act at cell level affecting from the cell wall or finely affecting the DNA. They offer a viable alternative to more traditional methods for the bacterial control. Once bacterial control is achieved using silver nanoparticles, continual bacterial monitoring becomes a critical component of the supply chain. For this, flow cytometry offers an accurate, novel and versatile technology through which to survey bacterial viability in assays of various bacterial control strategies.

Acknowledgements

We are indebted to Carolina Alvarado Gonzalez for the artwork.

Conflicts of interest

The authors declare there is no conflict of interest regarding the publication of this chapter. This chapter has not previously been published and is not being considered for publication elsewhere. The authors certify that neither the manuscript nor its main contents have already been published or submitted for publication in a scientific journal.

Author details

Graciela Dolores Avila-Quezada[1*] and Gerardo Pavel Espino-Solis[2]

1 Faculty of Agrotechnological Sciences, Autonomous University of Chihuahua, Chihuahua, Mexico

2 Faculty of Medicine and Biomedical Sciences, Autonomous University of Chihuahua, Chihuahua, Mexico

*Address all correspondence to: gavilaq@gmail.com

References

[1] Brady CL, Goszczynska T, Venter SN, Cleenwerck I, De Vos P, Gitaitis RD, et al. *Pantoea allii* sp. nov., isolated from onion plants and seed. International Journal of Systematic and Evolutionary Microbiology. 2011;**61**(4):932-937. DOI: 10.1099/ijs.0.022921-0

[2] Bautista-Baños S, Sivakumar D, Bello-Pérez A, Villanueva-Arce R, Hernández-López M. A review of the management alternatives for controlling fungi on papaya fruit during the postharvest supply chain. Crop Protection. 2013;**49**:8-20. DOI: 10.1016/j.cropro.2013.02.011

[3] Liu S, Tang Y, Wang DC, Lin NQ, Zhou JN. Identification and characterization of a new Enterobacter onion bulb decay caused by *Lelliottia amnigena* in China. Applied Microbiology: Open Access. 2016;**2**:1000114. DOI: 10.4172/2471-9315.1000114

[4] Vahling-Armstrong C, Dung JKS, Humann JL, Schroeder BK. Effects of postharvest onion curing parameters on bulb rot caused by *Pantoea agglomerans, Pantoea ananatis* and *Pantoea allii* in storage. Plant Pathology. 2016;**65**(4):536-544. DOI: 10.1111/ppa.12438

[5] García-ÁvilaCDJ,Valenzuela-TiradoGA, Florencio-Anastasio JG, Ruiz-Galván I, Moreno-VelázquezM,Hernández-MacíasB, et al. Organisms associated with damage to post-harvest potato tubers. Revista Mexicana de Fitopatología. 2018;**36**(2):308-320. DOI: 10.18781/R.MEX.FIT.1801-1

[6] Campa-Siqueiros P, Vallejo-Cohen S, Corrales-Maldonado C, Martínez-Téllez MÁ, Vargas-Arispuro I, Ávila-Quezada G. Reduction in the incidence of grey mold in table grapes due to the volatile effect of a garlic extract. Revista Mexicana de Fitopatología. 2017;**35**(3):494-508. DOI: 10.18781/R.MEX.FIT.1707-1 [Accessed: 19 May 2019]

[7] Nunes CA. Biological control of postharvest diseases of fruit. European Journal of Plant Pathology. 2012;**133**(1):181-196. DOI: 10.1007/s10658-011-9919-7 [Accessed: 19 May 2019]

[8] Liu J, Sui Y, Wisniewski M, Droby S, Liu Y. Utilization of antagonistic yeasts to manage postharvest fungal diseases of fruit. International Journal of Food Microbiology. 2013;**167**(2):153-160. DOI: 10.1016/j.ijfoodmicro.2013.09.004 [Accessed: 19 May 2019]

[9] Combrinck S, Regnier T, Kamatou GPP. In vitro activity of eighteen essential oils and some major components against common postharvest fungal pathogens of fruit. Industrial Crops and Products. 2011;**33**(2):344-349. DOI: 10.1016/j.indcrop.2010.11.011 [Accessed: 19 May 2019]

[10] Miranda-GómezB,García-HernándezA, Muñoz-Castellanos L, Ojeda-Barrios DL, Avila-Quezada GD. Pectate lyase production at high and low pH by *Colletotrichum gloeosporioides* and *Colletotrichum acutatum*. African Journal of Microbiology Research. 2014;**8**(19):1948-1954. DOI: 10.5897/AJMR2014.6765 [Accessed: 19 May 2019]

[11] Guetsky R, Kobiler I, Wang X, Perlman N, Gollop N, Avila-Quezada G, et al. Metabolism of the flavonoid epicatechin by laccase of *Colletotrichum gloeosporioides* and its effect on pathogenicity on avocado fruits. Phytopathology. 2005;**95**(11):1341-1348. DOI: 10.1094/PHYTO-95-1341 [Accessed: 19 May 2019]

[12] Prusky D, Alkan N, Mengiste T, Fluhr R. Quiescent and necrotrophic

lifestyle choice during postharvest disease development. Annual Review of Phytopathology. 2013;**51**:155-176. DOI: 10.1146/annurev-phyto-082712-102349 [Accessed: 19 May 2019]

[13] Torres-Armendáriz V, Manjarrez-Domínguez CB, Acosta-Muñiz CH, Guerrero-Prieto VM, Parra-Quezada RÁ, Noriega-Orozco LO, et al. Interactions between *Escherichia coli* O157:H7 and food plants. Has this bacterium developed internalization mechanisms? Revista Mexicana de Fitopatología. 2016;**34**(1):64-83. DOI: 10.18781/R.MEX.FIT.1507-4 [Accessed: 19 May 2019]

[14] Makavana JM, Makwana PJ, Kukadiya VD, Joshi AM. Post-harvest losses of lemon fruits: An assessment of microbial floral strength during post-harvest handling. International Journal of Current Microbiology and Applied Sciences. 2018;**7**(5):1184-1188. DOI: 10.20546/ijcmas.2018.705.144 [Accessed: 19 May 2019]

[15] Flores-Sánchez JL, Mora-Aguilera G, Loeza-Kuk E, López-Arroyo JI, Domínguez-Monge S, Acevedo-Sánchez G, et al. Yield loss caused by Candidatus *Liberibacter asiaticus* in Persian lime, in Yucatan Mexico. Revista Mexicana de Fitopatología. 2015;**33**(2):195-210. Available from: http://www.scielo.org.mx/pdf/rmfi/v33n2/2007-8080-rmfi-33-02-00195-en.pdf [Accessed: 19 May 2019]

[16] Ávila-Quezada GD, Téliz-Ortiz D, González-Hernández H, Vaquera-Huerta H, Tijerina-Chávez L, Johansen-Naime R, et al. Dinámica espacio-temporal de la roña (*Elsinoe perseae*), el daño asociado a trips y antracnosis (*Glomerella cingulata*) del aguacate en Michoacán, México. Revista Mexicana de Fitopatología. 2002;**20**:77-87. Available from: https://www.researchgate.net/profile/G_D_Avila-Quezada/publication/236569648_Dinamica_espacio-temporal_de_rona_dano_asociado_a_trips_y_antracnosis_en_aguacate/links/568767ab08ae19758397e6a3/Dinamica-espacio-temporal-de-rona-dano-asociado-a-trips-y-antracnosis-en-aguacate.pdf [Accessed: 19 May 2019]

[17] Chudasama B, Vala AK, Andhariya N, Mehta RV, Upadhyay RV. Highly bacterial resistant silver nanoparticles: Synthesis and antibacterial activities. Journal of Nanoparticle Research. 2010;**12**(5):1677-1685. DOI: 10.1007/s11051-009-9845-1

[18] Kim SW, Jung JH, Lamsal K, Kim YS, Min JS, Lee YS. Antifungal effects of silver nanoparticles (AgNPs) against various plant pathogenic fungi. Mycobiology. 2012;**40**(1):53-58. DOI: 10.5941/MYCO.2012.40.1.053 [Accessed: 19 May 2019]

[19] Min JS, Kim KS, Kim SW, Jung JH, Lamsal K, Kim SB, et al. Effects of colloidal silver nanoparticles on sclerotium-forming phytopathogenic fungi. The Plant Pathology Journal. 2009;**25**(4):376-380. DOI: 10.5423/PPJ.2009.25.4.376 [Accessed: 19 May 2019]

[20] Rai MK, Deshmukh SD, Ingle AP, Gade AK. Silver nanoparticles: the powerful nanoweapon against multidrug-resistant bacteria. Journal of Applied Microbiology. 2012;**112**(5):841-852. DOI: 10.1111/j.1365-2672.2012.05253.x [Accessed: 19 May 2019]

[21] Biao L, Tan S, Wang Y, Guo X, Fu Y, Xu F, et al. Synthesis, characterization and antibacterial study on the chitosan-functionalized Ag nanoparticles. Materials Science and Engineering: C. 2017;**76**:73-80. DOI: 10.1016/j.msec.2017.02.154

[22] Mukherji S, Ruparelia JP, Agnihotri S. In: Cioffi N, Rai M, editors. Nano-Antimicrobials: Progress and

Prospects. Berlin Heidelberg: Springer Verlag; 2012. pp. 225-251. Available from: https://books.google.com.mx/books?hl=es&lr=&id=dRh1fgmnOP0C&oi=fnd&pg=PR3&dq=Mukherji+S,+Ruparelia+JP,+Agnihotri+S.++in+Nano-Antimicrobials:+Progress+and+Prospects,+ed.+N.+Cioffi+and+M.+Rai,+Springer+Verlag,+Berlin+Heidelberg,+2012,+pp.+225%E2%80%93251+&ots=LxBFeqbcCA&sig=c3JIl_O3kpI7jD8ALHvks3u0nwo#v=onepage&q&f=false [Accessed: 19 May 2019]

[23] Agnihotri S, Mukherji S, Mukherji S. Size-controlled silver nanoparticles synthesized over the range 5-100 nm using the same protocol and their antibacterial efficacy. RSC Advances. 2014;**4**:3974-3983. DOI: 10.1039/C3RA44507K [Accessed: 19 May 2019]

[24] Lok CN, Ho CM, Chen R, He QY, Yu WY, Sun H, et al. Journal of Biological Inorganic Chemistry. 2007;**12**:527-534. DOI: 10.1007/s00775-007-0208-z [Accessed: 19 May 2019]

[25] Kim JS, Kuk E, Yu KN, Kim JH, Park SJ, Lee HJ, et al. Nanomedicine: Nanotechnology. Biology and Medicine. 2007;**3**(1):95-101. DOI: 10.1016/j.nano.2006.12.001 [Accessed: 19 May 2019]

[26] Arokiyaraj S, Vincent S, Saravanan M, Lee Y, Oh YK, Kim KH. Green synthesis of silver nanoparticles using *Rheum palmatum* root extract and their antibacterial activity against *Staphylococcus aureus* and *Pseudomonas aeruginosa*. Artificial Cells, Nanomedicine, and Biotechnology. 2017;**45**(2):372-379. DOI: 10.3109/21691401.2016.1160403 [Accessed: 19 May 2019]

[27] Sondi I, Salopek-Sondi B. Silver nanoparticles as antimicrobial agent: A case study on *E. coli* as a model for gram-negative bacteria. Journal of Colloid and Interface Science. 2007;**275**:77-82. DOI: 10.1016/j.jcis.2004.02.012 [Accessed: 19 May 2019]

[28] Castellano JJ, Shafii SM, Ko F, Donate G, Wright TE, Mannari RJ, et al. Comparative evaluation of silver-containing antimicrobial dressings and drugs. International Wound Journal. 2007;**4**:14-22. DOI: 10.1111/j.1742-481X.2007.00316.x [Accessed: 19 May 2019]

[29] Eby DM, Luckarift HR, Johnson GR. Hybrid antimicrobial enzyme and silver nanoparticle coatings for medical instruments. ACS Applied Materials & Interfaces. 2009;**1**(7):1553-1560. DOI: 10.1021/am9002155

[30] Sharma VK, Yngard RA, Lin Y. Silver nanoparticles: green synthesis and their antimicrobial activities. Advances in Colloid and Interface Science. 2009;**145**(1-2):83-96. DOI: 10.1016/j.cis.2008.09.002 [Accessed: 19 May 2019]

[31] Radzig MA, Nadtochenko VA, Koksharova OA, Kiwi J, Lipasova VA, Khmel IA. Antibacterial effects of silver nanoparticles on Gram-negative bacteria: Influence on the growth and biofilms formation, mechanisms of action. Colloids and Surface B: Biointerfaces. 2013;**02**:300-306. DOI: 10.1016/j.colsurfb.2012.07.039 [Accessed: 19 May 2019]

[32] Franci G, Falanga A, Galdiero S, Palomba L, Rai M, Morelli G, et al. Silver nanoparticles as potential antibacterial agents. Molecules. 2015;**20**:8856-8874. DOI: 10.3390/molecules20058856 [Accessed: 04 Jul 2019]

[33] Ambriz-AviñaV,Contreras-GarduñoJA, Pedraza-Reyes M. Applications of flow cytometry to characterize bacterial physiological responses. BioMed Research International. 2014. DOI: 10.1155/2014/461941 Accessed: 04 Jul 2019

[34] Lukomska-Szymanska M, Konieczka M, Zarzycka B,

Lapinska B, Grzegorczyk J, Sokolowski J. Antibacterial activity of commercial dentine bonding systems against *E. faecalis*–flow cytometry study. Materials. 2017, 2017;**10**(5):481. DOI: 10.3390/ma10050481 [Accessed: 04 Jul 2019]

[35] Mason DJ, Power EGM, Talsania H, Phillips I, Gant VA. Antibacterial action of ciprofloxacin. Antimicrobial Agents and Chemotherapy. 1995;**39**(12): 2752-2758. DOI: 10.1128/AAC.39.12.2752 [Accessed: 04 Jul 2019]

[36] Roth BL, Poot M, Yue ST, Millard PJ. Bacterial viability and antibiotic susceptibility testing with SYTOX green nucleic acid stain. Applied and Environmental Microbiology. 1997;**63**(6):2421-2431. Available from: https://aem.asm.org/content/63/6/2421.short [Accessed: 04 Jul 2019]

[37] Suller MTE, Stark JM, Lloyd D. A flow cytometric study of antibiotic-induced damage and evaluation as a rapid antibiotic susceptibility test for methicillin-resistant *Staphylococcus aureus*. Journal of Antimicrobial Chemotherapy. 1997;**40**(1):77-83. DOI: 10.1093/jac/40.1.77 Accessed: 04 Jul 2019

[38] Mortimer FC, Mason DJ, Gant VA. Flow cytometric monitoring of antibiotic-induced injury in *Escherichia coli* using cell-impermeant fluorescent probes. Antimicrobial Agents and Chemotherapy. 2000;**44**(3):676-681. DOI: 10.1128/AAC.44.3.676-681.2000 [Accessed: 04 Jul 2019]

[39] Novo DJ, Perlmutter NG, Hunt RH, Shapiro HM. Multiparameter flow cytometric analysis of antibiotic effects on membrane potential, membrane permeability, and bacterial counts of *Staphylococcus aureus* and *Micrococcus luteus*. Antimicrobial Agents and Chemotherapy. 2000;**44**(4):827-834. DOI: 10.1128/AAC.44.4.827-834.2000 [Accessed: 04 Jul 2019]

[40] Wickens HJ, Pinney RJ, Mason DJ, Gant VA. Flow cytometric investigation of filamentation, membrane patency, and membrane potential in *Escherichia coli* following ciprofloxacin exposure. Antimicrobial Agents and Chemotherapy. 2000;**44**(3):682-687. DOI: 10.1128/AAC.44.3.682-687 [Accessed: 04 Jul 2019]

[41] Suller MT, Lloyd D. Fluorescence monitoring of antibiotic-induced bacterial damage using flow cytometry. Cytometry. 1999;**35**(3):235-241. DOI: 10.1002/(SICI)1097-0320(19990301)35:3<235::AID-CYTO6>3.0.CO;2-0 [Accessed: 04 Jul 2019]

[42] Laflamme C, Lavigne S, Ho J, Duchaine C. Assessment of bacterial endospore viability with fluorescent dyes. Journal of Applied Microbiology. 2004;**96**(4):684-692. DOI: 10.1111/j.1365-2672.2004.02184.x [Accessed: 04 Jul 2019]

[43] Hammes F, Egli T. Cytometric methods for measuring bacteria in water: Advantages, pitfalls and applications. Analytical and Bioanalytical Chemistry. 2010;**397**(3):1083-1095. DOI: 10.1007/s00216-010-3646-3 [Accessed: 19 May 2019]

[44] Tracy BP, Gaida SM, Papoutsakis ET. Flow cytometry for bacteria: enabling metabolic engineering, synthetic biology and the elucidation of complex phenotypes. Current Opinion in Biotechnology. 2010;**21**(1):85-99. DOI: 10.1016/j.copbio.2010.02.006

[45] Comas-Riu J, Rius N. Flow cytometry applications in the food industry. Journal of Industrial Microbiology & Biotechnology. 2009;**36**(8):999-1011. DOI: 10.1007/s10295-009-0608-x

[46] Williams AJ, Cooper WM, Summage-West CV, Sims LM,

Woodruff R, Christman J, et al. Level 2 validation of a flow cytometric method for detection of *Escherichia coli* O157:H7 in raw spinach. International Journal of Food Microbiology. 2015;**215**:1-6. DOI: 10.1016/j.ijfoodmicro.2015.08.011 [Accessed: 19 May 2019]

[47] Day JP, Kell DB, Griffith GW. Differentiation of *Phytophthora infestans* sporangia from other airborne biological particles by flow cytometry. Applied Environmental Microbiology. 2002;**68**:37-45. DOI: 10.1128/AEM.68.1.37-45

[48] Valdameri G, Kokot TB, Pedrosa FDO, de Souza EM. Rapid quantification of rice root-associated bacteria by flow cytometry. Letters in Applied Microbiology. 2015;**60**(3):237-241. DOI: 10.1111/lam.12351

[49] Golan A, Kerem Z, Tun OM, Luzzatto T, Lipsky A, Yedidia I. Combining flow cytometry and gfp reporter gene for quantitative evaluation of *Pectobacterium carotovorum* spp. *carotovorum* in *Ornithogalum dubium* plantlets. Journal of Applied Microbiology. 2010;**108**(4):1136-1144. DOI: 10.1111/j.1365-2672.2009.04517.x

[50] Hahn MA, Keng PC, Krauss TD. Flow cytometric analysis to detect pathogens in bacterial cell mixtures using semiconductor quantum dots. Analytical Chemistry. 2008;**80**(3):864-872. DOI: 10.1021/ac7018365

[51] Sträuber H, Müller S. Viability states of bacteria-specific mechanisms of selected probes. Cytometry Part A. 2010;**77**(7):623-634. DOI: 10.1002/cyto.a.20920

[52] Freitas C, Nobre B, Gouveia L, Roseiro J, Reis A, da Silva TL. New at-line flow cytometric protocols for determining carotenoid content and cell viability during *Rhodosporidium toruloides* NCYC 921 batch growth. Process Biochemistry. 2014;**49**(4):554-562. DOI: 10.1016/j.procbio.2014.01.022

[53] Linhová M, Branská B, Patáková P, Lipovský J, Fribert P, Rychtera M, et al. Rapid flow cytometric method for viability determination of solventogenic clostridia. Folia Microbiologica. 2012;**57**(4):307-311. DOI: 10.1007/s12223-012-0131-8

Chapter 13

Bacteriological Quality of Borehole and Sachet Water from a Community in Southeastern Nigeria

Ogueri Nwaiwu, Chiugo Claret Aduba and Oluyemisi Eniola Oni

Abstract

Water from boreholes and packaged commercial sachet water from different areas in a community in southern Nigeria was analyzed with membrane filtration for a snapshot of heterotrophic count and coliforms. Two boreholes out of the 20 analyzed had counts of over 500 Cfu/mL and 7 boreholes indicated the presence of coliforms. Sixteen samples out of 20 sachet water brands analyzed showed a regulatory product registration code, whereas 4 samples had no number or code indicating that they were not registered. The heterotrophic count of all sachet water was well within the limit for all samples analyzed, and coliform was detected in only two samples. The overall quality of borehole water in the community studied was rated D (65%), whereas the sachet water was rated C (90%) according to the World Health Organization (WHO) surveillance guidelines. Improvements in water quality structure in the community studied are required to help achieve WHO sustainable development goals on water sanitation. The etiology, virulence properties, epidemiology, and pathogenicity of bacteria associated with borehole and sachet water are also discussed.

Keywords: bacteria, borehole, sachet water, coliforms, heterotrophic count

1. Introduction

Up to 2.1 billion people worldwide lack access to safe, readily available water at home according to a WHO/UNICEF report [1]. The report emphasized that majority of the people without good quality water are from developing countries and the lives of millions of children are at risk every day, with many dying from preventable diseases caused by poor water supply. The importance of good quality water is the reason why clean water and sanitation have been included as goal number 6 out of the 17 proposed sustainable development goals (SDGs) of the United Nations [2]. The proposal is that the SDGs will be the blueprint to achieving a better and more sustainable future for humanity by 2030.

In Nigeria, the public water supply is in a state of comatose in most towns and villages and dry taps without any hope of water running through the taps soon affect millions of homes. This has forced individuals and institutions to resort

 IntechOpen

to self-help by using water from boreholes as the only source of water supply for drinking and general use. Use of borehole is a simple way of obtaining potable water from the aquifer below the ground, after which the water can be pumped into storage tanks before distribution.

Many people that went into borehole drilling business, which reduced the price of new boreholes, aided the proliferation of boreholes in Nigeria, and many citizens were ready to pay more money in rent for houses, which had boreholes. Furthermore, the dependence on groundwater, which is believed to be purified, is on the increase due to the increasing contamination of the surface water [3]. It is known that properly designed and constructed borehole both ensures the success of the borehole as an adequate supply of water and minimizes the risk of local pollution affecting the source [4]. If a borehole facility is not properly managed, contamination may occur in the process through the accumulation of physical, chemical, and biological agents in the pipelines and storage tanks of a distribution system or water packaging company. One direct use of boreholes is in the production and packaging of drinking water in sachets made from low-density polyethylene sheets. These products are popularly known as "pure water" in Nigeria. From the early 1990s, the production of sachet water increased exponentially and provided jobs for producers and sellers of the product. There is hardly any community in Nigeria without a sachet water facility. It is possibly the most widely consumed commercial liquid in Nigeria, and no sophistication is required for production. The quest for a cheap, readily available, and inexpensive source of potable water contributed to the emergence of sachet water [5], and it is far better and safer than the hand-filled, hand-tied packaged water in polyethylene bag [6] sold in Nigeria in the past. In developing countries, production and consumption of sachet water are rapidly on the rise [7], and many unregulated producers exist.

Packaged drinking water like the sachet water could be water from any potable source such as tap, well, and rain, which may be subjected to further treatments like decantation, filtration, demineralization, remineralization, and other methods to meet established drinking standards [8, 9]. Packaged water is susceptible to microbial and chemical contamination regardless of their source [10]. Researchers have previously performed microbial analysis of sachet water in Nigeria using different laboratory techniques and found different bacteria and fungi. Occurrence of bacteria could lead to different disease conditions such as gastroenteritis, typhoid fever, cholera, bacillary dysentery, and hepatitis [11]. It has been reported [12] that waterborne diseases account for 80% of illnesses and diseases in developing countries, which leads to the death of several children every 8 seconds. In Nigeria, like most developing countries, various factors predispose packaged sachet water to contamination, and these include poor sanitation and source of raw material for food or water production [13]. Long storage of sachet under unfavorable environmental conditions and lack of good manufacturing practices (GMP) in general also contribute to contamination.

It has been found that the microbiome dynamically changes during different stages of water treatment distribution and the main important group in the past and present are fecal-associated bacterial pathogens like *Escherichia coli* [14]. However, opportunistic bacteria like *Legionella* and process-related bacteria, which form biofilms, are also a cause for concern [15, 16]. A review [17] elucidated that drinking water comprises a complex microbiota that is influenced by disinfection and that members of the phylum *Proteobacteria* represent the most frequent bacteria in drinking water. It was also pointed out that their ubiquity has serious implications for human health and that the first step to address the persistent nature of bacteria in water would be to identify and characterize ubiquitous bacteria. The manifestation of bacterial contamination in drinking water can become known when outbreaks occur, and surveillance data provides insights on the microbial etiology of diseases and process failures that facilitated the outbreak [18]. Sometimes it can also be detected from laboratory results especially when water treatment facility is contaminated by bacterial biofilms [19, 20].

In Nigeria, regulatory oversight is inadequate due to limited resources. Surveillance of bacteria in drinking water from boreholes and sachet water is necessary for the benefit of public health; hence, periodic surveys can help establish trends and identify where water quality of boreholes and sachet water is deficient. This chapter reports a survey, explores reports of bacteria associated with water from borehole and sachet water in Nigeria, and compares data found with WHO water standards. The organisms associated with boreholes and sachet water are discussed.

2. Methods

Water samples from boreholes were collected on different days using Whirl-Pak sampling bags (Nasco, Wisconsin, USA) and analyzed within 2 hours after collection. Twenty private boreholes and 20 different brands of commercial sachet water sold in four areas of a community were analyzed on different days. Sachet water was purchased (five each) from the different areas and were inspected for the inscription of an approved product registration code from the National Agency for Food and Drug Administration and Control (NAFDAC), the Nigerian national regulatory body. It was ensured that the same brand was not purchased twice from one area. The human population of the community (all 4 areas) was estimated to be over 5000 but less than 100,000.

Heterotrophic plate and total coliform count of bacteria were carried out using standard membrane filtration performed previously [21]. A slight modification of the method was introduced. Instead of using factory-made ready to use nutrient media sets, plate count agar (Oxoid, United Kingdom, CM0325) and violet red bile lactose agar (Oxoid, CM0107) for coliforms were prepared and used according to manufacturer's instructions. Briefly, the filtration process involved placing of 100 ml of water sample in a sterile multibranched stainless steel manifold and filter holder system. A 0.45 μm membrane filter was fitted into the filter system after which water was drawn through to retain bacteria on the membrane. The membrane filter was placed on the media prepared and then incubated at 32°C over 48 h for membrane filters placed on plate count agar, whereas incubation at 30°C for 48 h was used for filters grown on violet red bile lactose agar. The heterotrophic count was noted, and estimated coliform results obtained for boreholes and sachet water were compared to WHO quality guidelines for drinking water [22].

3. Results

3.1 Heterotrophic and total coliform count of borehole samples

This survey was carried out to have an overview of the bacterial load in water quality of some boreholes in the community surveyed. The borehole owners were apprehensive and thought they were being investigated for possible closure. To allow sample collection, it was agreed that the name of borehole owners and their location should remain anonymous when the findings were published. Results showed that borehole samples from area "C2" had the highest heterotrophic aerobic count. Two boreholes had counts of over 500 Cfu/mL, which is above the recommended heterotrophic limit [21]. All the other samples were below 500 Cfu/mL. Seven boreholes indicated the presence of coliforms because purple-pink colonies, which were 1–2 mm in diameter surrounded by a purple zone, were formed on the plates after incubation. Samples C2a, C2b, C2c, C2d, and C2e had coliform count of 17, 15, 9, 6, and 5 Cfu/mL, respectively, whereas samples C3b and C4b had coliform count of 4 and 2 Cfu/mL. The rest of the samples had no coliform on the plate used after incubation. A definitive trend was that samples with the highest heterotrophic count had the most coliform count (**Figure 1**).

3.2 Heterotrophic and total coliform count of sachet water samples

Periodic analysis of sachet water is important to public health because millions of people in Nigeria consume it. An ideal situation would be to analyze every borehole water from which sachet water is produced to establish water treatment effectiveness. Enquiries made to sachet water producers for access to their source of water for production were not successful. To refuse access some companies gave information and advice that they do not have a borehole and their water for production is sourced from the supply by water tankers. Hence, commercial samples of sachet water were purchased from different locations with unknown source of initial water for production of sachet water on sale. Sixteen samples out of the 20 analyzed showed a NAFDAC product registration code, whereas 4 samples had no number or code indicating that they were not registered. The heterotrophic count was well within the limit for all samples analyzed, and coliform was detected in only two samples. Sample SC1c and SC3c had a coliform count of 2 Cfu/mL each (**Figure 2**).

3.3 Comparisons with WHO guidelines

The WHO standards and guidelines are usually used to monitor water quality. The WHO categorizes drinking water systems based on population size and quality rating to prioritize actions. A quality score from A to D is awarded (quality decreases A to D) based on the proportion (%) of samples negative for *E. coli*. However, the samples under study were assessed for total coliforms and not *E.coli*; the scoring was carried out with the presumption that samples with high coliform count may contain *E. coli*. Total coliforms serve as a parameter to provide basic

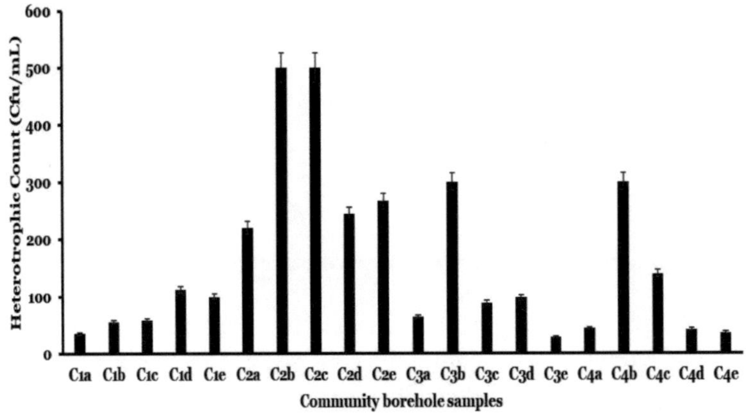

Figure 1.
Heterotrophic plate count of borehole water sourced from different areas of the community studied (C1–C4). The letters a to e represent different samples.

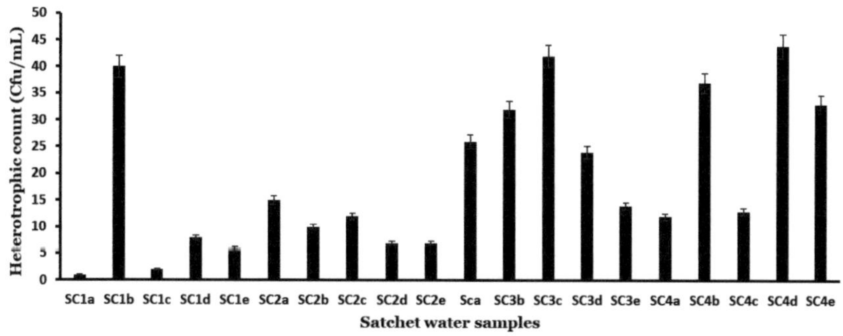

Figure 2.
Heterotrophic plate count of sachet water (S) sourced from different areas of the community studied (C1–C4). Letters a to e represent different samples.

information on water quality [23]. On this basis, the overall quality of borehole water in the community studied (all areas combined) was rated D (proportion of samples negative for coliform =13; 65%), whereas the sachet water was rated C (18 = 90%).

4. Discussion

4.1 Bacteria associated with boreholes in Nigeria

Pathogenic bacteria often occur in borehole water systems especially in developing nations [24–26]. Coliforms found in this study and other Gram-negative bacteria have been isolated from boreholes in different parts of Nigeria by many investigators [27–34]. The organisms mentioned in these studies include *Enterobacter aerogenes*, *Escherichia coli*, *Klebsiella aerogenes*, *Klebsiella* sp., *Klebsiella pneumoniae*, *Klebsiella variicola*, *Proteus* sp., and *Proteus vulgaris*. Other bacteria isolated are *Providencia sneebia*, *Pseudomonas aeruginosa*, *Salmonella paratyphi*, *Salmonella* sp., *Salmonella typhi*, *Staphylococcus aureus*, and *Vibrio cholera*.

The prevalence of the aforementioned species and genera may be due to the classical microbiological methods used for isolation. In most cases, MacConkey media was used for *E.coli* and coliform identification with no molecular studies that included 16S or whole-genome sequencing essential for establishing the actual prevalent bacteria species and strains in boreholes. An opportunity exists for regular molecular characterization of bacteria found in boreholes to help differentiate between harmless coliforms, fecal coliforms, and the deadly *E. coli* strain O157: H7. Borehole operators are required to deliver safe and reliable drinking water to their customers. If a community consistently consumes contaminated water, they may become unwell. Hence, regular monitoring and assessment of borehole water sources help maintain quality and provide data on groundwater management [35–38].

4.1.1 Bacteria contamination of groundwater

In Africa, many people rely on water from a borehole, but the purity of the drinking water from this source remains questionable [39, 40]. The high heterotrophic count found in Area "2" of the community studied suggests that the groundwater of that area may be contaminated. The corresponding increased coliform count observed is consistent with the findings of Amanidaz et al. [41], which showed that when the concentration of coliforms and fecal *Streptococci* bacteria increased in a water network system, there was also an increased concentration of heterotrophic bacteria. These contrasts with the work of others [42] where it was shown that high heterotrophic count inhibits coliform proliferation. Despite increased heterotrophic count and coliforms in the study of Amanidaz et al. [41], it was concluded that no correlation exists, and increased numbers could be due to variability in nutrient composition [43]. Another factor could be biofilm formation because it has been shown that attached bacteria in biofilms of a water system are more metabolically active than the ones that are free-living [44]. Groundwater is susceptible to contamination by both organic and inorganic contaminants [45–48]. Contamination could happen through natural processes, such as geological weathering and dissolution of numerous minerals beneath the earth's surface, which results in low natural concentrations of contaminants in groundwater [49]. Anthropogenic sources, such as seepages from agricultural wastewaters, domestic sewages, mining activities, and industrial effluents, can also affect the quality of groundwater in many parts of the world [50–52]. Other reports showed that borehole contamination may occur through domestic wastewater and livestock manure [53] industrialization and urbanization [54] and leakages from septic tanks [55] or pit latrines [56]. Seasonal environmental conditions may also contribute to increased bacteria count from borehole water because other investigators [57, 58] have demonstrated that higher bacterial count in borehole water occurs during the rainy season. This has been attributed to flooding which may allow floodwater to get into borehole systems that are not properly constructed.

4.2 Cases of sachet water contamination in Nigeria

Postproduction improper handling [59] and compromising safety and quality for profit during production [60] are factors that can affect sachet water contamination in Nigeria. Sachet water producers are expected to be food safety conscious in order not to jeopardize the health of the public. A large number of sachet water-producing companies in Nigeria are not registered and do not practice good manufacturing practices or follow international quality standards of water treatment [61] despite the efforts of NAFDAC to improve standards. Up to 25% of samples analyzed in this study had no regulation or expiration date code as recommended previously [62]. However, the fact that 75% of sachet water analyzed had date codes is a remarkable improvement from what was the norm (0%) when sachet water production started in the country. Unlike a previous study with larger sample size [11], which reported isolation of bacterial species in 54 out of 720 (7.5%) from 6 different brands of sachet water in northern Nigeria, all the samples in this study (100%) showed heterotrophic growth that were within permissible limits (<500 Cfu/mL).

Sachet water analysis from other parts of Nigeria has shown different levels of contamination. In this study, 10% (2 out of 20) of samples contained coliforms. In other studies carried out on samples sourced from Aba in the southeast, an analysis of 20 sachet water samples showed that 32% of the samples reportedly tested positive for *Staphylococcus* spp., 23% for *Pseudomonas*, 20% for *Klebsiella* spp., 15% for *Proteus*, and 10% for *Enterobacter* [59]. Another study in the same region reported a contamination in 8 out of the 10 sachet water samples analyzed, isolated microorganisms included *E. coli*, *Klebsiella* spp., *Pseudomonas* spp., *Bacillus* spp., *Proteus* spp., and *Staphylococcus* spp. [5]. Also 66% and 73% prevalence of pathogens have been reported [63] in this region after two batches of 30 sachet water samples were analyzed. In Oyo, which is situated in the southwest of Nigeria, *E. coli* (13.3%), *Pseudomonas aeruginosa* (39.9%), and *Enterobacter aerogenes* (53.3%) were isolated from commercially sold sachet water [64]. Another report in this region [26] highlighted that all brands of sachet water (100%) analyzed had the presence of coliforms.

4.3 Compliance with world standards

A recent SDGs progress report [3] shows that between 2000 and 2017, the proportion of the global population using safely managed drinking water increased from 61 to 71%. The report highlighted that despite the increase, water stress affects people on every continent, requiring immediate and accelerated collective action to provide billions of people with safely managed drinking water. The quality score for the boreholes and sachet water from the community studied showed that the water needs improvement to achieve the desired "A" rating. In this study, the borehole water quality in Area "2" is a source of concern, and the owners in that area were advised to boil and filter the water before drinking. It is common knowledge in Nigeria that some boreholes are not deep enough to produce clean water from the aquifer; hence, such boreholes are used for other domestic purposes but not for cooking food or drinking. Owners of such boreholes normally boil and filter the water for drinking.

Water quality specifications may depend on the particular use, but the presence of coliforms in drinking water indicates that disease-causing organisms could be in the water system and may pose an immediate health risk to the water consumers. When coliforms and other bacteria are found, it is always recommended [65] that an investigation should be carried out to establish the sources of contamination. This confirmation will enable risk assessment and identification of solutions that will eliminate or reduce the risk of waterborne disease within a large population [66].

4.4 Etiology, virulence, epidemiology, and pathogenicity of bacteria associated with borehole and sachet water

From the studies reviewed, the organisms found in borehole water are well-known food- and waterborne bacteria that are constantly monitored by regulatory authorities in many parts of the world. Outbreaks can occur in a community and cause fatalities and economic losses. Hence, a constant review of the growth conditions that enable the bacteria to proliferate, the features that enable survival in different environments, infection mode, and prevalence pattern of these bacteria is important to reduce outbreaks.

4.4.1 Staphylococcus

The bacterium *Staphylococcus aureus* from the genus *Staphylococcus* is known for methicillin resistance of some strains. The bacterium is a major environmental contaminant of food and water, and the human skin and nose are known to be major sources of the organism. Nasal colonization [67, 68] and atopic dermatitis of the skin [69, 70] are considered risk factors. Environmental contamination may be the source of contamination in borehole water analyzed in this study, whereas humans or personnel involved in sachet water production are likely to be contributors to contamination. In Nigeria, sachet water producers are known to lack resources; hence, it is possible that respiratory protective equipment like nose masks are not worn during production in some facilities. Since it is possible to distinguish community-associated MRSA from healthcare-associated MRSA based on genetic, epidemiologic, or microbiological profiles [71], it would be beneficial to screen the strains found in this study to determine if they are methicillin resistant and community-related.

The pathogenicity, epidemiology, and virulence factors of *Staphylococcus* have been comprehensively reviewed [72]. It was highlighted that colonization is aided by biofilm formation that is housed in extracellular polymeric substance (EPS) found in many bacteria and that virulence factors are expressed with accessory gene regulator (agr) system in response to cell density [73]. To avoid formation of biofilms and EPS in the sachet water-producing environment, adequate personnel hygiene and good manufacturing practices that meet food safety standards must be implemented.

4.4.2 Pseudomonas

The genus *Pseudomonas* especially *P. aeruginosa* is known globally as endemic [74] and an opportunistic pathogen that causes several infections [75]. They are often isolated in clinics [76], and other sources may include residential, recreational, or surface water [77]. The colonies are usually heavily mucoid on solid media. It has been reported that mechanisms of antimicrobial resistance in *Pseudomonas* strains and most bacteria include multidrug efflux pumps and down-regulation of outer membrane porins, whereas virulence may include secretion of toxins and the ability to form biofilms [78, 79]. A natural property of *Pseudomonas* is the possession of multiple mechanisms for different forms of antibiotic resistance [80], and this may have facilitated its occurrence in boreholes and sachet water.

4.4.3 Klebsiella

Klebsiella causes many infections, which includes urinary tract infections, pneumonia, bacteremia, and liver abscesses [81]. The genus is associated with water, and this may be why it has been isolated in both borehole and sachet water. The organism is found in drinking water [82], rivers [83], and sewage water [84], which may encourage environmental spread. It has been reported that the organism has a variety of virulence and immune evasive factors, which contribute to uptake of genes associated with antimicrobial resistance and pathogenicity [85]. A report [86] suggested that the species *K. pneumoniae* acquired antimicrobial resistance genes independently and their population is highly diverse. An analysis of strains from human and animal isolates spanning four continents has shown convergence of virulence and resistance genes, which may lead to untreatable invasive *K. pneumoniae* infections [87].

4.4.4 Escherichia

The most studied species of the *Escherichia* genus is *E. coli*, a coliform bacteria used to verify hygiene status in food and water. Usually, the presence of various strains of pathogenic or nonpathogenic *E. coli* in food or water samples indicates fecal contamination [88]. It has been reported that [89] a comparative analysis show that avian and human *E. coli* isolates contain similar sets of genes encoding virulence factors and that they belong to the same phylogenetic groups, which may indicate the zoonotic origin of extraintestinal pathogenic *E. coli*.

A study of the prevalence of *E. coli* strain O157:H7 in England and Scotland showed that it has a seasonal dependency, with greater fecal shedding of the organism in the warmer months together with increased reporting of *E. coli* O157:H7 infection among hospitalized patients [90]. This finding is very worrying because it suggests that there could be high prevalence when applied to Nigeria because the country has a warm climate all year round. However, good manufacturing practices irrespective of the climate appear to be the key factor in producing packaged water free of coliforms. It has been shown that levels of coliform bacteria and *E. coli* detected in sachet water samples in Ghana, a country with similar climate to Nigeria, were statistically and significantly lower than levels detected from several water sources including public taps [91].

4.4.5 Enterobacter

The genus *Enterobacter* consists of coliforms that are known to be of non-fecal origin. It is believed [92] that many *Enterobacter* species, which could act as pathogens, are widely encountered in nature but are most frequently isolated in human clinical specimens possibly because phenotypic identification of all species belonging to this taxon is usually difficult and not always reliable. Therefore, the identification of this genus in borehole and sachet water may need a revisit since molecular methods were not used. The organism is known as a ubiquitous and persistent Gram-negative bacterium in drinking water [17], but there are few studies of its occurrence or prevalence in borehole and sachet water or other water sources in Nigeria.

To understand the carbapenemase-producing *Enterobacter* spp. and the development of molecular diagnostics, Chavda et al. [93] used genomic analysis of 447 sequenced strains to establish diverse mechanisms underlying the molecular evolutionary trajectory of drug-resistant *Enterobacter* spp. Their findings showed the acquisition of an antibiotic resistance plasmid, followed by clonal spread and horizontal transfer of *blaKPC*-harboring plasmids between different phylogenomic groups. The report also showed repeated transposition of the *blaKPC* gene among different plasmid backbones.

4.4.6 Proteus

Proteus species are Gram-negative opportunistic rod-shaped bacteria known for its swarming motility and contamination of agar plates. Furthermore, on agar plates, the bacteria undergoes a morphological conversion to a filamentous swarmer cell expressing hundreds of flagella, and during infection, histological damage is caused by cytotoxins including hemolysin and a variety of proteases [94]. The organism is reported to have negative and positive advantages. According to Drzewiecka [95], *Proteus* species may be indicators of fecal pollution, which may cause food poisoning when the contaminated water or seafood is consumed, and it could be used for bioremediation activity due to its tolerance and ability to utilize polluting compounds as sources of energy.

Virulence factors may include fimbriae, flagella, outer membrane proteins, lipopolysaccharide, capsule antigen, urease, immunoglobulin A, proteases, hemolysins, and amino acid deaminases [96]. The ability to swarm and survive is facilitated by the upregulation of FlhD(2)C(2) transcription activator, which activates the flagellar regulon [97]. The prevalence of *Proteus* spp. in borehole or sachet water may be aided by its ability to swarm and colonize the production environment.

4.4.7 Vibrio

In Nigeria, the most reported species among the *Vibrio* species that cause water-related infection is *Vibrio cholerae*. The organism causes cholera, which is an infection that is characterized by watery stooling. The disease has killed hundreds of people in Nigeria in the last decade. According to Faruque et al. [98], a lysogenic bacteriophage designated CTXΦ encodes the Cholera toxin (CT), which is strongly influenced by environmental conditions [99]. The organism is responsible for the profuse diarrhea, and molecular epidemiological surveillance has revealed clonal diversity among toxigenic *V. cholerae* strains with continuous emergence of new epidemic clones. It has not been established if the strains found in boreholes and sachet water are the *V. cholerae* O1 or O139 strains that cause cholera [100]. There is a possibility that they could be non-O1 or non-O139 strains that are common in the environment.

In 2017, the WHO launched a global strategy on cholera control with a target to reduce cholera deaths worldwide by 90% [101]. The strategy is to use safe oral cholera vaccines in conjunction with improvements in water and sanitation to control cholera outbreaks and for prevention in areas known to be high risk for cholera. Nigeria can be classified as a high-risk area, and the occurrence of *Vibrio* species in borehole or sachet water suggests that they could transmit cholera. Outbreaks occur regularly in Nigeria, and it is always difficult to bring it under control. An outbreak in 2018 was characterized by four epidemiological waves and led to 836 deaths out of 43,996 cases [102], whereas that of 2010 killed a total of 1716 out of 41, 787 cases [103]. In both cases, the case fatality rate was over 1% recommended by WHO.

4.4.8 Bacillus

Bacillus cereus is a food safety concern among several species of *Bacillus*. It is naturally widely distributed in nature, and it is known as a Gram-positive rod bacterium that is responsible for food poisoning [104]. It can proliferate because of unhygienic practices [105] and can attach to drinking water infrastructure [106]. This suggests that the ubiquity of the organism, poor hygiene, and attachment to equipment may be why *Bacillus* has been repeatedly isolated from boreholes and sachet water by previous investigators.

Bacillus growth is sometimes considered an insignificant contaminant. Some strains like *B. subtilis* is used for probiotics [107], whereas a strain like *B. cereus* which secrets toxins like hemolysins, phospholipases, an emesis-inducing toxin, and proteases [108] is not used due to obvious reasons. Toxin production in *B. cereus* requires the transcription factor *PlcR*, which controls expression of virulence factors [109]. Virulence-associated gene profiles have been used to evaluate the genetic backgrounds and relationships of food poisoning cases among other isolates from the environment, and it was concluded that both molecular and epidemiological surveillance studies could be used effectively to estimate virulence [110].

4.4.9 Salmonella

The species *Salmonella typhi* and *Salmonella paratyphi* cause typhoid fever and remain a major public health concern in Asia and Africa [111] due to antimicrobial resistance. For developed countries, it is believed that some non-typhoidal strains are zoonotic in origin and acquire their resistance in the food animal host before onward transmission to humans through the food chain [112]. It has been reported that the overall global burden of *Salmonella* infections is high and this may be the reason why in 2017, the WHO listed fluoroquinolone-resistant *Salmonella* spp. as priority pathogens for which new antibiotics were urgently needed [113].

The bacterium can survive in aquatic environments by a number of mechanisms, including entry into the viable but non-culturable state or residence within free-living protozoa [114]. Survival in water may have contributed to the isolation from borehole and sachet water in studies by others. It is not certain if the isolates encountered in this study cause typhoid fever or are the non-typhoid causing strains. Hence, additional studies are required to establish the prevalent type of *Salmonella* in water-producing facilities in Nigeria. A recent report found that typhoid fever still poses a serious health challenge in Nigeria and is a major health security issue [115]. It was recommended that a combined approach that includes the use of typhoid vaccines, improvements in sanitation, and safe water supply is essential.

5. Conclusions

The overall bacteria quality of the borehole and sachet water in the community studied needs improvement. An improvement can be achieved by focusing on areas with coliform contamination. Boreholes should be sited where pollutants will not easily contaminate them. Regular water testing should be carried out to ensure the attainment of WHO guidelines always. Where deviations are found, corrective actions should be undertaken. The literature on bacteria from boreholes and sachet water in Nigeria shows that not much molecular characterization has been carried out; hence an opportunity exists for more investigations. Regulatory oversight for sachet water production and the use of boreholes by large community populations requires improvement. It is recommended that universities should carry out periodic surveillance of boreholes and sachet water sold near them to support the SDG targets of the WHO.

Conflict of interest

The authors declare no conflict of interest.

Author details

Ogueri Nwaiwu[1*], Chiugo Claret Aduba[2] and Oluyemisi Eniola Oni[3]

1 School of Biosciences, University of Nottingham, Sutton Bonington Campus, United Kingdom

2 Department of Science Laboratory Technology, University of Nigeria, Nsukka, Nigeria

3 Department of Microbiology, Federal University of Agriculture, Abeokuta, Nigeria

*Address all correspondence to: guerinwaiwu@yahoo.co.uk

References

[1] United Nations International Children's Emergency Fund (UNICEF). Water, sanitation and hygiene [Internet]. Available from: https://www.unicef.org/wash/ [Accessed: 03 November 2019]

[2] United Nations. About the sustainable development goals [Internet]. Available from: https://www.un.org/sustainabledevelopment/sustainable-development-goals/ [Accessed: 03 November 2019]

[3] Agwu A, Avoaja AG, Kalu AU. The assessment of drinking water sources in aba metropolis, Abia state, Nigeria. Resources and Environment. 2013;**3**(4):72-76

[4] Ensure availability and sustainable management of water and sanitation for all [Internet]. Available from: https://unstats.un.org/sdgs/report/2019/goal-06/ [Accessed: 02 November 2019]

[5] Anyamene NC, Ojiagu DK. Bacteriological analysis of sachet water sold in Awka Metropolis, Nigeria. International Journal of Agriculture and Biosciences. 2014;**3**(3):120-122

[6] Manjaya D, Tilley E, Marks SJ. Informally vended sachet water: Handling practices and microbial water quality. Water. 2019;**11**(4):800. DOI: 10.3390/w11040800

[7] Gangil R, Tripachi R, Patyal A, Dutta P, Mathur KN. Bacteriological evaluation of packaged bottled water sold at Jaipur city and its public health significance. Veterinary World. 2013;**6**:27-30

[8] Aroh KN, Eze EM, Ukaji D, Wachuku CK, Gobo AE, Abbe SD, et al. Health and environmental components of sachet water consumption and trade in Aba and Port Harcourt, Nigeria. Journal of Chemical Engineering and Material Science. 2013;**4**:13-22

[9] Isikwue MO, Chikezie A. Quality assessment of various sachet water brands marketed in Bauchi metropolis of Nigeria. International Journal of Advances in Engineering and Technology. 2014;**6**(6):2489-2495

[10] Sudhakar MR, Mnatha P. Water quality in sustainable water management. Current Science. 2004;**87**:942-947

[11] Thliza LA, Khan AU, Dangora DB, Yahaya A. Study of some bacterial load of some brands of sachet water sold in Ahmadu Bello University (Main campus), Zaria, Nigeria. International Journal of Current Science. 2015;**14**:91-97

[12] Hughes JM, Koplan JP. Saving lives through global safe water. Journal of Infectious Diseases. 2011;**10**:1636-1637

[13] Ashbolt NJ. Microbial contamination of drinking water and disease outcomes in developing regions. Toxicology. 2004;**198**:229-238

[14] Proctor CR, Hammes F. Drinking water microbiology--from measurement to management. Current Opinion in Biotechnology. 2015;**33**:87-94. DOI: 10.1016/j.copbio.2014.12.014

[15] Berry D, Xi C, Raskin L. Microbial ecology of drinking water distribution systems. Current Opinion in Biotechnology. 2006;**17**:297-302

[16] Benner J, Helbling DE, Kohler HP, Wittebol J, Kaiser E, Prasse C, et al. Is biological treatment a viable alternative for micropollutant removal in drinking water treatment processes? Water Research. 2013;**47**(16):5955-5976

[17] Vaz-Moreira I, Nunes OC, Manaia CM. Ubiquitous and persistent proteobacteria and other gram-negative bacteria in drinking water. Science of the Total Environment.

2017;**586**:1141-1149. DOI: 10.1016/j.scitotenv.2017.02.104

[18] Hunter PR, Anderson Y, Von Bonsdorff CH, Chalmers RM, Cifuentes E, Deere D, et al. Surveillance and investigation of contamination incidents and water borne outbreaks. In: Dufour A, Snozzi M, Koster W, Bartram J, Ronchi E, Fewtrell L, editors. Microbial Safety of Drinking Water: Improving Approaches and Methods. Cornwall: IWA; 2003. pp. 205-236

[19] Nwaiwu O, Nwachukwu MI. Detection and molecular identification of persistent water vessel colonizing bacteria in a table water factory in Nigeria. British Microbiology Research Journal. 2016;**13**(5):1-12

[20] Nwaiwu O. Data on evolutionary relationships of *Aeromonas hydrophila* and *Serratia proteamaculans* that attach to water tanks. Data in Brief. 2018;**16**: 10-14. DOI: 10.1016/j.dib.2017.10.073

[21] Nwaiwu O, Ibekwe VI. Prevalence and risk of heterotrophic bacteria in a carbonated soft drink factory. African Journal of Microbiology Research. 2017;**11**(6):245-253

[22] Guidelines for drinking-water quality: Fourth edition incorporating the first Addendum, Surveillance [Internet]. 2017. World Health Organization, pp. 89-91. Available from: https://www.who.int/water_sanitation_health/publications/drinking-water-quality-guidelines-4-including-1st-addendum/en/ [Accessed: 01 October 2019]

[23] Shibata T, Solo-Gabriele HM, Fleming LE, Elmir S. Monitoring marine recreational water quality using multiple microbial indicators in an urban tropical environment. Water Research. 2004;**38**(13):3119-3131. DOI: 10.1016/j.watres.2004.04.004

[24] Oludairo O, Aiyedun J. Contamination of commercially packaged sachet water and the public health implications: An overview. Bangladesh Journal of Veterinary Medicine. 2016;**13**(2):73-81

[25] Aroh KN, Eze EM, Ukaji D, Wachukwu CK, Gobo AE, Abbe SD, et al. Health and environmental components of sachet water consumption and trade in aba and Port Harcourt, Nigeria. Journal of Chemical Engineering and Materials Science. 2013;**4**(2):13-22

[26] Oyedeji O, Olutiola PO, Moninuola MA. Microbiological quality of packaged drinking water brands marketed in Ibadan metropolis and Ile-Ife city in South Western Nigeria. African Journal of Microbiology Research. 2010;**4**(1):096-102

[27] Onuorah S, Nwoke J, Odibo F. Bacteriological assessment of the public hand-pump borehole water in Onueke, Ezza south local government area, Ebonyi state, Nigeria. International Journal of Photochemistry and Photobiology. 2018;**2**(2):39-48

[28] Abdullahi M, Saidu BT, Salihu BA, Mohammed SA. Bacteriological and physicochemical properties of borehole water in Niger state polytechnic, Zungeru Campus. Indian Journal of Science Research. 2013;**4**(1):1-6

[29] Akinola OT, Ogunbode TO, Akintunde EO. Borehole water quality characteristics and its potability in Iwo, Osun state Nigeria. Journal of Scientific Research and Reports. 2018;**18**(1):1-8

[30] Ngele SO, Itumoh EJ, Onwa NC, Alobu F. Quality assessment of selected ground water samples in Amike-Aba, Abakaliki, Ebonyi State, Nigeria. Canadian Journal of Pure & Applied Science. 2014;**8**(1):2801-2805

[31] Josiah JS, Nwangwu CO, Omage K, Akpanyung OE, Amaka DD. Physicochemical and microbiological properties of water samples used for domestic purposes in Okada Town, Edo

State, Nigeria. International Journal of Current Microbiology and Applied Sciences. 2014;**3**(6):886-894

[32] Ibe SN, Okpalenye JI. Bacteriological analysis of borehole water in Uli, Nigeria. African Journal of Applied Zoology and Environmental Biology. 2005;**7**:116-119

[33] Uhuo CA, Uneke BI, Okereke CN, Nwele DE, Ogbanshi ME. The bacteriological survey of borehole waters in Peri-Urban areas of Abakaliki, Ebonyi State, Nigeria. International Journal of Bacteriology Research. 2014;**2**(2):28-31

[34] Ukpong EC, Okon BB. Comparative analysis of public and private borehole water supply sources in Uruan local government area of AkwaIbom state, Nigeria. International Journal of Applied Science and Technology. 2013;**3**(1):76-88

[35] Howard G, Bartram J, Pedley S, Schmoll O, Choros I, Berger P. Groundwater and public health. In: Schmoll O, Howard G, Chilton J, Chorus I, editors. Protecting Groundwater for Health: Managing the Quality of Drinking Water Sources. London: IWA Publishing; 2006. p. 17

[36] Mogheir Y, Singh VP. Application of information theory to groundwater quality monitoring networks. Water Resources Management. 2002;**16**(1):37-49

[37] Sundaram B, Feitz A, Caritat PD, Plazinska A, Brodie R, Coram J, et al. Groundwater sampling and analysis- A field guide, Geoscience Australia. Record: 2009/27. p. 95. Canberra [Internet]. Available from: http://www.cffet.net/env/uploads/gsa/BOOK-Groundwater-sampling-%26-analysis-A-field-guide.pdf [Accessed: 04 November 2019]

[38] Suthar S, Chhimpa V, Singh S. Bacterial contamination in drinking water: A case study in rural areas of northern Rajasthan, India. Environmental Monitoring and Assessment. 2009;**159**:43 [Internet]. Available from: http://doi.org10.1007/s10661-008-0611-0 [Accessed: 05 November 2019]

[39] Olalekan A, Abubakar B, Abdul-Mumini K. Physico-chemical characteristics of borehole water quality in Gassol, Taraba, State, Nigeria. African Journal of Environmental Science and Technology. 2015;**9**(2):143-154

[40] Ncube EJ, Schutte CF. The occurrence of fluoride in south African groundwater: A water quality and health problem. Water SA. 2005;**31**(1):35-40

[41] Amanidaz N, Zafarzadeh A, Mahvi AH. The interaction between heterotrophic bacteria and coliform, fecal coliform, Fecal streptococci bacteria in the water supply networks. Iranian Journal of Public Health. 2015;**44**(12):1685-1692

[42] Clark JA. The influence of increasing numbers of nonindicator organisms upon the detection of indicator organisms by the membrane filter and presence-absence tests. Canadian Journal of Microbiology. 1980;**26**:827-832

[43] LeChevallier MW, Schulz W, Lee RG. Bacterial nutrients in drinking water. Applied and Environmental Microbiology. 1991;**57**:857-862

[44] Møller S, Kristensen CS, Poulsen L, Carstensen JM, Molin S. Bacterial growth on surfaces: Automated image analysis for quantification of growth rate-related parameters. Applied and Environmental Microbiology. 1995;**61**:741-748

[45] Sun L, Peng W. Heavy metals in shallow groundwater of the urban area in Suzhou, northern Anhui Province, China. Water Practice and Technology. 2014;**9**(2):197-205

[46] Abduljameel A, Sirajudeen J, Abdulvahith R. Studies on heavy metal pollution of groundwater sources between Tamilnadu and Pondicherry, India. Advances in Applied Science Research. 2012;**3**(1):424-429

[47] Leung CM, Jiao JJ. Heavy metal and trace element distributions in groundwater in natural slopes and highly urbanized spaces in mid-levels areas, Hong Kong. Water Research. 2006;**40**(4):753-767

[48] Nouri J, Mahvi AH, Babaei AA, Jahed GR, Ahmadpour E. Investigation of heavy metals in groundwater. Pakistan Journal of Biological Sciences. 2006;**9**(3):377-384

[49] Rivett M, Drewes J, Barett M, Chilton J, Appleyard S, Dieter HH, et al. Chemicals: Health relevance, transport and attenuation. In: Schmoll O, Howard G, Chilton J, Chorus I, editors. Protecting Groundwater for Health: Managing the Quality of Drinking Water Sources. London: IWA Publishing; 2006. p. 123

[50] Stamatis G, Voudouris K, Karefilakis F. Groundwater pollution by heavy metals in historical mining area of Kavrio, Attical, Greece. Water, Air, and Soil Pollution. 2001;**128**(2):61-83

[51] Mahiknecht J, Steinich B, Navarro de Leon I. Groundwater chemistry and mass transfers in the independent aquifer, Central Mexico by using multivariate statistics and mass-balance models. Environmental Geology. 2004;**48**(6):781-795

[52] Rajasekaran R, Abinaya M. Heavy metal pollution in groundwater: A review. International Journal of ChemTech Research. 2014;**6**(14): 5661-5664

[53] Obi CN, Okocha CO. Microbiological and physico-chemical of selected bore-hole waters in world bank housing estate, Umuahia, Abia state, Nigeria. Journal of Engineering and Applied Sciences. 2007;**2**(5):920-929

[54] Longe EO, Balogun MR. Groundwater quality assessment near a municipal landfill, Lagos, Nigeria. Research Journal of Applied Sciences, Engineering and Technology. 2010;**2**(1):39-44

[55] Fubara-Manuel I, Jumbo RB. The effects of septic tank locations on borehole water quality in Port Harcourt, Nigeria. International Journal of Engineering and Technology. 2014;**4**(5):236-264

[56] Al-Sabahi E, Abdul RS, Wan Z, Al-Nozaily WY, Alshaebi F. The characteristics of leachate and ground water pollution at municipal solid waste landfill of Ibb city, Yemen. American Journal of Environmental Sciences. 2009;**5**(3):256-266

[57] Onuorah S, Igwemadu N, Odibo F. Bacteriological quality assessment of borehole water in Ogbaru communities, Anambra state, Nigeria. Universal Journal of Clinical Medicine. 2019;**7**(1):1-10. DOI: 10.13189/ujcm.2019.070101

[58] Adeyemi O, Oloyede OB, Olajidi AT. Physiochemical and microbial characteristics of leachate-contaminated groundwater. Asian Journal of Biochemistry. 2007;**2**(5):343-348

[59] Adegoke OO, Bamigbowu EO, Oni ES, Ugbaja KN. Microbiological examination of sachet water sold in Aba, Abia State, Nigeria. Global Research Journal of Microbiology. 2012;**2**:62-66

[60] Ackah-Arthur M, Amin AK, Gyamfi ET, Acquah J, Nyarko ES, Kpattah SE, et al. Assessment of the quality of sachet water consumed

in urban townships of Ghana using physico-chemical indicators: A preliminary study. Advances in Applied Science Research. 2012;**3**:2120-2127

[61] Falegan CR, Odeyemi AT, Ogunjobi LP. Bacteriological and Physico-chemicals studies of sachet water in locations at Ado- Ekiti, Ekiti State. Aquatic Biology. 2014;**2**:55-61

[62] Adewoye AO, Adewoye SO, Opasola OA. Microbiological examination of sachet water experimentally exposed to sunlight. International Journal of Pure and Applied Sciences and Technology. 2013;**18**:36-42

[63] Mgbakor C, Ojiegbe GC, Okonko IO, Odu NN, Alli JA, Nwaze JC, et al. Bacteriological evaluation of some sachet water on sales in Owerri metropolis, Imo State, Nigeria. Malaysian Journal of Microbiology. 2011;**7**:217-225

[64] Onilude AA, Adesina FC, Oluboyede OA, Adeyemi BI. Micro-biological quality of sachet packaged water vended in three local government of Oyo State, Nigeria. African Journal of Food Science and Technology. 2013;**4**:195-200

[65] Washington State Department of Health: Coliform bacteria in drinking water. 2019. Available from: https://www.doh.wa.gov/CommunityandEnvironment/DrinkingWater/Contaminants/Coliform [Accessed: 01 November 2019]

[66] Edema MO, Omemu AM. Microbiology and physiochemical analysis of different water sources. International Journal of Microbiology. 2001;**15**(1):57-61

[67] Botelho-Nevers E, Gagnaire J, Verhoeven PO, Cazorla C, Grattard F, Pozzetto B, et al. Decolonization of Staphylococcus aureus carriage. Médecine et Maladies Infectieuses. 2017;**47**(5):305-310

[68] Verhoeven PO, Gagnaire J, Botelho-Nevers E, Grattard F, Carricajo A, Lucht F, et al. Detection and clinical relevance of Staphylococcus aureus nasal carriage: An update. Expert Review of Anti-Infective Therapy. 2014;**12**(1):75-89

[69] Williams MR, Gallo RL. The role of the skin microbiome in atopic dermatitis. Current Allergy and Asthma Reports. 2015;**15**(11):65. DOI: 10.1007/s11882-015-0567-4

[70] Alexander H, Paller AS, Traidl-Hoffmann C, Beck LA, De Benedetto A, Dhar S, et al. The role of bacterial skin infections in atopic dermatitis: Expert statement and review from the International Eczema Council Skin Infection Group. Brazilian Journal of Dermatology. 2019;**2019**:1-12. DOI: 10.1111/bjd.18643

[71] Loewen K, Schreiber Y, Kirlew M, Bocking N, Kelly L. Community-associated methicillin-resistant *Staphylococcus aureus* infection: Literature review and clinical update. Canadian Family Physician. 2017;**63**(7):512-520

[72] Kırmusaoğlu S. Staphylococcal biofilms: Pathogenicity, mechanism and regulation of biofilm formation by quorum-sensing system and antibiotic resistance mechanisms of biofilm-embedded microorganisms. In: Dhanasekaran D, Thajuddin N, editors. Microbial Biofilms - Importance and Applications. Rijeka: IntechOpen;

[73] Arciola CR, Campoccia D, Ravaioli S, Montanaro L. Polysaccharide intercellular adhesin in biofilm: Structural and regulatory aspects. Frontiers in Cellular and Infection Microbiology. 2015;**5**:1-10

[74] Karampatakis T, Antachopoulos C, Tsakris A, Roilides E. Molecular

epidemiology of carbapenem-resistant *Pseudomonas aeruginosa* in an endemic area: Comparison with global data. European Journal of Clinical Microbiology and Infectious Diseases. 2018;**37**(7):1211-1220

[75] Rehman A, Patrick WM, Lamont IL. Mechanisms of ciprofloxacin resistance in *Pseudomonas aeruginosa*: New approaches to an old problem. Journal of Medical Microbiology. 2019;**68**(1):1-10

[76] Feng W, Sun F, Wang Q, Xiong W, Qiu X, Dai X, et al. Epidemiology and resistance characteristics of *Pseudomonas aeruginosa* isolates from the respiratory department of a hospital in China. Journal of Global Antimicrobial Resistance. 2017;**8**:142-147

[77] English EL, Schutz KC, Willsey GG, Wargo MJ. Transcriptional responses of *Pseudomonas aeruginosa* to potable water and freshwater. Applied and Environmental Microbiology. 2018;**84**(6):1-12. DOI: 10.1128/AEM.02350-17

[78] Driscoll JA, Brody SL, Kollef MH. The epidemiology, pathogenesis and treatment of *Pseudomonas aeruginosa* infections. Drugs. 2007;**67**(3):351-368

[79] Kamali E, Jamali A, Ardebili A, Ezadi F, Mohebbi A. Evaluation of antimicrobial resistance, biofilm forming potential, and the presence of biofilm-related genes among clinical isolates of *Pseudomonas aeruginosa*. BMC Research Notes. 2020;**13**(1):27. DOI: 10.1186/s13104-020-4890-z

[80] Moradali MF, Ghods S, Rehm BH. *Pseudomonas aeruginosa* lifestyle: A paradigm for adaptation, survival, and persistence. Frontiers in Cellular and Infection Microbiology. 2017;7:39. DOI: 10.3389/fcimb.2017.00039

[81] Hennequin C, Robin F. Correlation between antimicrobial resistance and virulence in *Klebsiella pneumoniae*. European Journal of Clinical Microbiology and Infectious Diseases. 2016;**35**(3):333-341

[82] Hamza E, Dorgham SM, Hamza DA. Carbapenemase-producing *Klebsiella pneumoniae* in broiler poultry farming in Egypt. Journal of Global Antimicrobial Resistance. 2016;**7**:8-10

[83] Jelić M, Hrenović J, Dekić S, Goić-Barišić I, Tambić AA. First evidence of KPC-producing ST258 *Klebsiella pneumoniae* in river water. Journal of Hospital Infection. 2019;**103**(2):147-150

[84] Ovejero CM, Delgado-Blas JF, Calero-Caceres W, Muniesa M, Gonzalez-Zorn B. Spread of mcr-1-carrying Enterobacteriaceae in sewage water from Spain. Journal of Antimicrobial Chemotherapy. 2017;**72**(4):1050-1053

[85] Gomez-Simmonds A, Uhlemann AC. Clinical implications of genomic adaptation and evolution of carbapenem-resistant *Klebsiella pneumoniae*. Journal of Infectious Diseases. 2017;**215**(Suppl 1):S18-S27

[86] Moradigaravand D, Martin V, Peacock SJ, Parkhill J. Evolution and epidemiology of multidrug-resistant *Klebsiella pneumoniae* in the United Kingdom and Ireland. MBio. 2017;**8**(1). pii: e01976-16. DOI: 10.1128/mBio.01976-16

[87] Holt KE, Wertheim H, Zadoks RN, Baker S, Whitehouse CA, Dance D, et al. Genomic analysis of diversity, population structure, virulence, and antimicrobial resistance in *Klebsiella pneumoniae*, an urgent threat to public health. Proceedings of the National Academy of Sciences USA;**112**(27):E3574-E3581. DOI: 10.1073/pnas.1501049112

[88] Carvalho F, George J, Sheikh HMA, Selvin R. Advances in screening, detection and enumeration of

Escherichia coli using nanotechnology-based methods: A review. Journal of Biomedical Nanotechnology. 2018;**14**(5):829-846. DOI: 10.1166/jbn.2018.2549

[89] Sarowska J, Futoma-Koloch B, Jama-Kmiecik A, Frej-Madrzak M, Ksiazczyk M, Bugla-Ploskonska G, et al. Virulence factors, prevalence and potential transmission of extraintestinal pathogenic *Escherichia coli* isolated from different sources: Recent reports. Gut Pathogenes. 2019;**11**:10. DOI: 10.1186/s13099-019-0290-0

[90] Money P, Kelly AF, Gould SW, Denholm-Price J, Threlfall EJ, Fielder MD. Cattle, weather and water: Mapping *Escherichia coli* O157:H7 infections in humans in England and Scotland. Environmental Microbiology. 2010;**12**(10):2633-2644

[91] Guzmán D, Stoler J. An evolving choice in a diverse water market: A quality comparison of sachet water with community and household water sources in Ghana. American Journal of Tropical Medicine and Hygiene. 2018;**99**(2):526-533

[92] Mezzatesta ML, Gona F, Stefani S. *Enterobacter cloacae* complex: Clinical impact and emerging antibiotic resistance. Future Microbiology. 2012;**7**(7):887-902

[93] Chavda KD, Chen L, Fouts DE, Sutton G, Brinkac L, Jenkins SG, et al. Comprehensive genome analysis of carbapenemase-producing *Enterobacter* spp.: New insights into phylogeny, population structure, and resistance mechanisms. mBio;**7**(6): e02093-16. DOI: 10.1128/mBio.02093-16

[94] Armbruster CE, Mobley HLT, Pearson MM. Pathogenesis of *Proteus mirabilis* infection. EcoSal Plus. 2018;**8**(1). DOI: 10.1128/ecosalplus.ESP-0009-2017

[95] Drzewiecka D. Significance and roles of Proteus spp. bacteria in natural environments. Microbial Ecology. 2016;**72**(4):741-758

[96] Różalski A, Sidorczyk Z, Kotełko K. Potential virulence factors of *Proteus bacilli*. Microbiology and Molecular Biology Reviews. 1997;**61**(1):65-89

[97] Morgenstein RM, Szostek B, Rather PN. Regulation of gene expression during swarmer cell differentiation in *Proteus mirabilis*. FEMS Microbiology Reviews. 2010;**34**(5):753-763

[98] Faruque SM, Albert MJ, Mekalanos JJ. Epidemiology, genetics, and ecology of toxigenic *Vibrio cholerae*. Microbiology and Molecular Biology Reviews. 1998;**62**:1301-1314

[99] Zhu J, Miller MB, Vance RE, Dziejman M, Bassler BL, Mekalanos JJ. Quorum-sensing regulators control virulence gene expression in *Vibrio cholerae*. Proceedings of the National Academy of Sciences USA. 2002;**99**(5):3129-3134

[100] Cabral JP. Water microbiology. Bacterial pathogens and water. International Journal of Environmental Research and Public Health. 2010;**7**(10):3657-3703

[101] WHO cholera key facts 2019 [Internet]. Available from: https://www.who.int/en/news-room/fact-sheets/detail/cholera[Accessed: 02 January 2020]

[102] Elimian KO, Musah A, Mezue S, Oyebanji O, Yennan S, Jinadu A, et al. Descriptive epidemiology of cholera outbreak in Nigeria, January–November, 2018: Implications for the global roadmap strategy. BMC Public Health. 2019;**19**:1264. DOI: 10.1186/s12889-019-7559-6

[103] Dalhat MM, Isa AN, Nguku P, Nasir S-G, Urban K, Abdulaziz M, et al.

Descriptive characterization of the 2010 cholera outbreak in Nigeria. BMC Public Health. 2014;**14**:1167. DOI: 10.1186/1471-2458-14-1167

[104] Ishida R, Ueda K, Kitano T, Yamamoto T, Mizutani Y, Tsutsumi Y, et al. Fatal community-acquired *Bacillus cereus* pneumonia in an immunocompetent adult man: A case report. BMC Infectious Diseases. 2019;**19**(1):197. DOI: 10.1186/s12879-019-3836-3

[105] Chen D, Li Y, Lv J, Liu X, Gao P, Zhen G, et al. A foodborne outbreak of gastroenteritis caused by Norovirus and *Bacillus cereus* at a university in the Shunyi District of Beijing, China 2018: A retrospective cohort study. BMC Infectious Diseases. 2019;**19**(1):910. DOI: 10.1186/s12879-019-4570-6

[106] Szabo JG, Meiners G, Heckman L, Rice EW, Hall J. Decontamination of bacillus spores adhered to iron and cement-mortar drinking water infrastructure in a model system using disinfectants. Journal of Environmental Management. 2017;**187**:1-7

[107] Jeżewska-Frąckowiak J, Seroczyńska K, Banaszczyk J, Jedrzejczak G, Żylicz-Stachula A, Skowron PM. The promises and risks of probiotic bacillus species. Acta Biochimica Polonica. 2018;**65**(4):509-519

[108] Bottone EJ. Bacillus cereus, a volatile human pathogen. Clinical Microbiology Reviews. 2010;**23**(2):382-398

[109] Slamti L, Lereclus D. A cell–cell signaling peptide activates the PlcR virulence regulon in bacteria of the Bacillus cereus group. The EMBO Journal. 2002;**21**:4550-4559

[110] Okutani A, Inoue S, Noguchi A, Kaku Y, Morikawa S. Whole-genome sequence-based comparison and profiling of virulence-associated genes of Bacillus cereus group isolates from diverse sources in Japan. BMC Microbiology. 2019;**19**(1):296. DOI: 10.1186/s12866-019-1678-1

[111] Kariuki S, Gordon MA, Feasey N, Parry CM. Antimicrobial resistance and management of invasive salmonella disease. Vaccine. 2015;**33**(Suppl 3):C21-C29. DOI: 10.1016/j.vaccine.2015.03.102

[112] Threlfall EJ. Antimicrobial drug resistance in Salmonella: Problems and perspectives in food- and water-borne infections. FEMS Microbiological Reviews. 2002;**26**(2):141-148

[113] Cuypers WL, Jacobs J, Wong V, Klemm EJ, Deborggraeve S, Van Puyvelde S. Fluoroquinolone resistance in Salmonella: Insights by whole-genome sequencing. Microbial Genomics. 2018:**4**(7). DOI: 10.1099/mgen.0.000195

[114] Liu H, Whitehouse CA, Li B. Presence and persistence of Salmonella in water: The impact on microbial quality of water and food safety. Frontiers in Public Health. 30 May 2018;**6**:159. DOI: 10.3389/fpubh.2018.00159

[115] Akinyemi KO, Oyefolu AOB, Mutiu WB, Iwalokun BA, Ayeni ES, Ajose SO, et al. Typhoid fever: Tracking the trend in Nigeria. The American Journal of Tropical Medicine and Hygiene. 2018;**99**(Suppl 3):41-47